ROCK ART AND POSTERITY

Conserving, managing and recording rock art

Proceedings of Symposium M,
'CONSERVATION AND SITE MANAGEMENT'

and Symposium E,
'RECORDING AND STANDARDISATION IN ROCK ART STUDIES'

of the

First AURA Congress,
held in Darwin in 1988

Edited by

COLIN PEARSON

National Centre for Cultural Heritage Science Studies, University of Canberra, Australia

and

B. K. SWARTZ, Jr

Department of Anthropology, Ball State University, U.S.A.

Occasional AURA Publication No. 4
Australian Rock Art Research Association
Melbourne 1991

OCCASIONAL AURA PUBLICATION No. 4
General Editor: Robert G. Bednarik

The printing costs of this volume have been met by equal grants from the Australian Institute of Aboriginal and Torres Strait Islander Studies and the Australian National Commission for Unesco.

All money recouped from the sales of this volume is directed into the AURA Publications Fund, which will meet the cost of publishing the proceedings of future AURA congresses.

ISBN 0 646 03751 X

Published by Archaeological Publications, Melbourne.
Format: Robert G. Bednarik

Australian Rock Art Research Association, P.O. Box 216, Caulfield South, Victoria 3162, Australia

Foreword and acknowledgments

The First AURA Congress, held in Darwin in August/September 1988, has proved to be a watershed in international studies of pre-Historic art. It has led to the establishment of a new scientific discipline, the adherents of which, drawn from numerous other disciplines, have spontaneously formed a global community of researchers. The proceedings of this event form a record of developments that will clearly determine future directions in this area of scientific endeavour.

The academic program of the First AURA Congress consisted of eleven subject-oriented symposia. In the present volume appear most of the papers given in two of these symposia. Symposium M, chaired by Associate Professor Colin Pearson and Alan Watchman, dealt with aspects of rock art site management and conservation, while Symposium E, chaired by Professor B. K. Swartz, Jr, addressed the subjects of recording and standardisation. Thus two of the most pressing aspects of rock art, those dealing with questions of the survival of this earliest record of the cognitive, intellectual and cultural development of humanity, are appropriately combined in a single volume, and edited by two of the world's foremost scholars in the respective fields. This volume is a milestone in the endeavours to ensure the survival of rock art, in one form or another, for posterity.

Several points deserve to be mentioned here. Firstly, this is a welcome opportunity to thank, once again, the sponsors of the First AURA Congress, without whose help the congress would not have been able to subsidise the attendance costs of thirty participants, most of whom came from developing countries. The principal sponsors of the congress were the Australian National Parks and Wildlife Service, the International Seminars Support Scheme of the International Development Assistance Bureau, Qantas Airways Ltd, the Australian Institute of Aboriginal Studies (now Australian Institute of Aboriginal and Torres Strait Islander Studies), the Cultural Relations and Visits Branch of the Department of Foreign Affairs and Trade, Ansett Airlines of Australia, Energy Resources of Australia Ltd, the French Embassy in Australia, the British Council, the International Centre for the Study of the Preservation and the Restoration of Cultural Property, Argyle Diamond Mines Pty Ltd, the Australia - China Council, Pancontinental Mining Limited, Woodside Offshore Petroleum Pty Ltd, the Northern Territory Museum of Arts and Sciences, and the Northern Territory Government Tourist Bureau.

Secondly, the costs of printing this present volume of the proceedings of the First AURA Congress have been met by two grants of equal size, from the Australian Institute of Aboriginal and Torres Strait Islander Studies in Canberra, and the Australian National Commission for Unesco, also in Canberra. I express my sincere gratitude to the two underwriting organisations, who have in this way contributed to establishing a permanent basis for publishing quality work in this discipline. All money recouped from the sales of this volume will be returned into the AURA Publications Fund, the exclusive purpose of which it is to meet the cost of future Occasional AURA Publications, most specifically the proceedings of future AURA congresses. Under the supervision and guidance of the Editorial Committee, this perpetual fund will thus provide a solid basis for the future publication of high-quality work in a field which so far has not had the considerable benefit of possessing such an established publication base, dedicated wholly to the development of this discipline. It is hoped that in this way the dynamic scientific development the discipline experienced during the 1980s can be sustained into the next millennium. Seen in this context, the two sponsors of this volume have truly made a lasting contribution to the emerging discipline.

More specifically, this volume, which is one of a series of several volumes resulting from the Darwin congress, will be of great value to those concerned with the survival of rock art for posterity. I thank the Australian Institute of Aboriginal and Torres Strait Islander Studies and the Australian National Commission for Unesco for their generous support, on behalf of the editors and authors, of AURA, and on behalf of the International Federation of Rock Art Organisations.

Robert G. Bednarik
Founder of AURA
AURA Congress Chairman

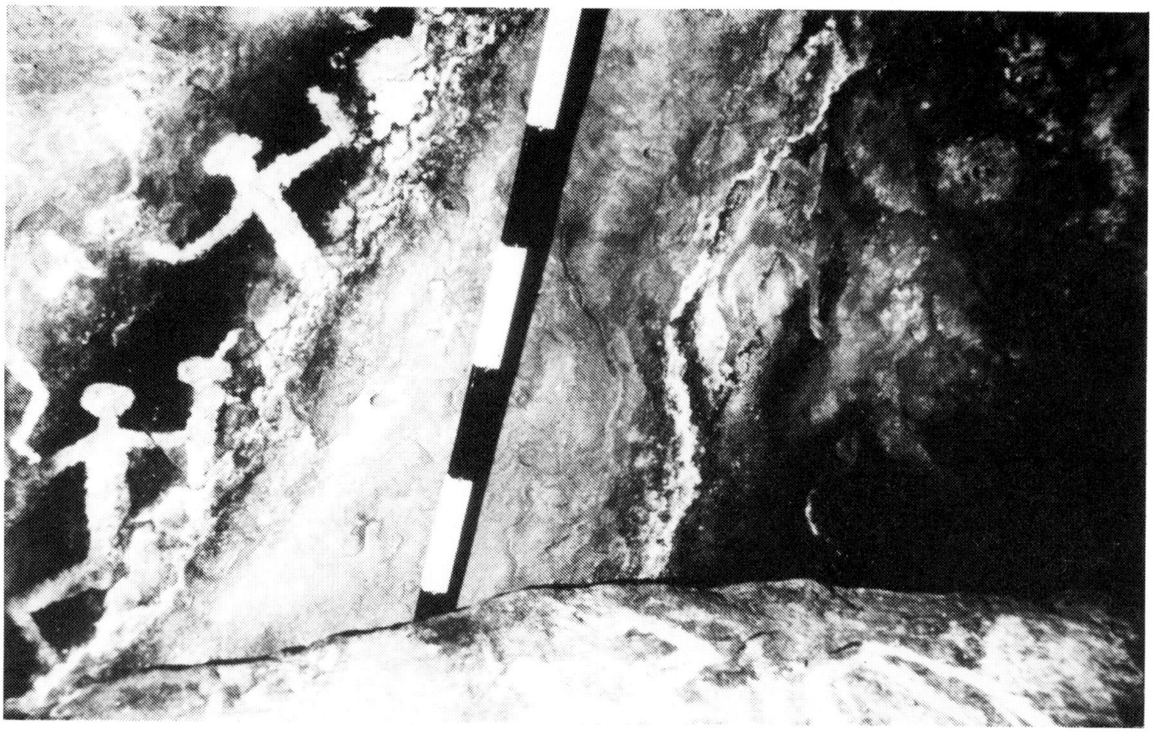

Natural rock art deterioration: the effects of intermittent waterwash on rock paintings at Mount Grenfell, New South Wales, Australia. Photograph taken in 1971.

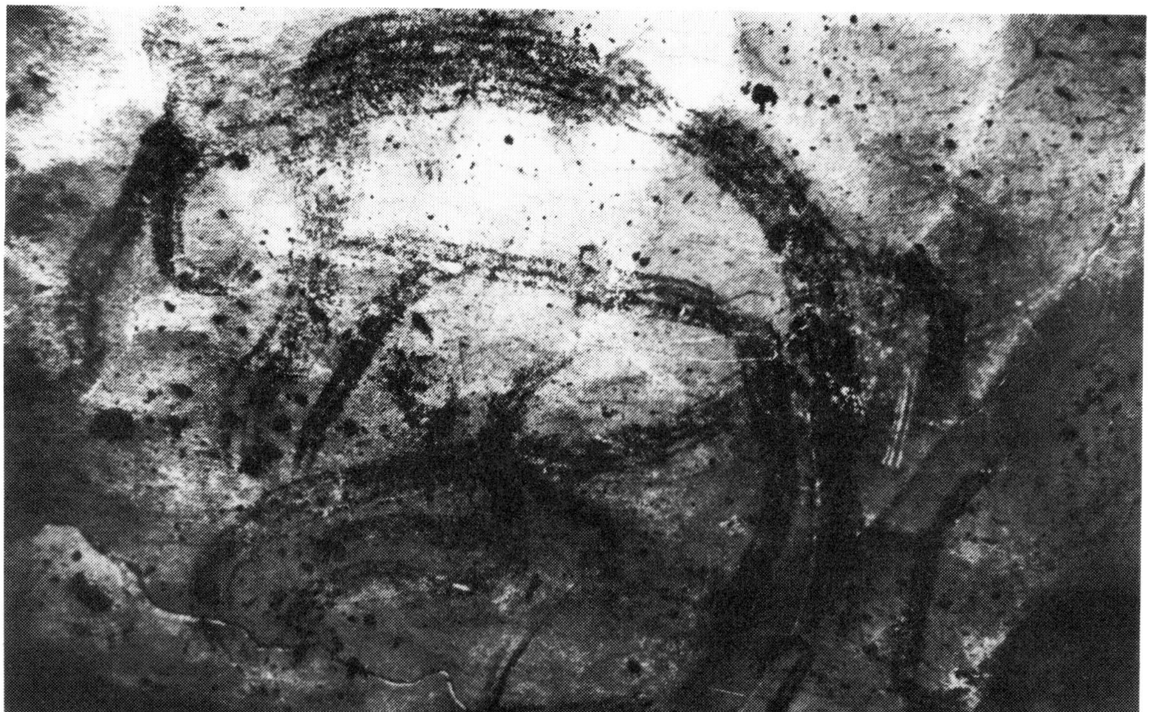

Rock art vandalism: clay balls thrown onto decorated cave ceiling, Baume Latrone, Gard, France. Photograph taken in 1981, prior to restoration by J. Brunet. (Both photographs by R. G. Bednarik.)

CONTENTS

Front cover: Rainbow Serpent design from Arnhem Land, Australia.
By George Chaloupka, Darwin.

PART ONE

ROCK ART CONSERVATION AND SITE MANAGEMENT

Proceedings of Symposium M

First AURA Congress

29 August to 2 September 1988

Darwin, Australia

Edited by

COLIN PEARSON

With contributions by Robert G. Bednarik, Tony Blanks, Steve Brown, J. Brunet, Elizabeth Caldicott, J. Clarke, Paul Faulstich, J. Flood, Ivan P. Haskovec, P. Haydock, D. Lambert, Colin Pearson, N. North, Peter Randolph, Eric W. Ritter, Nicholas P. Stanley Price, Hilary Sullivan, Andrew Thorn, Gajendra S. Tyagi, P. Vidal, J. Vouvé, Joseph Wallam and Grahame L. Walsh.

1

SITE MANAGEMENT IN KAKADU AND ULURU NATIONAL PARKS

Hilary Sullivan

Abstract. This paper deals with the strategies of site management in Kakadu National Park, where thousands of sites are located in remote and inaccessible areas. In particular, the attitude of the local Aboriginal community, who owns the land, to conservation and visitor management is examined. Methods of visitor management in a large park with high visitor numbers are also considered.

Introduction

Kakadu and Uluru National Parks, located in the Northern Territory of Australia, have much in common. Both are rich in rock art, are owned by Aboriginal people and jointly managed by them and the Australian National Parks and Wildlife Service (ANPWS); both have high cultural and biological values and are entered on the World Heritage List; both are high profile tourist destinations, visited by many thousands of people every year. Many groups have interests in these parks and may at times have very different expectations in relation to site management because of different cultural perceptions of what is important about sites and what is an appropriate form of management. In this paper some of these conflicts will be outlined and their effects on management practices discussed.

Kakadu National Park

Kakadu National Park is located in the Northern Territory some 250 km east of Darwin. It was established in 1979 and is managed by the Australian National Parks and Wildlife Service. Kakadu has increased in size over the years and is now a substantial park of some 18 000^2 km.

The park has a resident Aboriginal population of about 300 people who live in settlements throughout the Park. Much of Kakadu is either freehold Aboriginal land or currently under land claim. The landowners have leased the parts which they own to the Director of the ANPWS.

The Alligator Rivers Region, in which Kakadu is located, is highly mineralised and the mining industry has interests in the area. In particular, uranium deposits occur and Ranger Uranium Mine, which is excluded from, but surrounded by, the Park has been operating for some years. A town to cater for mine personnel and their families (some 1200 people) is located in the Park.

The Park has very high biological and cultural values and Stages I and II of the Park are entered on the World Heritage List.

The cultural resources of the Park encompass a range of site types formed between at least 23 000 years ago and the present. There are an estimated 5000-7000 rock art sites in the Park. Three complexes of art sites are open to the public, two of which have been developed for intensive visitor use while the third has been left undeveloped.

One of the effects of this high profile is very high visitor numbers. In 1987 around 185 000 people visited Kakadu which is the equivalent of 666 000 visitor days. About 80% of these people travelled privately and 20% were on commercial tours. Between 1986 and 1987 there was a 104% rise in our tour patrons and this is expected to continue to increase.

Uluru National Park

Uluru National Park (1325^2 km) is a relatively small Park located east of Alice Springs in central Australia. An Aboriginal population of approximately 200 people live in the Park. As the Park contains Ayers Rock (Uluru) it is very popular with tourists, especially 1-2 day package tours which are accommodated at the Yulara tourist centre located nearby but outside the Park. Some 250 000 people visit the Park each year. Rock art is prolific around the base of the rock but the majority of visitors concentrate on two groups of sites adjacent to other areas of visitor interest. Protective measures for and interpretation of the sites is gradually being developed. As most of the art is composed of tracks and symbols, visitor appreciation and understanding of the art depends greatly on interpretation. Four registered Sacred Sites around the base of the rock are closed to the public.

The importance of the art

While rock art issues have a relatively low profile, there are a range of people such as the owners, researchers, tourists and managers with quite different interests in rock art. Each have their own view of the significance of the art and consequently have different expectations from site management.

Aboriginal people own the sites in both a legal and cultural sense and they value sites for a variety of reasons. They see art sites as a storehouse of traditional knowledge. The art depicts objects, animals and activities familiar to Aboriginal people today. Mythological heros from the very strong oral tradition are portrayed. Many paintings have a deeper 'inside' meaning related to religious and ceremonial life. Some of the art can be attributed to individual artists and is part of the life experience of people now living. The older art, often not recognised as human art by Aboriginal people today, is however incorporated into Aboriginal cosmology and is tangible proof of beliefs about the formation of the landscape. Aboriginal interests in rock art then are diverse, but essentially rock art is viewed as one aspect of a range of things that make up a sense of place. It is preserving this sense of place that Aboriginal custodians are most concerned about.

Researchers and academics working in Kakadu see rock art as documents of *human* history. While they obviously look at ethnographic information their emphasis is not in the personalised sense of place of local perception but in larger scale internal and external comparisons. Their main concern about the management of rock art is in its preservation and conservation so that sites are available to provide the information that is the lifeblood of their research. Occasionally, however, researchers sacrifice conservation for knowledge such as when they excavate a site. Increasingly this may become the case with rock art studies, for example, as new dating techniques are developed requiring the removal of surface layers of salt-impregnated paint.

Visitors value rock art for the pleasure, entertainment and the insight into Aboriginal culture that it gives them. Their management expectations are oriented to the provision of access and the availability of interpretation. Aboriginal culture is very sophisticated but many people do not appreciate this because it is based on knowledge passed between individuals without ostentatious physical trappings. Rock art is an exception, especially in Kakadu where the art is plentiful and is often colourful and spectacular. In particular visitors find x-ray art surprising and exciting. Because visitors expect that human influence will be slight, Kakadu is often seen as a wilderness, whereas the rock art is testimony to the thousands of generations of Aboriginal history.

As a management authority the Park Service has some responsibility to all these groups although under the lease agreements the prime responsibility is to the Aboriginal land owners. The wishes of the owners must be foremost in the development of management policies for rock art and in the day to day decisions on site development and interpretation.

Researchers provide a great deal of information that is of use in site interpretation. It is ironic that while many of these findings are contrary to Aboriginal perceptions of the past they often legitimise Aboriginal claims for ownership of land to the wider Australian community. Research also provides management authorities with vital information about conservation and the significance of sites, from a research perspective. This knowledge helps in the formation of management policies.

Visitors interests and demands often provide the rationale for site management at least in terms of justifying the allocation of funds and staffing. Visitor satisfaction is of considerable concern to managers. Park managers need to balance the wishes of the owners, the protection of the sites, the need to educate and inform the public and to provide a satisfactory experience to visitors.

At times it is difficult for managers to accommodate all these different and at times conflicting interests and compromises are sometimes necessary. Of prime importance is the need for sites to maintain their cultural integrity. Subject to the qualification and given goodwill on all sides, most groups expectations can at least be partly met. When this is not the case the Park Service unfortunately has to accept criticism from groups which may feel they have been disadvantaged to some extent.

Management and research in two cultures

It is interesting to look at a couple of examples where the interests of these groups may not coincide.

At a large site in Kakadu the Park Service was concerned about the fire risk posed by large trees growing close, within centimetres, of the painted surface. In the fire-prone environment of the North, a hot fire may have caused spalling of the painted surface. One option that we canvassed with traditional owners of the site was to remove some of these large trees, but this was found to be unacceptable to them. The reduction of shade and the possible alteration to the water table which may have reduced the availability of surface water in the vicinity of the site would have diminished its value as a good place to camp. Although people do not camp at this site any more, its value was perceived in terms of its traditional use pattern. Alternative fire protection was given to the site.

Interestingly a proposal to construct an elaborate roof over this site was not a problem for traditional owners as it would have enhanced its value as a good place to camp. In contrast, this was viewed by white Australians as very intrusive and destructive of the site environment.

A second example relates to the research interest in investigating how Aboriginal society has changed over time. From an Aboriginal viewpoint such questions about the past are meaningless and inappropriate. A recent leaflet produced by the Australian National Parks and Wildlife Service and the local Aboriginal community at Uluru states:

> Anangu assert that the cave art was done by their ancestors a long time ago. The specific significance of sites around Uluru has been as it is since the beginning of time. The possibility that some sites have been used differently by Aboriginal people in the past is unthinkable; because, according to Aboriginal law, it would be *wrong* (ANPWS: 1987).

It is obvious that while Aboriginal custodians value sites as the places of 'the old people', their interest in the protection of the deposit or the art surface is on a different basis to the attitudes of researchers. This is not to say that the two groups are in constant conflict.

In northern Australia there has been a history of Aboriginal co-operation and interest in the many research projects that have been undertaken. However, while Aboriginal participants have been keenly interested in the practical results of research such as the types of artefacts found and the identification of species of animals hunted in the past, the broader conclusions researchers draw from this data have not been so readily grasped across the language and cultural barriers. Certainly the implications of this research on how their past is perceived by the Australian community is not understood. That their past can be contemplated without reference to the creation era and all that that concept entails comes as a shock to many Aboriginal people. Some communities are reasonably relaxed about telling 'that whitefella story about those old people' as long as the Aboriginal view is also put. Others however take the view that on Aboriginal land the Aboriginal view should prevail.

The World Heritage listing of both Kakadu and Uluru National Parks require the ANPWS to manage the Parks in a way that is acceptable by world standards. Aboriginal communities in the north do have diverse views about the effects of World Heritage listing. Some landowners see World Heritage listing as a means to protect the status of their land against the vagaries of government policy. Other Aboriginal people however are suspicious of a step that may reduce their control and do not like the high profile and increased tourism that results from World Heritage Listing.

Generally speaking Aboriginal people in Kakadu and Uluru National Parks have been quite amenable to new techniques of site protection. Physical conservation of painted surfaces is not a major concern. Indeed it is of much less concern than the proposal to repaint sites as a conservation measure. To the present generation of Aboriginal custodians rock paintings now belong to the realms of the past. It is something that their fathers and grandfathers did, something to be preserved if possible but not to be changed or added to. Aboriginal people are concerned much more with the proper use of country. Perhaps the major concern of Aboriginal custodians is access to sites and country and appropriate behaviour at these places. This is a recurring theme in any discussion between park management and landowners.

At the Kakadu workshop on visitors and site management in 1983, Aboriginal speakers emphasised this concern:

> Also I have been worried about *balanda* (white people) going into places where only certain Aboriginal men should go ... I think as long as people don't go where they shouldn't ... then visitors in moderate numbers are alright (Neiijie 1974: 57).
>
> Tourists can come up here and enjoy themselves - have a good look at bird life, go to Jim Jim Falls, look at cave paintings but they can't go to ceremony places. They've got a lot of places to go but they can't go to sacred sites, dreaming places and burial places ... We've got to have a big National Park so there is place for everybody to go (Gangale 1974: 57).
>
> I know what's a good place and can tell people it's all right to go there. It's good for those visitors to go to places like Cooinda and Yellow Water, but they can't go everywhere. They would get lost or go to a dangerous place (sacred site) and get hurt (Gabirrigi 1984: 59).

Art as recreation

Consideration needs to be given to the situation that arises when rock art becomes associated with holidays, with relaxation and recreation.

Three major rock art complexes are available for visitor access in Kakadu and between 50% and 60% of all Park visitors go there. They are one of the major attractions of Kakadu and the Top End generally. While the art sites do not have such a high profile at Uluru they are nevertheless visited by the majority of visitors. They feature on videos, posters and travel brochures. They are marketed as holiday attractions down south, and viewing rock art has become a feature of the tourist industry.

Access

A consequence of the promotion of rock art is that many people want to come and look at sites and indeed, as it is a part of their holiday itinerary, they demand access. This means that some sites must be available to show people, i.e. sites that traditional owners approve for this purpose. Aboriginal people are often very keen for visitors to have access to some rock art sites. They feel that it will help visitors to appreciate and understand Aboriginal culture. Indeed one traditional owner at Kakadu was worried that visitors experience at rock art sites was too fast and superficial.

> I like visitors to come to Ubirr to look at the paintings and learn the stories about the paintings. The outside (or public stories) are good to tell visitors and are very good for children to know. Its good for *balanda* (white people) to learn about the Aboriginal way of life.
>
> Some *balanda* who come to Ubirr don't really understand what they see. They rush to the sites and then back to their buses. They don't stay long enough at the sites. How can they understand? (ANPWS: 41).

Some sites present problems for use as a visitor destination. One rock art site in Kakadu is adjacent to an area where, until recent times, many people have been buried in rock cavities. Before the Park was declared, these burials were disturbed and in some cases removed by visitors. After the Park was declared, traditional owners were reticent to disrupt visitor access, but insisted on proper supervision of visitors. As a result the Park opens the site to visitors only when there are enough staff available to supervise and give information. In Stage III, recently included in Kakadu National Park, there is a frequently visited site to which Aboriginal custodians attach great religious significance and do not believe is an appropriate place for visitors to go. This site has been closed.

One result of large numbers of tourists at one place is that local Aboriginal people increasingly feel uncomfortable at such art sites and generally stop visiting them. Aboriginal people are aware of this trend and are prepared to trade this for the positive aspects of allowing visitors to learn about Aboriginal culture through rock art. However, while Aboriginal people have been very generous in agreeing to the opening up of their sites, the numbers of sites open to the public is unlikely to rise significantly.

Sites available to the public have to be protected from the damage associated with access by large numbers of people. This can cost a great deal of money because, not only does the immediate site environment such as deposits, art faces and the local site environment need physical protection, but access has to be rationalised, car parks and toilet facilities built, information provided and staff allocated to carry out surveillance of the site and maintain all these new assets. This of course costs a great deal of money. The ANPWS spent close to half a million dollars on the Nourlangie Development in Kakadu. So from a management point of view it is only possible to have a few sites open to the public.

Generally, the Kakadu experience has shown that large numbers of visitors can be catered for without undue site damage. Visitors generally find these developments as a positive contribution to their enjoyment of the site. The short term visitor with only one or two hours to spend at art sites appreciate ease of access and the information provided. A minority however, hanker after a more natural, untouched experience. This is especially true of permanent residents who have been to the developed sites three or four times and are looking for something 'new'. This is in fact a critical factor in determining which sites become known to the public and which remain unknown. Along this continuum decisions about site development at one end and site closure at the other end are made.

Interpretation

Because management authorities are usually staffed with university trained managers, interpretation at sites is usually focused on academic concerns such as: How old is the art? What style is it? etc. At Kakadu an attempt is made to link this information with the Aboriginal viewpoint by using appropriate information about whose country the site is on, some traditional stories about the art, etc.

However, the tested truths of the archaeological record can be a political statement and in conflict with Aboriginal perceptions of the past. Continually stating that Aboriginal people have been in Australia for at least 40 000 years when Aboriginal people know they have been here forever is an obvious example. Archaeologists are always speculating about the different environmental or cultural conditions of the past. As in the Uluru example quoted previously, this is often not an appropriate way to approach the past from an Aboriginal perspective.

Whilst this approach is in conflict with the basic premise of modern Western scientific philosophy, it is essential that this view of the past be understood by visitors if they are to appreciate rock art in its cultural context and not as a disembodied relic. Achieving this end is notoriously difficult. A recent Uluru Park Note states:

> Traditional owners of Uluru-Kata Tjuta National Park feel very strongly about their role in the interpretation of the Park to visitors. They feel particularly strongly about their responsibilities for interpreting the traditional aspects of the Park to the public in appropriate ways. Their first task, therefore, has been less to provide yet more information than to attempt to turn people's heads around so they can think 'straight' and appreciate the problems of cultural interpretation from an Anangu perspective.
>
> Anangu view many visitors as being 'greedy for the Tjukurpa', (creation era) wanting to hear 'all' about it without learning how to think about it.
>
> 'They hear a little about this place and about that place', said a senior Mutitjulu man, 'and they put it all together in one bucket and shake it up. Everything gets broken and mixed up, and when they pour it out in their own country to try to remember they don't know what pieces go together. They should take it home in their hearts. Then they'd remember' (ANPWS: 1987).

A positive response of visitors to rock art in some ways depends on whether people perceive it as 'good' art. Fortunately at Kakadu x-ray art does appeal to most visitors. This may be one of the reasons for the very low incidence of graffiti in the Park although there is a less positive response to the older monochrome art.

It is interesting that some people feel disappointed and confused when confronted with the non-figurative art of central Australia. Some visitors use such comparisons to judge the local Aboriginal people as 'less sophisticated' etc. Explaining the cultural context of this art would give visitors a new understanding.

ANPWS provide free guided tours to sites by specially employed and trained seasonal rangers. In particular at Kakadu this is a large undertaking with the ANPWS running 2-3 tours per day at art sites during the high visitor season. These tours are very popular and from the visitor response on survey forms they add a great deal to peoples understanding of a wide range of issues.

Increasingly visitors are wanting to talk about wider issues. Rock art is often the stepping off point for questions not only about Aboriginal culture but about land rights and other political issues relevant to the interaction between Aboriginal people and other Australians. Rock art tours these days can become the forum of political debate.

Appropriate behaviour

Where visitors are travelling together in a group on holiday, the tour captain's job is often seen not only as information giver but as entertainer. This is a dangerous situation since the distinction between fact and fiction and good and bad taste often becomes blurred. It is easy to pander to a groups' desire for entertainment and titillation at the expense of the truth.

This is especially so at sites where there are few detailed stories about the art such as at Uluru National Park. There the art sites are largely composed of non-figurative designs and tracks. These designs are common throughout central Australia and many can be identified as having a generalised meaning such as the stumpy boomerang shape often symbolising a seated person. Beyond this however, Aboriginal people say that only the artist, who are all now deceased, knew the specific story he was depicting. However, visitors and tour guides in particular do tell stories about the art. This speculation is considered nonsense by Aboriginal people but where it involves guesses about ceremonial or religious life it becomes offensive to their perception of what is right. At Uluru the Aboriginal community live nearby and visit the rock regularly. As they are often confronted with these discussions their concern that visitors be given appropriate and correct information is understandable. The often outright racist nature of some commentaries tentatively linked to rock art create further disenchantment in Aboriginal communities with rock art interpretation by tour captains.

Tour captains generally find it very difficult to admit that they, and in fact no one, knows the answers to many questions about rock art. To a certain extent they have to justify the cost of the tour and their position as guide by demonstrating some knowledge.

The ANPWS has led the way in this area by designing and running courses for tour operators and providing written material. These courses have been successful in raising the standard of commentaries at art sites, although a lot of improvement is still necessary.

Aboriginal people do not expect overly serious behaviour at art sites but they do expect visitors to attempt to appreciate the art from an Aboriginal perspective and that information given at sites both written and verbal accurately portrays their beliefs about the art.

Rock art, and especially its presentation to visitors, at times presents dilemmas to those responsible for its management. In north Australia rock art is increasingly an element in the race relations and land ownership issues. Site managers in this situation have a great opportunity to provide an experience for visitors which will broaden the quality of this debate.

Acknowledgments

I would like to thank the Traditional Owners of Kakadu and Uluru National Parks for sharing their ideas about the issues raised in this paper with me. I would also like to thank Sharon Sullivan, Ivan Haskovec and Professor Ovington for their comments and suggestions on the first draft of this paper.

REFERENCES

ANPWS 1987a. Park notes on Uluru (Ayers Rock - Mount Olga) National Park. 'Some common misapprehensions'.

ANPWS 1987b. Park notes on Uluru (Ayers Rock - Mount Olga) National Park. 'Some Anangu perceptions of Uluru'.

GABIRRIGI, N. 1984. Back to my country. In Sullivan, H. (ed.), *Visitors to Aboriginal sites: Access, control and management*, p. 59. ANPWS, Canberra.

GANGAL, T. 1984. Still living on the land. In Sullivan, H. (ed.), *Visitors to Aboriginal sites: Access, control and management*, p. 57. ANPWS, Canberra.

NEIJIE, B. 1984. Visitors to sites: a traditional owners perspective. In Sullivan, H. (ed.), *Visitors to Aboriginal sites: Access, control and management*, pp. 41-42. ANPWS, Canberra.

2

CONSERVATION AND PROTECTION WITHIN THE AMERICAN ROCK ART RESEARCH ASSOCIATION

Eric W. Ritter

Abstract. The American Rock Art Research Association (ARARA) has been a key agent in the conservation/protection (C/P) of rock art within North America. The organisation has served as a focal point for advice on C/P matters to corporate, government and academic sectors. C/P attention has been directed to the development of guidelines, public education, symposia, expert and volunteer referral lists, and reference listings. The C/P future promises increased participation in co-operative agreements, education utilising diverse media, greater integration of rock art undertakings with ethnic groups whose ancestors made the art, contributions in the technical and ethical treatment of rock art (and its environment), and directed advice.

Introduction

Rock art, its anthropological and humanistic study, its conservation and protection has come of age in the last few decades in North America. But there remain problems.

Beginning in the late 19th century there was recognition in the United States of the deterioration and destruction of cultural resources, primarily ruins and trash deposits within the Southwest. Such attention in these early days resulted in passage of the 1906 American Antiquity Act. This Federal act not only protected (on paper) heritage values on public lands but authorised the establishment of various National Monuments, such as Death Valley, Lava Beds, Homolovi, El Moro and others which contain important rock art complexes. There has been subsequent legislation for the protection of cultural resources, almost exclusively at the Federal and State levels. Still, human caused losses of rock art (Fig. 1) and other cultural resources, to say nothing of natural deterioration, persists, irrespective of ownership, owing largely to ignorance and indifference, greed, absence of law enforcement, weaknesses in laws, and inattention by professionals.

Archaeology in North America was well-established by 1900 and has increased in popularity and influence since. Rock art studies, on the other hand, until quite recently have remained on the periphery of archaeology, largely within the domain of knowledgeable avocationalists. Because of rock art's popularity within a wide spectrum of interests beyond archaeology, as in the humanities, arts and other social sciences, and since increasing losses were apparent, a group of concerned enthusiastic individuals from various perspectives - and largely avocational - formed the non-profit American Rock Art Research Association (ARARA) in 1974, a time of heightened environmental awareness among many Americans. The purposes of this organisation are straightforward, to: (1) advance research, (2) work for the preservation and protection of rock art sites in North America through co-operation with land owners and managers, (3) promote the non-destructive utilisation of rock art for scientific, educational and artistic purposes; and, (4) inform and educate people with regard to the rock art heritage of the United States and to a lesser extent Canada and Latin America. In informing and educating members, the public agencies and corporations, there is a quarterly newsletter, *La Pintura*, a publication series, and a yearly symposium.

It soon became apparent that a number of issues, among them rock art conservation and protection, could be best served within the organisation by a committee. Thus, in 1977 a Conservation and Preservation, later Conservation and Protection Committee was formed. ARARA president Klaus Wellmann stated (1977: 3) 'what it (the committee) can, and should, do is to make its voice heard, decisively and persistently, in all cases in which the survival of a rock art site is at stake. I have no illusions it will be an easy task.' Furthermore, in the association bylaws it is written 'It's chairman is obligated to initiate action by writing letters to appropriate authorities, or by taking other steps deemed suitable ... It will be very important to be persistent and to follow up each case to its eventual solution.'

This paper is intended to provide a historical summary from the author's perspective of ARARA's involvement in conservation and protection matters (detailed Committee reports are contained in *La Pintura*), to point out strong points and weaknesses in efforts to date, to present some opinions and recommendations of certain ARARA members and a group of non-ARARA archaeologists regarding various rock art conservation and protection issues, to forecast the direction of conservation and protection in ARARA and the United States and beyond, and, finally, to offer some recommendations for future action on a broad base building upon ARARA's and others' work (cf. Van Tilburg and Meighan 1981; Anati 1983; Wainwright 1985).

A history of conservation and protection action

ARARA's involvement with rock art conservation and protection has largely been individualistic and on a crisis basis. Volunteers are widely separated. Recognition and awareness of the group's capabilities and limitations has been slow in coming. Early conservation and protection actions were largely educational, geared to members to increase their awareness of issues by discussing such problems as rock art theft in the California Desert, the removal of graffiti from pictographs at Hueco Tanks State

Park in Texas, discussing problems in pin-pointing rock art site locations in publications, and the feasibility of restoring petroglyphs. Dr Richard Wright was appointed first committee chairman in 1977, a position he held until 1980. He effectively led an informal lobbying effect against several highway projects that threatened rock art (Halloran Springs, California, and the Waterflow Site, New Mexico), and the publication of an ill-advised brochure leading visitors to an unprotected site in Georgia. One of the major accomplishments of the committee during this time period was the publication and distribution of a rock art brochure stressing that 'American Indian rock art is ours to enjoy, to protect, to preserve'.

This author was appointed committee chairman in 1980. Actions and accomplishments since can be categorised as follows: (i) visitor control/vandalism, (2) planning and co-ordination, (3) conservation, (4) education, (5) development, (6) inquiries and information, and (7) miscellaneous.

Visitor control/vandalism

Visitor control and vandalism problems have largely been agency oriented and focussed in the western United States where the vast majority of rock art is located (Fig. 2). Most ARARA members reside in the West as well. Problems brought to the committee's attention concerning vandalism and visitor control have been ameliorated through co-ordinated or influenced actions like the placement of a grate at a pictograph site to prevent campfire smoke damage (Snow Creek, California). Individual problems reported to ARARA have been investigated through solicitations for action or field inspections by ARARA members, generally leading to a satisfactory solution. Thousands of volunteer hours have been invested by members in this regard. Dr Georgia Lee, for one, has been vigorously working with the Chilean government on visitor control/vandalism difficulties on Easter Island (Lee 1987), and more recently has focused attention on rock art problems on Hawaii (Lee 1988). Dr Frank Bock and A. J. Bock have documented a number of endangered sites in California and Arizona (Bock and Bock 1988). Occasionally committee efforts have been stymied by a complete lack of response from agencies.

Largely related to visitor/control vandalism is the planning issue; broad land use planning, specific management planning for a rock art site or complex of sites, and general planning for rock art management. ARARA has informally lobbied high government officials in Washington DC for proper attention to rock art sites. Members have contributed to rock art planning and management by providing planning input and volunteer efforts towards site mapping, recording, signing, fencing, and monitoring. Examples include a plan for Canyonlands National Park in Utah, the California Desert Plan, California's State Office of Historic Preservation county cultural resource planning, and volunteer efforts towards plan implementation at locations in Arizona, California, Colorado, New Mexico, Nevada, Ohio, Utah, and Wyoming and elsewhere. ARARA has entered into co-operative management plans for rock art recording and other efforts with such agencies as the Department of Interior's Bureau of Land Management.

Planning and co-ordination

Co-ordination efforts, as pointed out by Anati (1983: 16),

> serve to develop standards and procedures in the research, documentation and conservation of rock art and to lay the foundation of a world strategy for the safeguarding, evaluating and promoting of the knowledge and information in the field of rock art.

In this regard ARARA began its still on-going co-operation with the Canadian Rock Art Research Association; the American Society for Conservation Archaeology; the Getty Conservation Institute; ICOMOS-CAR; ICOM Committee for Conservation, rock art conservation working group; the American Land Resource Association; the Sierra Club; and the Coalition for Applied Technology. Many ARARA members are also members of other rock art groups.

Conservation

Only a few members of ARARA are rock art conservators (and, in fact, there are very few in the United States), and rock art conservation work through the auspices of ARARA has been less than protection oriented tasks. Yet there have been some noteworthy contributions despite risks to the rock art and workers that such efforts can entail. Much of the work has been at the Federal level, in response to vandalism and natural deterioration (cf. U.S. Congress, Office of Technology Assessment 1986). A recent major rock art publication (Connor et al. 1987) was found to contain photographs of panels illuminated by bonfires, a practice undoubtedly detrimental to both the art and any cultural deposits. ARARA has mounted a concerted campaign to prevent such photographic techniques in the future and to alert the publisher of our concern. As of 1989 the publisher has declined to remove the book from distribution or to attach any cautionary notes.

One of the more humorous responses to an inquiry came from the National Park Service regarding construction of lava flow diversionary structures around petroglyph fields in Hawaii. They stated that 'should a lava flow override a petroglyph field, the underlying rock is not melted, just covered. So, rather than destroying the petroglyphs, it is the ultimate form of protection.'

Education

Education, at all levels, is certainly recognised by ARARA as one of the cornerstones in protecting and conserving rock art. Many members are involved in presenting rock art classes and lectures, undertaking recording projects with groups, conducting tours, etc. Education and rock art was set up with its own committee within ARARA in 1986. A major effort of this committee is the development of educational packets, video tapes, brochures, and the like for schools and various groups. The ARARA brochure, publications and meetings, and individual efforts have been of considerable educational value. A rock art code of ethics has been developed for members and interested or targeted individuals and groups.

Development

The issue most involving the committee to date centres on proposed developments; those related primarily to energy development, highway construction, recreation, and new housing. Since 1981, following on the heels of the Arab oil embargo, large projects related to fossil fuel extraction and nuclear waste dump siting in Utah, and oil exploration and coal fired and thermal plant construction in California have been proposed in rock art rich areas. Noxon and Marcus (1986) present a good analysis of one such project. On a state and Federal level, existing laws and regulations generally allow for adequate consideration of direct impacts. But the most significant factor with respect to the rock art, a factor usually not well considered, is the creation of indirect impacts such as vandalism from increased visitation. The creation of acid rain and its deteriorating effects on some rock art is another problem. An important and all too obvious observation is that political action on even a global scale can have a trickle-down effect on the remotest of rock art sites.

Other significant development conflicts addressed by the committee relate to several highway and road projects where letter lobbying was successful in one case (Utah). In another case, a letter was sent providing information to a local (California) planning department (whose road crews had destroyed a site) with the hopes of changing their procedures. Dam construction or upgrading and reservoir development present recurring problems for rock art, as has happened in many cases in the past. In a number of situations, like the Los Vaqueros, Bear, and Burns dams in California, the Salt Caves Dam in Oregon, the Muddy River Dam in Nevada, and dams along the Churchill River in Manitoba, Canada, the committee has initiated action. Indirect impacts are once again one of the major areas of concern. These interests were brought to the responsible agency's attention. Of particular concern in these instances were impacts that could result from construction tremors and lake edge erosion.

Other developments to draw committee action included proposed new housing (California, New Mexico), timber harvesting (California), a solid waste disposal facility where concentrated exhaust fumes could affect pictographs (California), quarrying (Wisconsin), and military bombing and access problems (Chocolate Mountains, California). In most cases co-operation was promised, in others either no response was received, little substantive attention paid to the inquiry, or the project was eventually dropped.

Inquiries and information

Agency and public inquiries are routinely received by ARARA requesting the names of available rock art conservators, volunteer assistance in recording, recording standards, management recommendations regarding visitor control, and questions regarding chalking, latex molding, rubbing etc. Such inquiries and requests are becoming more numerous as ARARA becomes more visible.

Miscellaneous

There is a potpourri of other items related to rock art conservation and protection handled by the committee. Important activities include the presentation of a symposium in 1987 devoted to conservation and protection issues, and the publication of guidelines in 1988 for public and private agencies, groups, and individuals for assisting in the recognition and reporting of threatened examples of the rock art heritage.

Other committee actions include initial work on creation of a rock art conservation and protection bibliography, listings of rock art conservation and protection specialists and volunteers, a register of government agencies and archaeological groups, and establishment of a conservation-protection fund from voluntary contributions. Committee members have been active in writing support letters, conservation projects, monitoring efforts, planning review, and limited communication with Native American groups where conflicts of values are present between researchers/artists and Native Americans. This listing covers many of ARARA's past activities involving conservation and protection of rock art and its environs. There are *numerous* individual actions not cited.

A questionnaire for action

In order to assess the achievements and shortcomings of the Conservation and Protection Committee of ARARA, to garner opinions on some current issues regarding conservation and protection, and to gain recommendations and guidance for future activities by the committee and others, especially fledgling rock art groups, a questionnaire was developed. This was distributed to ARARA members attending the 1988 symposium business meeting who wished to fill one out, to officers and advisory board members, to a non-random sample of other members not in attendance known to be knowledgeable on rock art conservation/protection issues, and to 23 archaeologists with the United States Department of Agriculture's Forest Service in Arizona and New Mexico. (One ARARA member passed the questionnaire along to this group).

Forty-nine questionnaires were returned, 33 from professionals or rock art authorities (by their own account), the remainder from knowledgeable avocationalists (not all of whom are knowledgeable in rock art matters). This represents approximately a 20% return. The results are quite informative and directional, but biased due to the nature of the non-random sample of rock art devotees and ARARA members, and because of the relatively high number of government archaeologists from one region of the United States (albeit one of the areas with high numbers of rock art sites). Other biases include little representation of corporate archaeologists or managers/administrators, art historians, conservators, Native Americans, specialists outside the United States, and so on. Still, the results are considered informative and directional and will be used cautiously by ARARA - and hopefully others - to formulate future activities. There are 13 questions, 10 of which are yes-no-don't know with room for comment/recommendations, and three which are open-ended.

Question 1

Question 1 asked whether conservation and protection of rock art had been enhanced through ARARA's efforts. Seventy-three percent of respondents thought in the affirmative, eight per cent in the negative, and the rest did not know. Proportionately within each group twice as many professional/authority individuals didn't know compared to others. Comments included positive responses on the committee achievements chronicled earlier. A number of respondents noted that the committee has had a positive role in the recruitment and leadership in preservation, that it has raised the awareness, sensitivity and consciousness of people and agencies regarding research goals and preservation efforts. One individual praised the increasing enlightenment brought to members on safe recording measures. A few noted the increased awareness of conservation and protection of rock art through the survey and recording projects of the organisation. Several others noted that there has been some negative aspects such as large ARARA field trips to sensitive areas, and limited influence of the organisation in problem solving. Some projects were deemed relevant by one respondent, others were listed as questionable in value.

Question 2

The second, related question asked about the shortcomings of ARARA's conservation and protection efforts. While there were a number of 'no comments' or 'none' responses, there were numerous informative statements. One individual made note of the lack of administrative infrastructure and permanent staff. Another person believes ARARA cannot do more than general education. A third sees a need for a budget to initiate protective measures at critical sites. Several individuals commented on an absence of time and people to work on the many problems; the absence of ARARA field projects. Another individual suggests the membership is too concentrated (California and the Southwest). One authority comments on the excessive talking and writing (undoubtedly this paper aside!) with not enough action on the ground. Other respondents want to see more political action, more input to enforcement of antiquity laws, bonding for protection, acquisition, more involvement of Native Americans, more internal participation by members, more information on how to participate and how to correctly record rock art, more marketing and publicity of the organisation, especially to land managers and archaeologists, more public respect, more publicity of problem areas and the need to put out more literature. Several individuals noted that the committee is making strides but that it will take time, and that more membership participation is needed.

Question 3

The next query asked about the shortcomings or achievements to date of Federal and none-Federal governmental efforts directed at the conservation and protection of rock art. Of the 39 people who responded with answers to the question, the main theme was the absence of financial and labour support and planning for inventory, protection and conservation; apathy and ignorance, 'too little, too late', and misplaced priorities, as in tourism and development because of revenue and pressure. One individual noted the lack of consistent policy and direction and differences in policy by and within each agency. Another commented on the weaknesses in Federal and state laws, and the lack of antiquity violation prosecutions. Other responses on the negative side focused on too little attention paid to conservators, conservation training and conservation technology, the absence of technology development and transfer, and the lack of a holistic approach to conservation of sites. Several recent examples of poor government action were cited including paving a road into rock art rich Chaco Canyon in New Mexico where development now threatens thousands of petroglyphs (cf. McDowell 1987: 621), and several other perceived agency miscues.

On the positive side several respondents made note of the recent passage of legislation favourable to cultural site protection, to the increasing recognition among some agencies and archaeologists of the value and fragility of rock art and its rampant vandalism and abuse, and the need for protection, such as with more visitors; the steward program of heritage protection started in many states; the co-operative management between Federal land managing agencies and ARARA; the involvement of amateurs by agencies in survey and recording, several examples of well-developed local and Federal parks with excellent rock art protection, and site stabilisation efforts such as at Council Rocks in Arizona and the Grimes Site in Nevada.

Question 4

Are Native Americans being sufficiently involved in rock art conservation and protection efforts at governmental levels was the fourth question. Only three respondents indicated yes, 25 said no, and the remainder didn't know. Comments varied considerably from encouragement to consternation and were clearly regionally dependent. Several members noted that few efforts had been made to involve Native Americans on sites or this was discouraged, but that they should be actively encouraged and invited to do so. One professional stated that except for Puebloans, it was a tricky proposition. Another respondent sees a trend in this direction but that not all Native people value conservation, that there are cultural preferences for deterioration. One individual stated that 'Native Americans are the least qualified to help in conservation of rock art. They destroy more rock art than they preserve'. In a related answer a respondent suggests that if one considers vandalism the chief evil, then Native Americans would probably be of little help. Another person indicated that 'recorders stab around in the dark and if we had more of the people (Native Americans) that know what's going on ... we could do a better job'. Still another individual proclaimed that research designs and site management plans need knowledgeable Native American review, by tribal elders directly, traditionally connected to sites, that there was too much input from 'professional Indians'. Similarly, a member noted the problem is imitation Indians. 'Use of self-proclaimed and politically-appointed Indians is a travesty'. An interesting comment was that 'archaeologists are not aware of rock art, let alone anyone else in government. How can we get a third party involved?' Others related that generally the rock art is a mystery to Native Americans, but that there is interest and a desire to participate at some level. Finally, one person stated that knowledge, inventory and protection of rock art on tribal lands is worse that on land administered by such agencies as the Bureau of Land Management and Forest Service.

Question 5

A related question, number five, centres on whether there is sufficient use of Native Americans at non-governmental levels, such as in private projects including ARARA members. In total only three indicated yes, all professionals; 19 said no, and 24 didn't know, with no difference between professionals and others.

Many comments were the same or similar to those for the previous question. One professional stated that Indians should be involved when they take the initiative to express an interest, otherwise it is merely political, not scholarly involvement. One person suggested that 'Native Americans are more interested in keeping Anglos out of their land and cultural heritage'. A different individual advises that not enough professionals want to contact Native Americans, 'that some comments by Native Americans four generations removed from their pre-contact days that are getting into the literature by avocationalists have little bearing on what rock art is all about'.

Several respondents pronounced that they were taking elders and children to visit sites, that they were actively involved with Native Americans. Another person expressed a need for the involvement of all concerned parties to achieve effective protection and respect of rock art.

Question 6

Question 6 deals with whether deteriorating Indian rock art should be repainted or repecked by professional conservators and/or Native Americans. Seven of the respondents indicated yes, 34 said no and eight didn't know or did not answer the question. A number of individuals qualified their answer indicating that such a practice by Native Americans should occur only when a direct connection to the tradition still exists (Fig. 3), 'otherwise professionals should work for replication, as in adjacent areas suitable for viewing and interpreting to the general public'. An archaeologist who has worked very closely with conservators indicated no to the question unless reintegration (not 'repainting' as such) can be fully justified by the conservator or if the site is still actively used by Native Americans. One individual thought that public access sites should be repaired.

A number of respondents pointed out the problems that repainting or repecking would have on trace element studies and dating, that there could be misinterpretation. Others emphasised the need to provide stabilisation and more protection at such sites, to let the natural processes take their course or to leave them alone as they are sacred to Native American groups who do not wish them touched.

Question 7

Question 7 has two parts. I asked if public access to rock art sites in the United States was too restrictive on (a) public lands and, (b) private lands.

Answers to both question sub-parts were nearly the same between those in both categories. For public lands only four persons thought access too restrictive, 38 thought not, and seven, all but one professional, didn't know. For private land, nine thought access too restrictive (Fig. 4), 28 did not think so, and 12 didn't know or failed to answer.

Responders noted the unique situation of each site, and variation among land managing agencies. A number of them thought there were not enough restrictions until better education and public awareness and adequate safeguards, such as fencing, explanatory signing, monitoring and even attendants, were in place. Then the public should be encouraged to use it, said one specialist. One individual stressed the need to inform private landowners with sites of the importance of rock art preservation. Another person indicated there should be at least one site protected and available in or close by each metropolitan area, that this would create interest. An example from Painted Rocks State Park near Gila Bend, Arizona was cited. One individual cautioned that excessive restrictions should be avoided, as might occur in wilderness areas (and does occur on some military reservations).

Question 8

The eighth question was specific to ARARA's conservation and protection committee. Should the organisation develop rock art tourist-visitor control guidelines? While one respondent has a problem with the word 'control', 38 of the responses were favourable, only three negative, and eight didn't know or did not respond. A good deal of advice and caution was offered. A negative response centred on the publicity the guidelines would bring to sites and the worsening of the situation. Several individuals wondered if local managers or others would use them. In related comments, one respondent suggested they be limited to public lands where monitoring/control is already available or can be constituted; another thought suggestions would be welcomed by administrators/managers, but that they would need to be realistic. A respondent stated that such guidelines should encourage visiting while indicating how a visit can be made without prejudice to the rock art. One respondent noted that many groups are looking for things like this; another suggested incorporating state preservation specialists (State Historic Preservation Officers). A few who were queried thought ARARA should do it only if the job could be done right utilising experts and research on guidelines from elsewhere, such as Australia and France. One board member stated that ARARA really needs funds to implement a strong policy in this direction, that nothing should be stated if it is too 'Mickey Mouse-ish'. This adviser suggested acquiring a few graduate internships for this effort.

Question 9

Should ARARA participate in rock art conservation/protection research through grant and investment solicitation, cost sharing or other means other than volunteerism was the ninth question. Once again the answer was predominantly yes, by 77% of the respondents. Only two, both professionals, stated no, and eight individuals didn't know or failed to answer.

One ARARA advisory board member noted that this would require individual commitment, that ARARA as an organisation is unlikely to manage these matters well. Another professional noted the potential problems in this direction and believes 'there should be a US Rock Art Institute stimulated into activity and funded (by foundation and Federal support) using ARARA initiative'. Several individuals wondered about the fund raising flexibility, who would do the research, administer the program, provide review and management? Another person indicated that fund-raising should be set up through an endowment committee.

On the other side of the coin, one member said we should be involved certainly as consultants but this person was uncertain whether ARARA should be involved in fund-raising. One conservator thought it worthwhile 'because it would help emphasise to the membership how fundamental conservation and protection is to the study of rock art'. Another responder noted that the only way to have in-depth research by professionals is through grants. A related statement by another member was that 'the (lack of) cash helps keep a lot of qualified people from getting involved'. One member thought that any method of building protection and public awareness should be pursued.

Question 10

Question 10 stated: Should ARARA publish or reproduce for limited distribution a bibliography of major works related to rock art conservation/protection? Once again the vast majority of respondents (81%) indicated yes, with only one no. While this is presently under completion by ARARA in terms of a computerised listing, its final form is undecided. Many constructive comments were received. Advice included: make it annotative; include American and non-American works; limit it for a few key sites; do not limit its distribution; and keep it short. One person doubted if it was practical; another thought it would be a key contribution but again it would depend on individuals to do the work. A responder suggested that State and Federal agencies be targeted first, then the public. One individual lamented on the cost. An added person thought such a work would help publicise ARARA and its efforts. An archaeologist cautioned that it should avoid disseminating information on 'how to' techniques to non-qualified personnel. One professional thought that rather than a bibliography it should be a good, short, easily read brochure on the subject for wide distribution. A recent work by Rosenfeld (1985) was suggested as a good example to follow.

Question 11

A split vote was tabulated for Question 11, whether ARARA should adopt an official policy to defer rock art site information requests to appropriate land management agencies. Of the professional/authority individuals, eight indicated yes, 10 said no, seven didn't know, and eight did not respond. Other individuals were more negative, with three indicating yes, ten saying no, and three who were unsure or didn't answer.

Several members who answered thought this should be handled on a case by case basis, that good judgement and flexibility (by either agency or ARARA) is needed in providing information. One advisory member indicated that if people come to ARARA for information they should be responsible enough to get it. One thought the question sounded like a threat to jurisdiction by a private organisation. Another stated that ARARA should leave legal management concerns to appropriate agencies since they are most familiar with particular site concerns. One person asked if land management agencies are always prepared to offer this information? Another respondent stated that government agencies in many cases lack knowledge concerning protection etc. of rock art. 'We can certainly lend our support and concern, offer guidelines, but that it is up to the appropriate agency to handle a problem'. One person thought we should refer individuals to the UCLA Rock Art Archive. Finally, one respondent wrote that education is the key as 'I'm sure some agencies don't know what rock art is'.

Question 12

The twelfth question is an opinion request regarding whether ARARA should become involved financially or otherwise with such American groups as The Archaeology Conservancy and Trust for Public Lands, to place threatened rock art sites under more protection. Almost everyone who answered (81%) indicated yes; there were four negative responses and four who didn't know.

A group of individuals qualified their answers with statements such as 'If the money is there', 'But how possible?', and 'But cautiously and after lots of conservation'. On the negative side, one person stated that ARARA doesn't have the resources. There should be volunteer work and publicity help (from ARARA). In a similar vein, an adviser stated flatly that 'ARARA does not have the managerial or fiscal resources to take on this responsibility'. Others suggested ARARA should seek co-operation of those organisations where their help can be useful. One individual stated 'Insofar as we may make recommendations to them only'. A respondent noted that 'The Archaeological Conservancy has already helped protect rock art sites'. But another who answered the query and is a member of the organisation wrote that he had seen internal policy problems, such as their acquisition of valuable (irreplaceable) archaeological resources and then using them as 'trading stock' rather than conserving them. One person suggested that we should also encourage site documentation as soon as possible, while another individual proclaimed that 'the more affiliations the greater the strength we can present as our goals are similar'.

Question 13

The last question is open-ended, and generally solicited any further comments, criticisms, recommendations, guidance, etc. regarding ARARA's conservation and protection efforts toward rock art sites and their environs. Twenty-six individuals provided no further comment, 17 in the professional/authority category. A number of respondents made a plea for the complete recording of all sites, that more members need to be involved in recording projects.

Of the varied comments received, the following are judged by the author to be most cogent. One person stressed that ARARA speaks from an emotional rather than educated viewpoint. This individual wished to see a professional versus avocational certification process. Several responders wished to see more educational involvement and increased communication internally, as with local chapters of ARARA (e.g. through a section of *La Pintura*). Another member stated that 'Native Americans have done more real damage than the general public to rock art. They have erased or defaced rock art so it can't be reconstructed. Public agencies and road builders are also guilty'. Furthermore, 'general public education is the best bet'. On other notes from reviewers, one specialist stated that there is a need to disseminate conservation-protection action information on specific areas widely and rapidly to the membership, news media and other conservation groups. One person wanted to see ARARA develop recommendations and guidelines for property owners with sites. The answers from Forest Service archaeologists indicated a need for increasing ARARA's visibility among professionals, especially those charged with managing rock art, and seeking their rock art needs.

Two comments from advisory board members are particularly thought-provoking. One indicated that 'education and research must be interrelated, that ARARA (by its efforts and publicity) must be given greater academic visibility ... if a constructive and dynamic role for ARARA is to be achieved'. The other adviser stated that 'It is very difficult to launch complex and time-consuming projects and expect them to be carried forward successfully using volunteer labour of the kind available to ARARA'. This individual suggested lining up the small number of educated individuals first before promising results, that there are questions regarding the realistic fulfilment of commitments.

Conclusion

Some concluding observations and action items are offered with this background of past ARARA conservation and protection operations and revealing tabulations and comments from respondents on a questionnaire. These are appropriate for both ARARA and the rock art community at large in order to understand some of the problems and possible solutions to the conservation and protection of rock art.

Rock art conservation and protection is on the upswing in North America, but it would appear that the balance is still tipped toward the loss side. This is due not only to inadequate protection and education but to population shifts and increases. More and more Americans are moving at least for part of the year to the southern sectors of the United States, to the so-called Sun Belt, and populations are rapidly increasing in many western cities, zones of concentrated rock art. There is also considerable developmental pressures, more uncaring people concomitant with more people, more leisure time, and increasing access to sites and areas coupled with continued development and utilisation of off-highway vehicles.

A number of individuals and groups, North American and international, have offered their observations and recommendations for rock art research, protection and conservation, to lessen resource stress (cf. Van Tilburg and Meighan 1981; and Anati 1983). These are quite relevant to ARARA's conservation and protection work. Wainwright (1985: 23) has stated for Canada that 'Successful rock art conservation programmes have usually been co-operative, involving provincial archaeologists and ministries responsible for cultural resource (sic) management, Indian Bands and other concerned parties including ARARA associates, universities, museums and private industry'. He has stated further (Wainwright 1985: 34) that 'The success of rock art research and conservation has never been a question simply of allocating funds from government revenues. Rather, it has depended on the enthusiasm, the interest, the expertise and the dedication of individuals'. These comments are echoed here. It is also encouraging to see the Getty Conservation Institute training qualified conservators in rock art conservation methods.

It is clear that ARARA has had some successes, but that there are deficiencies, mostly the result of its inherent structure, not its members. For ARARA to increase its effectiveness in rock art conservation and protection and as general guidance to more efficiency in grass roots efforts throughout the work, the following summaries of potential action items (as each situation warrants) are offered.

1. Establish a sound administrative infrastructure; delegate duties to competent volunteers or workers.
2. Ensure significant professional involvement and regional representation, commitment and communication.
3. Publicise aims and achievements through the media and other effective communication channels; market the rock art organisation and its aims and directions.
4. Develop or utilise well-researched visitor guidelines, co-operatively if possible with owners/managers.
5. Encourage agencies and individuals to limit access to sensitive, unprotected sites.
6. Explore grant and investment solicitation possibilities by special committee or sub-committee. If fruitful, establish sound means of professionally managing assets and resulting undertakings.
7. Establish an emergency conservation/protection fund.
8. Encourage and support better curation of archival materials.
9. Develop through volunteers a select international annotated rock art conservation and protection bibliography. Provide appropriate cautionary statements regarding its use.
10. Produce a short, easy to read brochure on conservation-protection of rock art for wide distribution.
11. Develop slide programs, videos, study kits, etc. geared toward both school children and adults on rock art values and conservation and protection. Participate in heritage protection programs.
12. Handle information requests regarding rock art locations on a case by case basis through professional judgement. Where there are doubts offer only referrals to administrative agencies and clearing houses of site information.
13. Maintain co-operation or affiliation with groups and agencies responsible or concerned with land management and/or cultural resource study and preservation.
14. Pursue rock art protection and conservation on the political front, seeking better enforcement of existing laws and regulations, or the legislation of enforcible laws.
15. Strive to bring rock art studies into the mainstream of academia.
16. Actively undertake organisation projects and enter into co-operative agreements for rock art work with government agencies and private corporations. Work in concert with academia where feasible.
17. Maximise internal and external communication through various educational and informational channels. Maintain close ties and interchange with sister groups.
18. Consider traditional Native American (or other ethnic) beliefs and practices when undertaking projects or encountering re-pecked or re-painted rock art, and consult with knowledgeable traditionalists and elders with perceived or known cultural ties to the rock art.
19. Work within organisational and individual limits.
20. Support a National Centre for Preservation Technology or similar facility.

ARARA is a critical cog in the American wheel of rock art conservation and protection. It is a volunteer group of individuals with various persuasions and opinions, with some shortcomings. Yet ARARA's efforts at improving the understanding and state of rock art in North America and elsewhere should not be overlooked. When coupled with the many increasingly co-ordinated efforts of individuals, groups, corporations and agencies, the waves of deterioration and loss of these heritage values are being diminished. No small task lies ahead but no one should question the value of conservation and protection efforts. Rock art brings from the past a small measure of hope for the present and future, and promise in its information, for what this can contribute to world understanding.

Acknowledgments

I wish to express my gratitude to Helen Crotty, A. J. and Frank Bock and Pat Ritter for review and comment on this paper. My gratitude is also extended to those who took the time to answer the questionnaire. This paper remains the author's responsibility, and the opinions, recommendations, and conclusions are not necessarily those of the American Rock Art Research Association.

REFERENCES

ANATI, E. (ed.) 1983. Preservation and presentation of rock art 1981-83. *Studies and documents on the cultural heritage 1*, UNESCO and Centro Camuno Di Studi Preistorici, Brescia, Italy.

BOCK, F. and A. J. BOCK 1988. ARARA research team aids Petrified Forest in rock art survey. *La Pintura* XV (1 & 2): 6.

CONNOR, L., R. DINGUS, S. FITCH, J. PFAHL and C. ROITZ 1987. *Marks in place: Contemporary responses to rock art.* The University of New Mexico Press, Albuquerque.

LEE, G. 1987. Conservation of petroglyphs on Isla de Pascua. Paper presented at the World Monuments Fund Meeting, Santiago.

LEE, G. 1988. Petroglyphs of Lana'i, Hawai'i. *La Pintura* XV (1 & 2): 10, 12.

McDOWELL, B, 1987. New Mexico: Between frontier and future. *National Geographic* 172(5): 602-33.

NOXON, J. S. and D. A. MARCUS 1986. Waste to create waste? Potential impact of high-level nuclear waste on rock art in Canyonlands National Park, southeastern Utah. In W. D. Hyder, H. Crotty, K. Sanger and F. Bock (eds), *American Indian Rock Art* 11: 19-32. American Rock Art Research Association, El Toro, California.

ROSENFELD, A. 1985. *Rock Art Conservation in Australia.* Australian Heritage Commission, Canberra.

US Congress, Office of Technology Assessment 1986. *Technologies for prehistoric and historic preservation.* United States Government Printing Office, Washington. DC.

VAN TILBURG, J. and C. W. MEIGHAN (eds) 1981. Prehistoric Indian rock art: Issues and concerns. *UCLA Institute of Archaeology Monograph* XIX. Los Angeles.

WAINWRIGHT, I. N. M. 1985. Rock art conservation research in Canada. *Bollettino del Centro Camuno di Studi Preistorici* 22: 15-46.

WELLMANN, K. F. 1977. From the President. *La Pintura* 3(4): 3.

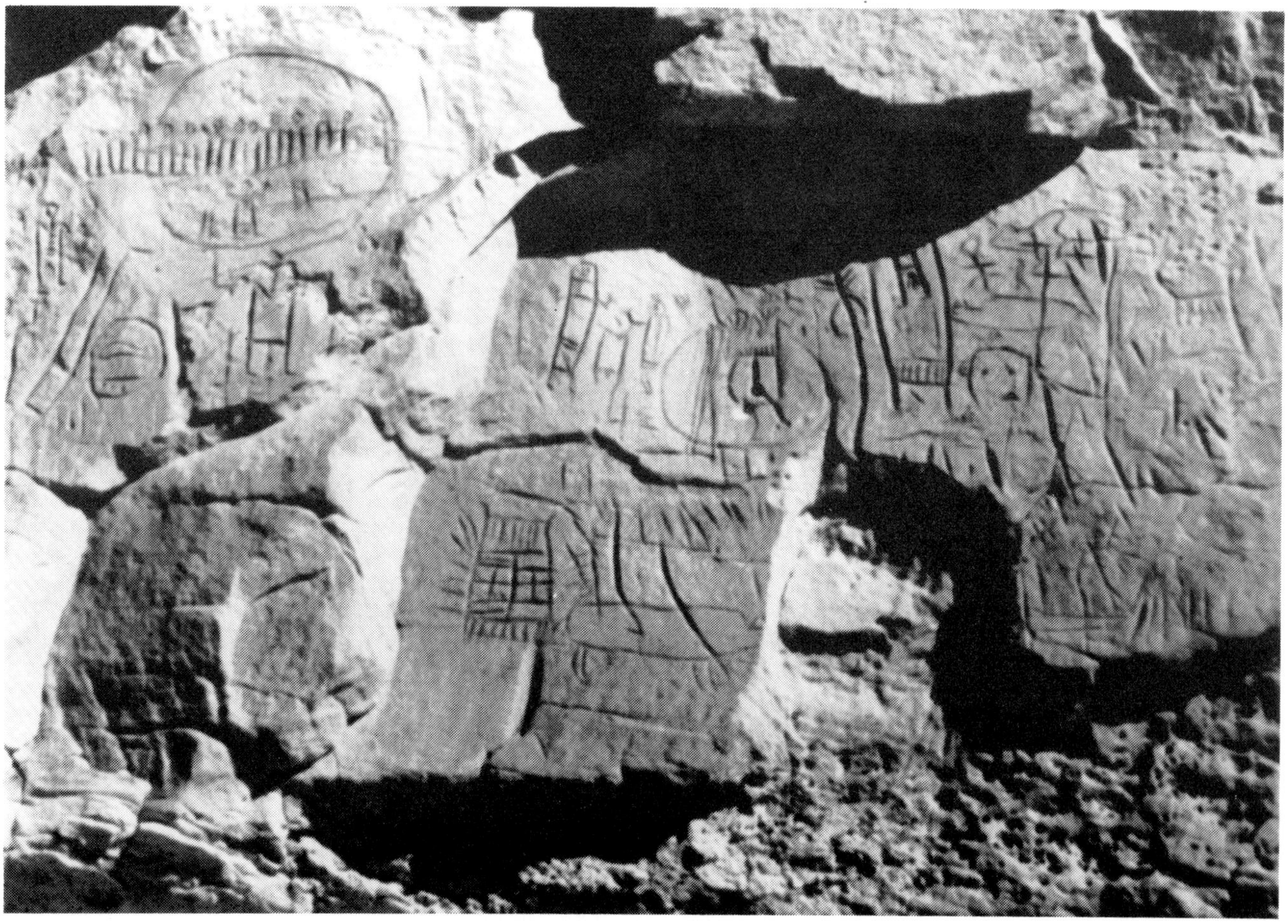

Figure 1. *Portion of petroglyph panel chiselled away at the Castle Garden site in Wyoming. Photo taken in 1961.*

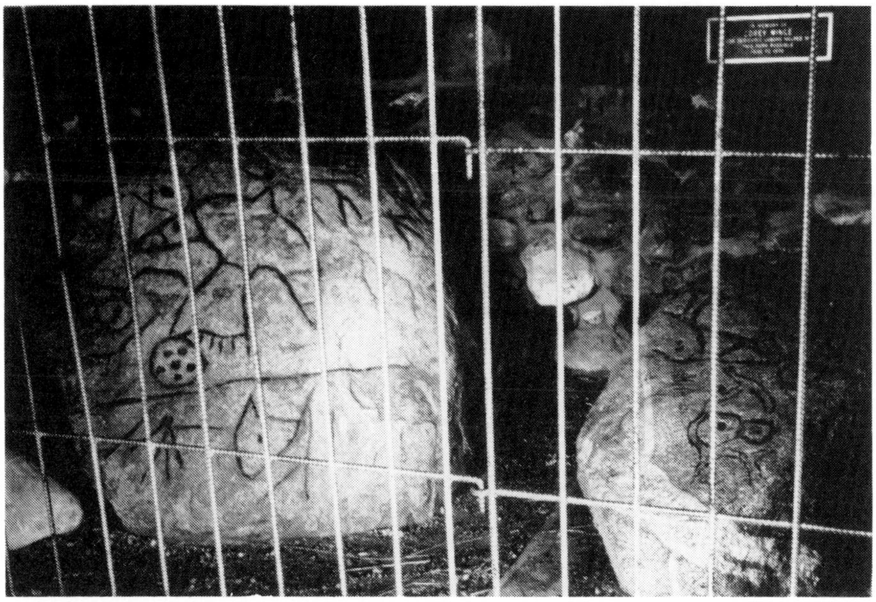

Figure 2. 'Jailed' petroglyphs at Writing Rock, North Dakota (1971). Photo illustrates drastic measures taken by local citizens in an attempt at protecting rock art while still allowing visitation.

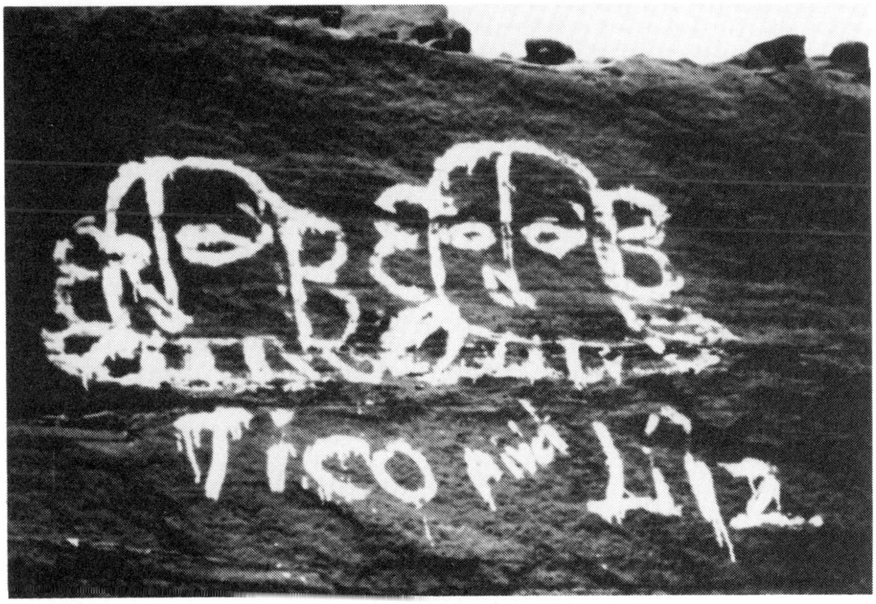

Figure 3. Modern (1963) Hopi pictograph from Oraibi, Arizona incorporating traditional (katchina mask) symbolism with a love duo statement in English.

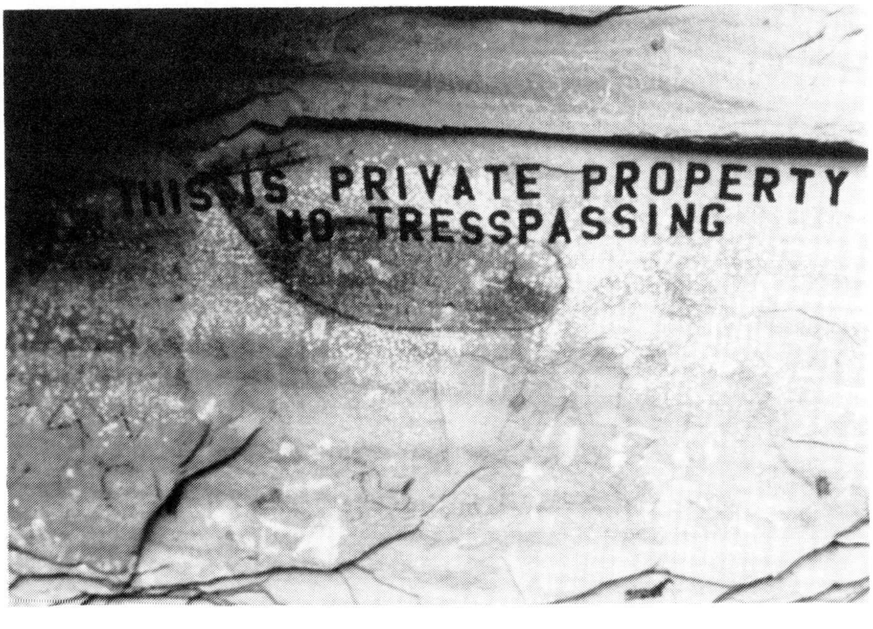

Figure 4. Red pictograph from Rasmussen's Cave at Nine Mile Canyon in Utah defaced by private landowner's statement regarding entry rights.

3

LA GROTTE DES COMBARELLES I AUX EYZIES;

Aménagement et contact direct du public avec les oeuvres pariétales

P. Vidal, J. Vouvé et J. Brunet

Abstract. Combarelles Cave is open to the public. The idea of tourist access to such places of cultural interest must be reconciled with their upkeep. With this in mind, Combarelles Cave was fitted out to facilitate the visits of tourists. This also meant organising the course of the visits, taking into account specific aspects (climatological, morphological etc.) of the decorated gallery. The site's future depends on finding an acceptable compromise between visits and preservation needs.

Introduction

La valeur archéologique et esthétique des gravures de la grotte des Combarelles aux Eyzies a justifié l'action de l'administration du Ministère de la Culture dès 1972 pour améliorer d'une part, les conditions de visite de la cavité et d'autre part, les conditions de conservation des oeuvres pariétales. Aux dispositions correspondant à une réfection des installations techniques, se sont ajoutées des recommandations concernant la fréquentation obtenues après l'analyse de mesures recueillies en période touristique et sur plusieurs années.

1) Travaux d'aménagement: récapitulation

Dès le début du siècle (précisons que la découverte et la reconnaissance des figurations préhistoriques datent de septembre 1901), le sol de l'entrée et de la galerie qui sera dénommée plus tard Combarelles I, fut abaissée pour faciliter la venue des spécialistes. A cette initiative, se sont ajoutées des améliorations du circuit de visite qui se sont poursuivies jusqu'en 1975, date à laquelle, la cavité réaménagée sous l'autorité du conservateur des Monuments Historiques peut être réouverte au public. Les différents aspects de ces travaux peuvent se récapituler ainsi (Sarradet 1979).

La description que firent les premières personnes à parcourir la galerie justifie les travaux mentionnés, relevons au passage

> '... à cette époque, aucun aménagement ne facilitait l'ascension des dénivellations de l'entrée; les premiers mètres étaient encombrés des débris du plancher stalagmitique ancien effondré; en maint endroit, le remplissage de graviers et les gours superposés, souvent élevés, rendaient la circulation pénible, et nécessitaient de cheminer courbés en deux ou sur les genoux, en garant sa tête des stalactites pointues et des angles rocheux retombant ...'.
> '... sa hauteur est très variable: de près de 4 m a l'entrée, elle tombe bientôt à 2 m, pour se maintenir ensuite longtemps à une moyenne de 1.60 m à 1.75 m sur une largeur de 1.50 m à 2 m. Des coudes à angles presque droits rompent la monotonie de ce long couloir ...' (Breuil 1952).

Jusqu'en 1972, la grotte des Combarelles I était, par sa morphologie, inondable pendant une partie de l'année; le sol présentait de nombreuses mares alimentées par les infiltrations qui se manifestaient essentiellement de l'automne au printemps. L'accumulation de ces eaux était donc facilitée par le profil irrégulier du sol; une partie de l'année, une humidité néfaste régnait dans la cavité. Par ailleurs, la cavité était souvent inaccessible au printemps, et les gardiens tant bien que mal évacuaient cette eau par va et vient de seaux ou par pompage. Tout cela n'était donc pas sans risques et inconvénients pour les parois en raison des dimensions réduites de la section des galeries. L'amélioration des conditions de visites demandait la mise en oeuvre d'un drain, la réfection du réseau électrique et le nettoyage général de la cavité.

La Figure 1 qui est une coupe longitudinale de l'entrée et des premiers mètres de la galerie illustre l'importance de l'abaissement des sols qui atteint 2 m au maximum à l'entrée et 1 m à 0.60 m vers le fond de la galerie. Mentionnons que dans certaines parties pour rendre la galerie accessible, il a fallu descendre jusqu'au niveau des calcaires dans laquelle la grotte s'est creusée. La pose d'un réseau de drainage dans le sol de la galerie (Fig. 2) a facilité l'écoulement des eaux d'infiltrations. La tranchée sert aussi pour le passage des câbles électriques. Le sol après tous ces aménagements a été comblé de galets et recouvert de panneaux métalliques en treillis.

Les parois ont pu être débarassées des gaines électriques et des interrupteurs; quelques uns avaient à l'époque été maladroitement scellés à proximité ou dans les zones à gravures au grand préjudice des figures préhistoriques (Photo 1, 2).

L'éclairage réaménagé, se fait sous tension de 24 volts et est commandé par une série de commutateurs échelonnés au long de la galerie. Deux modes d'éclairage existent donc, l'un comprend des lampes placées au ras du sol pour le cheminement des visiteurs, l'autre est mobile ('lampes balladeuses') et sert à révéler les gravures.

Au cours des années, les infiltrations d'eau avaient laissé des dépôts de limon sur les parois. De légers brossages suivis de lavages ont assuré l'élimination de ces dépôts. Ils ont permis de retrouver certains tracés décrits par les découvreurs (Breuil 1952) et depuis lors masqués sous un voile de calcite. Ce travail a notablement amélioré l'aspect général de la cavité.

2) Entretien

Les microorganismes peuvent trouver dans les grottes des conditions favorables à leur développenment; en effet, l'air est humide par suite d'infiltrations d'eau, des mécanismes de condensation et aussi par la vapeur d'eau dégagée par la respiration des visiteurs (il faut rappeler que chaque personne rejette en moyenne 40 g de vapeur d'eau par heure). Des quantités importantes de matières organiques sont également apportées par l'haleine, la sueur (Brunet et Vidal 1987; Quintos et al. 1987).

Si ces effets sont minimes au niveau de chaque individu, ils deviennent importants par accumulation à l'issue de la saison touristique.

En effet, les allées et venues de visiteurs contrarient ou accélèrent les courants d'air, facilitant ainsi le transport de spores végétales, de débris organiques qui, nous l'avons vu, peuvent favoriser le développement des microorganismes. L'essaimage de ces colonies sur les parois décorées est préjudiciable à la bonne conservation des oeuvres rupestres et il est nécessaire d'assurer un entretien régulier.

Jusqu'a ces dernièes années, dans le cadre de la Conservation des Monuments Historiques d'Aquitaine, un programme d'entretien a été mené et la cavité a fait l'objet de nettoyage et de pulvérisation de solutions biocides. Les résultats de ce traitement sont satisfaisants et il est très important que ce type d'action soit assuré chaque année.

L'entretien est un point à ne pas sous-estimé. Cette année encore (1988) grâce à l'appui des services des Bâtiments de France, la réfection des panneaux de treillis métalliques a pu être assurée. En effet, ceux-ci étaient exposés en permanence à l'humidité, la rouille avait produit de nombreux dégâts qu'il fallait réparer rapidement.

3) La cavité et les variations thermiques liées à la fréquentation

Depuis 1976 des mesures de température ont été effectuées à l'aide de thermomètres au mercure au 1/20° de ^{o}C de précision, disposés en permanence dans l'air, la roche et le sol et répartis en stations de l'entrée au fond de la galerie (Fig. 3).

a) Analyse de mesures en période touristique (stations A, B, C, D, E, H, I, K, L)

La comparaison des mesures faites à différentes reprises matin et soir - donc avant et après les visites - pour un même jour, montre l'élévation de température de la roche en liaison avec celle de l'air. Cette variation positive est fonction de la localisation des points de mesures, de la durée de présence des groupes à proximité des stations de thermomètres et du nombre de visiteurs de la journée (cf. Tableaux 1, 2, 3, 4).

La comparaison des mesures du matin et de la veille au soir, montre une diminution de l'élévation des températures citée précédemment (Tableau 1 - exemple: 24/8/87 : 9 heures et 23/8/87 : 18 heures).

La comparaison des mesures du soir (la veille du jour de fermeture) et du matin de réouverture (Tableau 1), met en évidence une baisse des températures. En valeur absolue, elle est supérieure à la différence observée entre le soir et le matin de deux jours consécutifs.

A partir de l'examen des variations journalières, nous pouvons différencier trois sections dans la galerie (Fig. 3):

- de l'entrée jusqu'au-delà de la station D - soit sur 100 m de long - les échanges naturels d'air se font facilement ce qui permet l'élimination des calories dûes aux visiteurs
- de la station D à l'escalier (station F) - soit sur environ une cinquantaine de m de long - une zone de transition
- de l'escalier au fond de la partie visitée - soit sur 45 m de long - une zone de confinement se dessine; nous y notons l'élévation de la température diurne entre mesures du matin et du soir et une tendance au rééquilibrage thermique durant la nuit. Ce rééquilibrage thermique est amplifié au cours des 36 heures consécutives de fermeture hebdomadaire.

Les valeurs des températures au niveau de la roche et de l'air, toutes très proches le matin avant le passage des visiteurs, dans les differentes stations de la 3éme section, soulignent que la surface pariétale de cette partie du réseau est plus soumise aux perturbations du volume d'air qu'à celles du contexte géologique et hydrogéologique.

b) Analyse d'ensemble des mesures thermiques

Une analyse graphique d'ensemble a pu être esquissée à partir de la restitution des mesures faites chaque semaine après le jour de fermeture, dans la zone à forte concentration de gravures (partie terminale de la galerie visitée). La courbe No. 1 correspondant aux températures de la roche (cf. Planche 1) est représentative de la section non influencée ou peu influencée par le va et vient des visiteurs. Là, en effet, les visiteurs progressent rapidement, le temps de présence est de 15 minutes avec de rares arrêts. Remarquons au passage sur un plan général que le découpage par zone de la galerie des Combarelles I ne coincide pas tout à fait d'une part, lors de l'interprétation des mesures faites en période touristique et donc naturellement influencées par les visiteurs et d'autre part lors de l'analyse d'ensemble. On notera aussi une fréquentation quasi-nulle de la cavité en saison hivernale de 1977 à 1981, ce qui se traduit pour les températures par un palier. Le détail de la fréquentation est fourni par la série d'histogrammes mensuels du nombre de visiteurs, placée en vis à vis des diagrammes de l'évolution thermique et par les Tableaux 3 et 4.

La courbe No. 2 est l'expression des températures de la roche des stations de la partie décorée où séjournent les visiteurs (35 mn) (stations I, J et K). Le détail des mesures est fourni par la figure B - Planche 1 - qui souligne le regroupement des mesures à un même moment.

La courbe No. 3 concerne un seul point de mesure (station L) situé au fond de la partie visitée dans la roche à la voûte; du fait de sa localisation (fin de visite, à côté de la partie non aménagée et du réseau profond) ce point de mesure traduit par la courbe No. 3 une amplitude thermique plus restreinte que les courbes No. 1 et No. 2 mentionnées.

4) Fréquentation touristique

L'étude de la fréquentation de la cavité porte sur 7 années de 1979 à 1986; elle permet de suivre l'évolution et l'augmentation du nombre de visiteurs qui va de pair avec l'augmentation des gardiens.

Globalement, la fréquentation touristique nous apparaît trop élevée au cours du mois de juillet et août (Tableau 5); les moyennes journalières par mois traduisent bien ce phénomène. A titre indicatif, mentionnons les valeurs de cette moyenne pour juillet à août, soit 239 et 254 pour l'année 1983, 186 et 247 pour l'année 1984 (Tableau 6).

La baisse de fréquentation de juillet et août 1985 ne correspond pas à la désaffection des visiteurs, mais à l'utilisation différente du personnel. Il y avait cette année là, deux personnes pour l'encadrement de chaque groupe. De meilleures conditions de visites de la cavité se traduisaient par une moindre fréquentation, ce n'était pas sans répercussions financières. En 1986, où il n'y avait plus qu'un responsable par groupe, la fréquentation est de nouveau en hausse.

L'intérêt d'un grand nombre de touristes pour ces manifestations esthétiques datant de la préhistoire, peut faire redouter avec juste raison, des effets néfastes pour la conservation des oeuvres pariétales. Ce sujet de préoccupation n'est pas l'apanage des cavités françaises en général et des Combarelles en particulier; et existe aussi pour les sites d'art rupestre de plein air du Sahara et les parcs nationaux australiens (Gale 1985; Gale and Jacobs 1987; Flood 1986; Dragovich 1986; Lorblanchet 1986). Pour les cavités françaises, comme pour les autres monuments (églises, chateaux ...) (Saint Victor 1987; Vouvé 1987) il faudra parler à l'avenir de gestion du patrimoine, de 'management'.

5) Conclusions: Intérêt d'une limitation des visites et propositions

La fréquentation journalière de la cavité sur les bases de 140 à 150 personnes par jour (valeurs du 23 et 27/8/87) provoque des élévations diurnes des températures de la roche évoluant entre 0.6^{o} et 0.8^{o}C dans la partie ornée et confinée de la cavité des Combarelles I. A cette élévation de température s'ajoutent un accroissement de l'hygrométrie de l'air (les valeurs mesurées se rapprochent de la saturation) de la concentration en gaz carbonique et de la contamination (cf. entretiens, Ch. 2). D'où l'intérêt de restreindre la fréquentation estivale pour diminuer les possibilités de déstabilisation de l'interface air-roche qui joue un rôle actif dans les phénomènes de condensation, dissolution, concrétionnement ...

En prenant la saison estivale de l'année 1977 comme référence pratique concrète (juillet 6050 visiteurs et août 5400 visiteurs), nous remarquons l'élévation rapide et permanente des températures sur la période concernée. La référence prévisionnelle de la fréquentation que nous proposons serait ramenée à 3500 personnes par mois au maximum, soit environ 133 personnes par jour.

Aussi compte tenu de notre connaissance de la grotte des Combarelles I, et des observations faites en période estivale, il convient de recommander pour chaque saison touristique dans le même esprit que pour Font de Gaume (Brunet et Vidal 1984)

- le départ régulier toutes les 20 minutes d'un groupe de visiteurs
- la durée de visite de l'ordre de 40 minutes
- la présence simultanée de trois groupes au maximum dans la cavité
- l'effectif de groupe ne dépassant pas 7.

Ces aménagements de la fréquentation de la cavité correspondent à 19 groupes au maximum par jour, dont 8 le matin et 11 l'après-midi. Ils doivent se traduire par l'amélioration des conditions de conservation. Si l'amélioration était insuffisante, il faudrait restreindre encore plus la fréquentation.

Remerciements.

Les auteurs remercient Monsieur et Madame Claude Archambeau, guides de la grotte des Combarelles I, de l'aide apportée pour cette étude, et Madame Gisèle Tudela.

REFERENCES

BREUIL, H. 1952. Les Combarelles. In *400 siècles d'art pariétal*. Editions Max Fourny Art et Industrie, pp. 91-106.

BRUNET, J. et P. VIDAL 1984. Le problème du seuil de fréquentation. In *Actes des colloques de la Direction du Patrimoine. L'art pariétal paléolithique*, pp. 229-39. Le Thot, Périgueux.

BRUNET, J. et P. VIDAL 1987. Les cavités préhistoriques: conservation et accès du public; remarques. In *Actes des 3éme Rencontres Internationales pour la protection du patrimoine culturel*, Avignon, pp. 97-112.

DRAGOVICH, D. 1986. A plague of locusts or manna from heaven? Tourists and conservation of cave art in southern France. *Rock Art Research* 3: 140-7.

FLOOD, J. 1986. Visitors to Aboriginal sites: Access, control and management. *Heritage Newsletter* 7(2): 3-6.

GALE, F. 1985. Monitoring visitor behaviour at rock art sites. *Rock Art Research* 2: 112-8.

GALE, F. et J. JACOBS 1987. *Tourists and the National Estate*. Special Australian Heritage Publication Series No. 6. Australian Government Publishing Service, Canberra.

LORBLANCHET, M. 1986. Comment on D. Dragovich, 'A plague of locusts or manna from heaven?' *Rock Art Research* 3: 152-5.

QUINTOS, L. S., A. BONET, N. DIAZ-CANEJA, P. L. FERNANDEZ, I. GUTTIEREZ, J. R. SOLANA, J. SOHO et D. VILLAR 1987. Studies of the environmental variables affecting the national preservation of the Altamira cave paintings located at Santillana del Mar, Spain. *Atmospheric Environment* 21: 551-60.

SAINT VICTOR, B. de, 1987. L'accueil du public dans les grands monuments nationaux. In *Actes des 3éme Rencontres Internationales pour la protection du patrimoine culturel*, Avignon, pp. 131-9.

SARRADET, M. 1979. Travaux de conservation dans la grotte des Combarelles (Les Eyzies). In *Publicacions del Patronato de las cuevas prehistoricas de la provincia de Santander*, pp. 135-47.

VOUVE, J. 1987. Essai sur les risques touristiques et de société face à la surconsommation culturelle, cas de Lascaux. Lascaux II. In *Actes des 3éme Rencontres Internationales pour la protection du patrimoine culturel*, Avignon, pp. 81-96.

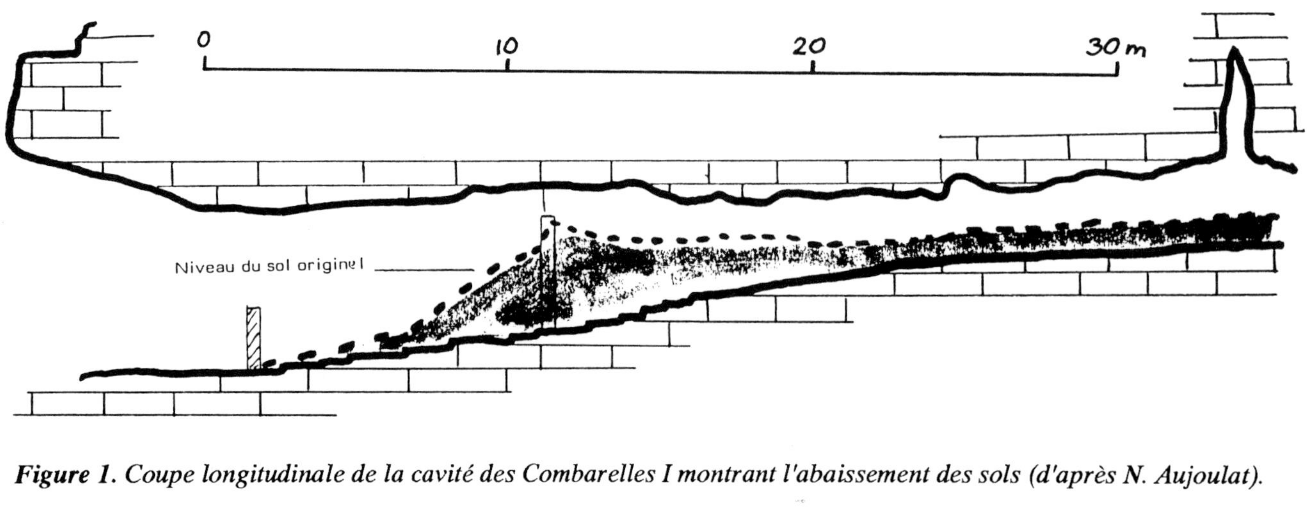

***Figure 1.** Coupe longitudinale de la cavité des Combarelles I montrant l'abaissement des sols (d'après N. Aujoulat).*

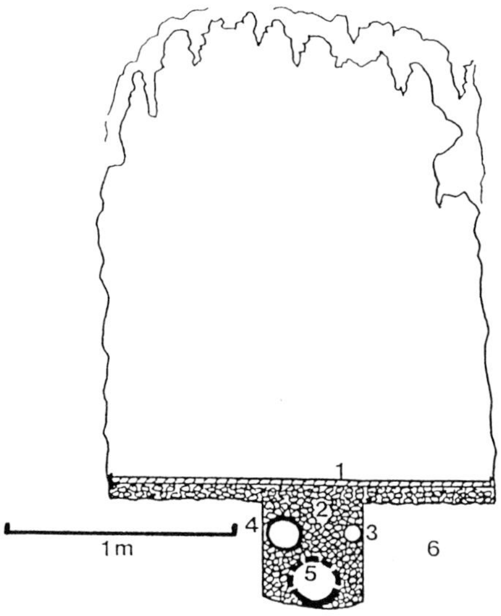

Figure 2.

Section sur la galerie de Combarelles

1. Cheminement en "métal déployé" sur chassis cornières
2. Remplissage de galets de rivière
3. Canalisation électrique
4. Buse en attente (extraction du CO_2, ou autres usages)
5. Drain pour l'évacuation des eaux d'infiltration
6. Rocher en place

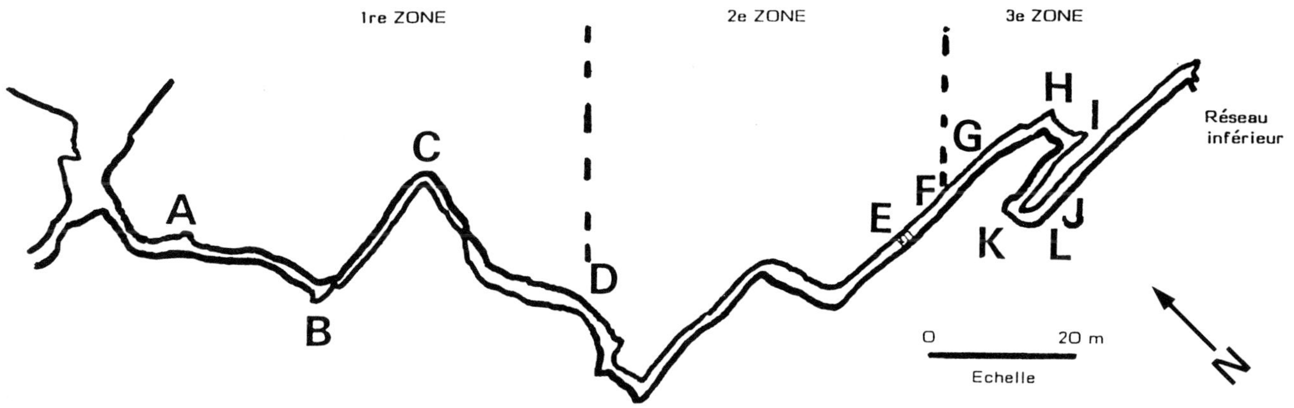

Figure 3*. Plan de la cavité des Combarelles I (d'après N. Aujoulat). Localisation des stations de mesure des températures.*

DATE DES MESURES	STATIONS A	B	C	D	E	F	G	H	I	Nombre de visiteurs
9/8/86 9 h	12,85	11,92	11,30	11,51	11,68	12,70	13,04	12,93	12,53	160
9/8/86 18 h	13,17	11,95	11,32	11,56	11,98	13,28	13,50	13,88	12,74	
10/8/86 9 h	13,18	11,97	11,28	11,52	11,70	12,75	13,12	12,98	12,56	138
10/8/86 18 h	13,60	12,04	11,36	11,60	12,02	13,30	13,56	13,83	12,75	
23/9/87 9 h	14,12	12,32	11,48	11,41	11,43	12,45	12,63	12,52	12,32	148
23/9/87 18 h	14,10	12,32	11,60	11,45	11,65	12,88	13,08	13,20	12,43	
24/8/87 9 h	13,82	12,26	11,48	11,36	11,44	12,42	12,72	12,52	12,32	112
24/8/87 18 h	13,76	12,25	11,62	11,44	11,66	12,82	13,02	13,15	12,42	
26/8/87 9 h	13,20	12,08	11,42	11,34	11?38	12,26	12,52	12,32	12,26	

Tableau 1*. Mesures thermiques de la roche à différentes stations.*

	A	B	C	D	E	F	G	H	I
DT_R1	+ 0,32	+ 0,03	+ 0,02	+ 0,05	+ 0,30	+ 0,58	+ 0,46	0,95	0,21
DT_R2	- 0,02	0,0	+ 0,12	+ 0,04	+ 0,22	0,43	+ 0,65	0,68	0,11
DT_R3	- 0,28	- 0,06	- 0,12	- 0,09	- 0,21	- 0,46	- 0,36	- 0,68	- 0,09
DT_R4	- 0,56	- 0,17	- 0,20	- 0,10	- 0,28	- 0,56	- 0,50	- 0,83	- 0,16

DT_R1 : Différences de températures entre 9 h et 18 h le même jour (9/8/86)

DT_R2 : Différences de températures entre 9 h et 18 h le même jour (23/8/87)

DT_R3 : Différences de températures entre 18 h et 9 h deux jours consécutifs (23/8 et 24/8)

DT_R4 : Différences de températures entre 18 h (veille de fermeture) et 9 h (lendemain de fermeture)

***Tableau 2.** Variations thermiques de la roche.*

DATES DES MESURES	STATIONS								
	A	B	C	D	E	F	G	H	I
9/8/86 9 h	12,94	12,06	11,49	11,48	12,52	12,89	12,83	12,88	12,56
9/8/86 18 h	14,04	12,25	11,80	11,52	13,25	14,91	14,32	14,70	13,56
10/8/86 9 h	13,60	12,20	11,55	11,38	12,12	12,95	12,92	12,98	12,61
10/8/86 18 h	14,61	12,32	11,88	11,65	13,47	14,08	14,40	14,22	13,55
23/8/87 9 h	14,82	12,62	11,90	11,40	12,68	13,09	12,60	12,82	12,40
23/8/87 18 h	14,80	12,80	12,21	11,61	12,76	13,50	13,40	13,12	12,80
24/8/87 9 h	14,13	12,52	11,90	11?30	11,96	12,58	12,56	12,57	12,26
24/8/87 18 h	14,20	12,55	12,18	11,44	12,82	13,64	13,55	14,20	13,01
26/8/87 9 h	13,24	12,23	11,77	11,25	11,78	12,38	12,40	12,38	12,20

***Tableau 3.** Mesures thermiques de l'air à différentes stations.*

	STATIONS A	B	C	D	E	F	G	H	I
DT_A1	1,1	0,19	0,31	0,04	0,73	2,03	1,49	1,82	1
DT_A2	0,02	0,18	0,31	0,21	0,08	0,51	0,80	0,30	0,40
DT_A3	- 0,67	- 0,28	- 0,31	- 0,31	- 0,80	- 1,02	- 0,84	- 0,55	- 0,54
DT_A4	- 0,42	- 0,32	- 0,41	- 0,19	- 1,04	- 1,26	- 1,15	- 1,82	- 0,81

DTA_1 : Différences de températures entre 9 h et 18 h le même jour (9/8/96)

DTA_2 : Différences de températures entre 9 h et 18 h le même jour (23/8/87)

DTA_3 : Différences de températures entre 18 h et 9 h deux jours consécutifs (23/8 et 24/8)

DTA_4 : Différences de températures entre 18 h (veille de fermeture) et 9 h (lendemain de fermeture)

***Tableau 4**. Variations thermiques de l'air.*

ANNEES	1979	1980	1981	1982	1983	1984	1985	1986
JANVIER	31	48	96	130	51	196	102	102
FEVRIER	184	164	134	226	183	228	172	123
MARS	169	321	162	380	713	526	538	609
AVRIL	840	815	1597	1666	1291	2298	1187	811
MAI	770	1339	1151	1514	1519	1397	1239	1320
JUIN	1258	1723	1883	2095	2278	2085	1462	1178
JUILLET	4312	5534	5936	6092	6237	4854	2559	4723
AOUT	6243	4680	5977	6693	6618	6424	2957	5358
SEPTEMBRE	2546	939	2260	2626	2324	2465	787	1296
OCTOBRE	431	543	644	841	778	1085	619	953
NOVEMBRE	181	294	337	185	178	419	339	163
DECEMBRE	76		61	41	318	302	175	135
TOTAL	17041	16400	20238	22588	22489	22279	12132	16769

***Tableau 5**. Fréquentation mensuelle de la cavité de 1979 à 1986.*

Années	1979	1980	1981	1982	1983	1984	1985	1986
JUIN	50	69	75	84	91	83	58	47
JUILLET	165	221	228	234	239	186	98	189
AOUT	240	212	229	257	254	247	113	198
SEPTEMBRE	101	68	90	105	92	98	91	51

Tableau 6. *Fréquentation moyenne journalière mensuelle des touristes.*

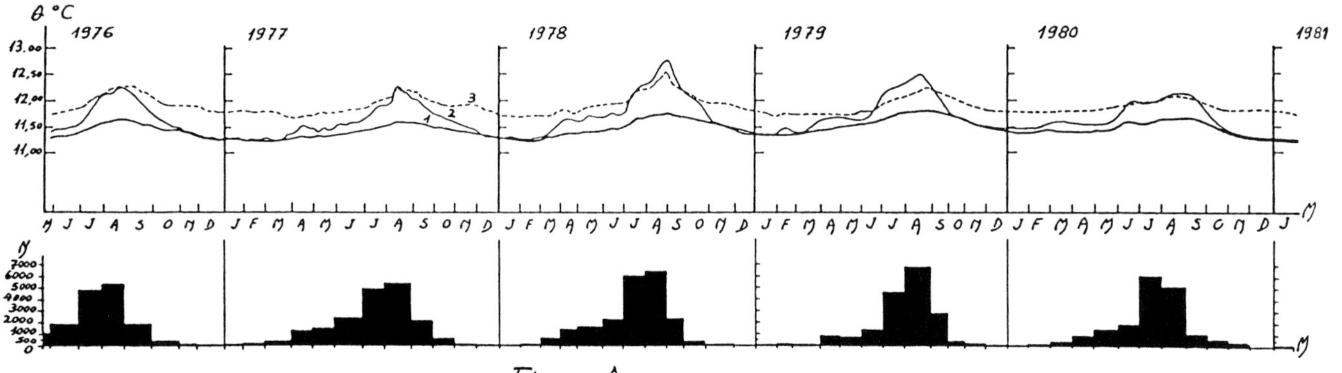

Figure A

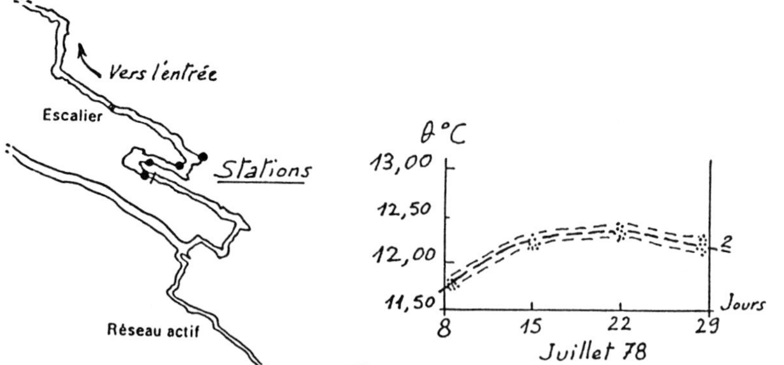

Figure B

Planche 1. *Analyse d'ensemble de mesures thermiques au niveau de la roche sur plusieurs années dans la partie finale de la galerie visitée.*

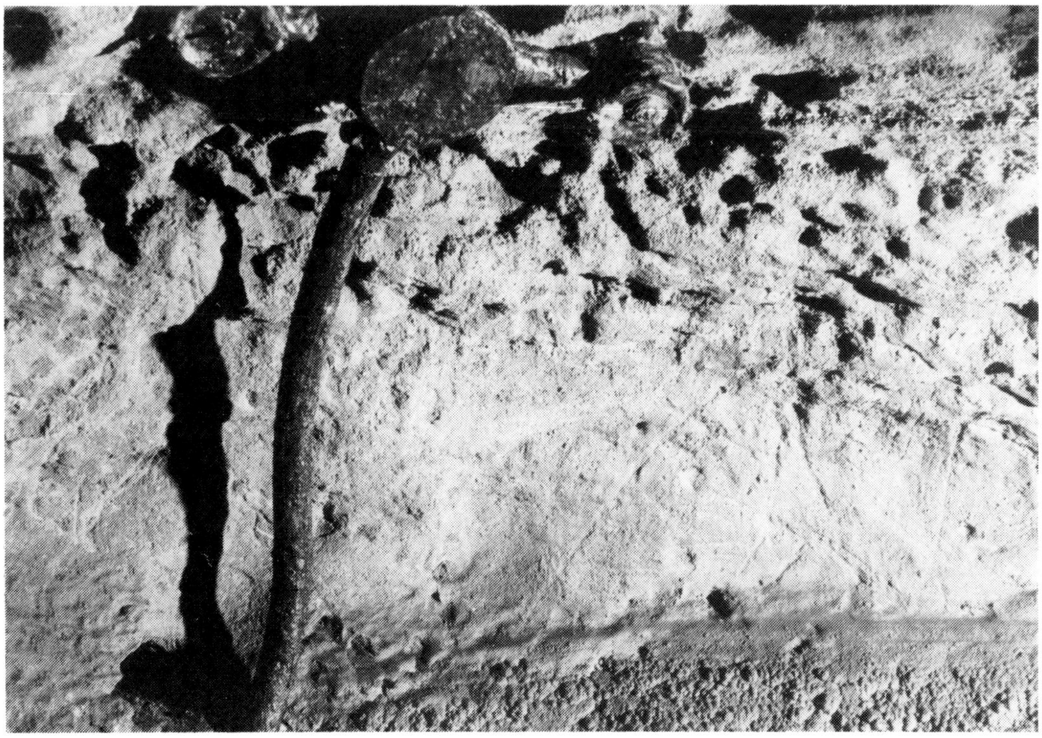

***Photo 1.** Grotte des Combarelles I: Etat de la cavité en 1974. Gaines électriques à proximité des zones gravées (cliché J. Brunet L.R.M.H.)*

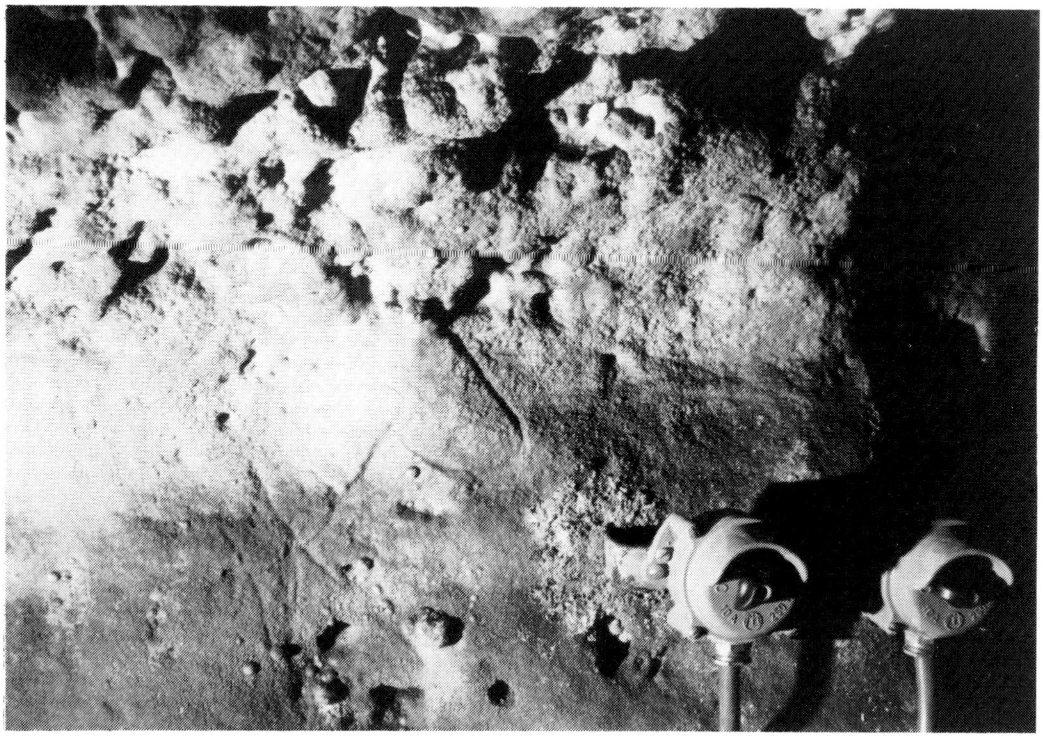

***Photo 2.** Grotte des Combarelles II: Etat de la cavité en 1974. Interrupteur électrique à proximité des gravures (cliché J. Brunet L.R.M.H.)*

4

FROM ASHES TO GRAVESTONES:

The Charcoal Drawings of Gua Badak, Malaysia

Paul Faulstich

Abstract. Gua Badak (Rhinoceros Cave) in Perak, Malaysia, was an occupation site of the Malay Negritos and had considerable archaeological significance. The walls of this shallow cave were decorated with abstract designs, humans and animal figures - most of which were drawn with charcoal. Some of the realistic motifs illustrated historical events, while the abstract drawings may have depicted aspects of mythologies. The author documented Gua Badak in 1984, just prior to its destruction by a marble quarrying company.

Introduction

Deep in the rainforest, a Negrito man ties a bamboo quiver of poisoned darts around his waist, grabs his blowpipe, and vanishes into the dense jungle foliage. As he disappears into the forest, the rustling of leaves gives way to the distant sound of bulldozers. Roads are being cut far into the jungle to facilitate logging. Development is encroaching on this jungle world and will soon destroy or forever change the lives of Malysia's indigenous peoples.

The Malay Peninsula lies in the tropics. Annual rainfall is high, and seasonal temperatures fluctuate very little, resulting in a consistently warm and humid environment with an average temperature of 28°C, and an average relative humidity of 83 per cent. Thick jungle and tropical rainforest once covered the land, and provided a rich habitat for a wide variety of flora and fauna. Sadly, most of the virgin forest has been destroyed, along with the tigers, rhinoceros and other animals that roam there. In some of the more isolated niches of jungle, however, the resonant calls of the lar gibbon can still be heard from high above in the jungle canopy, and small groups of hunting and gathering people struggle to maintain their traditional way of life in the face of encroaching modernisation.

Aspects of Malaysia's Aboriginal heritage, evident in archaeologically precious rock art sites, are threatened as well. As an expression of creativity, Malaysian rock art illustrates the vitality of its makers. It conveys a special quality of endurance and continuity, and beckons us to appreciate the rich and varied cultures of those who came before. These intellectual and emotional expressions of indigenous artists serve as reminders that various cultures have inhabited Malaysia for many thousands of years.

The prehistoric rock art of Malaysia was produced by the Orang Asli. The term 'Orang Asli' is a Malay phrase meaning 'Original People', and is synonymous with the English word 'aborigines'. The Orang Asli encompass the ethnic divisions of Negritos, Senoi and Proto-Malays. In some parts of Malaysia, small groups of Orang Asli are still living a life which follows many of the cultural traditions of their ancient ancestors. To our knowledge contemporary Orang Asli no longer produce rock art, but their ancestors left vivid images which can help us to better understand Malaysia's Aboriginal peoples. This rock art, however, like the Orang Asli themselves, faces an uncertain future. The rock art site that this paper is concerned with, Gua Badak (Rhinoceros Cave), has systematically been destroyed, and now nothing remains of this once striking example of Negrito art.

Gua Badak rock art site

Gua Badak was located in territory historically occupied by Negritos. Gua Badak was a deep shelter which was used as a temporary occupation site, and its walls were decorated with abundant charcoal drawings. The shelter was probably occupied intermittently, perhaps being visited seasonally to take advantage of the local ripening of fruit, and the corresponding movement of game animals.

The Negritos are traditionally a hunting and gathering people, ranging over their tribal territories utilising the available food resources. They hunt animals such as gibbon and birds, and they collect a variety of wild plant foods including tubers and various fruits. Their highly mobile lifestyle requires portability in their material objects, and their success in obtaining food is dependent on their knowledge of the natural environment.

As foragers, their possessions are few, but traditionally included objects such as fire-starting kits, digging sticks, hunting tools, and woven mats and baskets. In addition to making use of natural rock shelters, they construct temporary thatched dwellings made of poles and fronds. Modern examples of these shelters were visited by the author, and many were found to contain charcoal drawings strikingly similar to those at Gua Badak.

When Gua Badak was last surveyed in 1984, the walls of the cave were decorated with abstract designs, along with drawings of humans and animals. Most of the figures were drawn directly on the rock with charcoal. There were a few examples of where the Negritos had scraped off the ubiquitous parietal mold to produce a negative image. Virtually all of these unique 'pictoglyphs' had been re-covered by mold growth, and were faintly visible at best. The charcoal drawings were extremely fragile and susceptible to deterioration.

At Gua Badak there were pictures of monkeys, elephants, horses, people, bicycles, and even automobiles (Plates 1-5). The subject matter of many of the drawings was an obvious indication of their age, and they could not have been much more than one hundred years old.

The more realistic motifs appeared to illustrate some of the current events of the time. It has been suggested, for example, that the drawings of a man on a horse may have been of Hubert Berkeley, who was the District Judge of Hulu Perak in 1889 (Nasir 1977: 45). Berkeley was regarded as a very important person by the Negritos.

When he travelled to the Malay villages he normally rode a horse, and the route that he followed to the village of Lenggong took him directly past Gua Badak.

Other compositions in this shelter appeared to represent hunting scenes, and depict men with blowpipes (Plate 1). There was only one drawing of a man with a bow and arrow. The bow and arrow was the early hunting implement of the Negritos, but about one hundred years ago they began to favour the blowpipe, and the use of the bow and arrow ceased. These charcoal drawings were obviously made after the introduction of the blowpipe.

In 1928 the anthropologist Ivor Evans visited Gua Badak while a group of Negritos was living there. He was unable to obtain interpretations for several of the drawings, and recorded that many of the large abstract designs represented the sleeping mats of the ancestral beings held in esteem by the Negritos (Evans 1928: 106), see Plates 2 and 3.

The non-figurative motifs at Gua Badak were also very similar to the designs that are carved on pieces of bamboo by the Negritos of today. These so-called 'myth bamboos' symbolically relate aspects of cultural knowledge, and depict various themes in symbolic form. The myth bamboos portray special geographical features, as well as shamans (medicine men), animals and plants. The drawings at Gua Badak may have depicted these same subjects through their abstract designs. Ornamental hair combs, incised with similar designs, are among the most valued aesthetic possessions of contemporary Negrito women.

Destruction of Gua Badak

The cave walls of Gua Badak were formed of colourful banded marble. It is ironic that this beauty also encouraged the destruction of this valuable archaeological site (Plate 3). A marble quarrying operation has been mining the Gua Badak area for many years. Recently the entire Gua Badak rock art complex was demolished for its high quality marble, and all of the unique and irreplaceable drawings lost forever. These photographs are some of the only documentation of Gua Badak, a priceless but destroyed national treasure.

Being a decorated occupation site, Gua Badak had considerable archaeological and ethnographic significance. Cultural resources, like the art at Gua Badak, are irreplaceable, and unless strong measures are taken to protect and preserve them they will continue to be demolished. All too often, rapid development is at the expense of indigenous peoples and their unique heritages. The destruction of the rock drawings at Gua Badak exemplifies a general lack of interest and concern over small, relatively isolated societies around the world. Caught in the path of industrialisation, indigenous peoples and their cultural heritages are often exploited, or treated with little regard. There is a sad irony in the fact that some of the marble from Gua Badak has been transformed from cave walls decorated with charcoal drawings to gravestones in burial sites scattered throughout Malaysia. Rest in Peace Gua Badak.

Through future protection of cultural resources we can begin to foster a greater appreciation of both the unity and diversity of humankind. As a tool used by historians, philosophers and archaeologists, the remaining rock art of Malaysia can help us to better understand the events, concerns and cultures of the Orang Asli - Malaysia's 'Original People'.

Acknowledgments

Field Research for this report was carried out under a grant from the Malaysian Culture Study Group (Kumpulan Kebudayaan Malaysia). I wish to especially acknowledge the efforts of Kay Hashim; her gracious support and encouragement made the documentation of Gua Badak possible.

REFERENCES

EVANS, I. H. N. 1928. Negrito cave drawings at Lenggong, Upper Perak. *Journal of the Federated Malay States Museums* 12: 105-6.

NASIR, A. H. 1977. *Sejarah Perak: Pertama*. Jabatan Muzium, Kuala Lumpur.

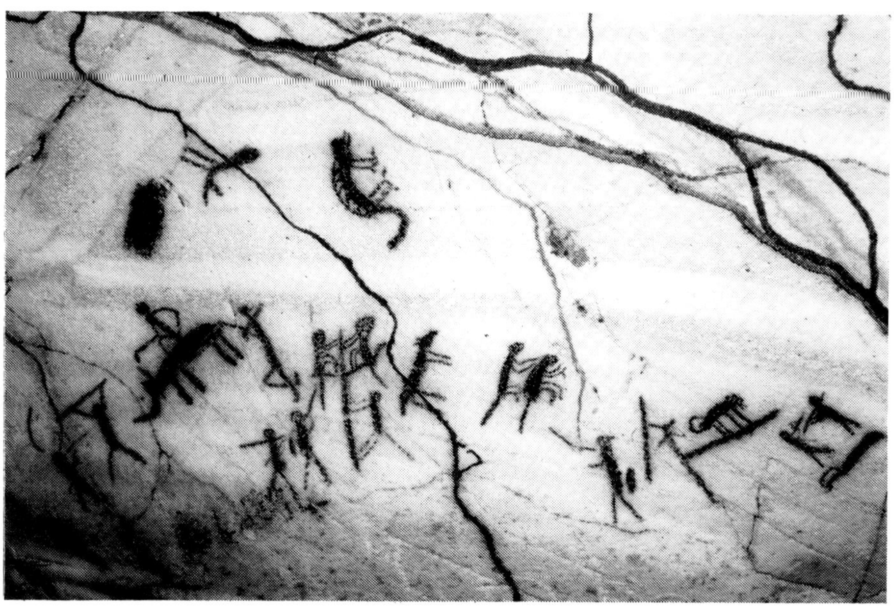

Plate 1. *These drawings from Gua Badak depict various scenarios, including a person riding an elephant and a man using a blowpipe to hunt monkeys.*

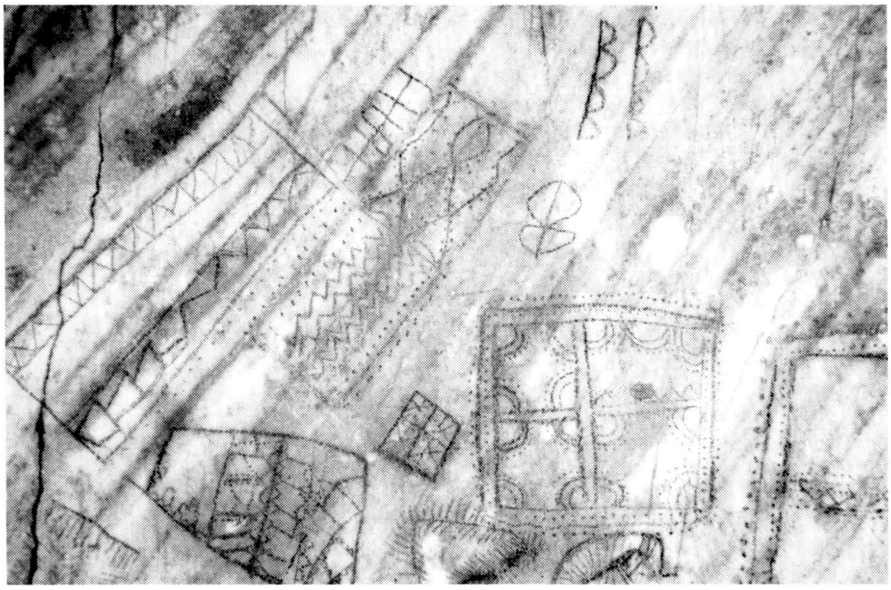

Plate 2. *The non-figurative motifs at Gua Badak might have represented the sleeping mats of the Negrito's ancestors and may illustrate various places and events in symbolic form.*

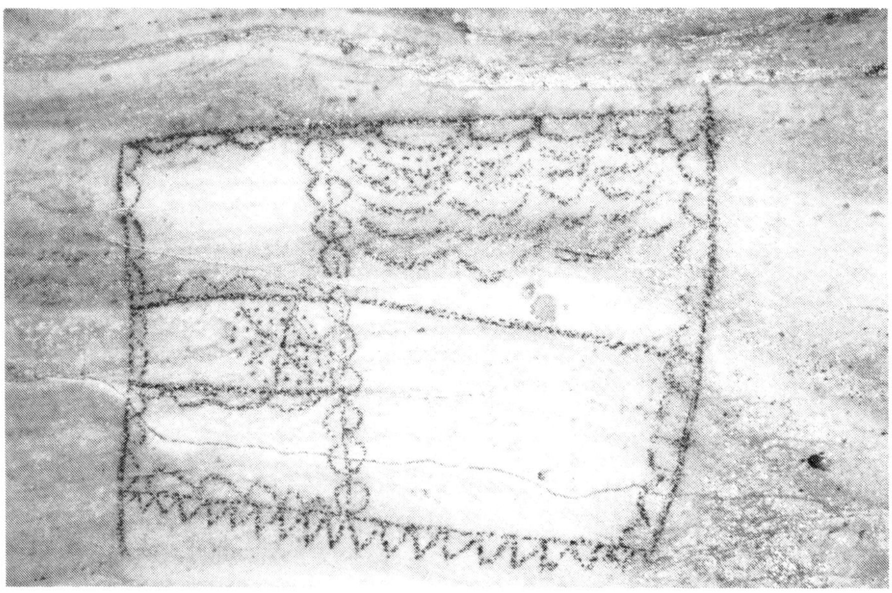

Plate 3. *The colourful marble of Gua Badak was considered more valuable than the unique rock drawings. Gua Badak was destroyed for the beautiful marble out of which this cave was formed.*

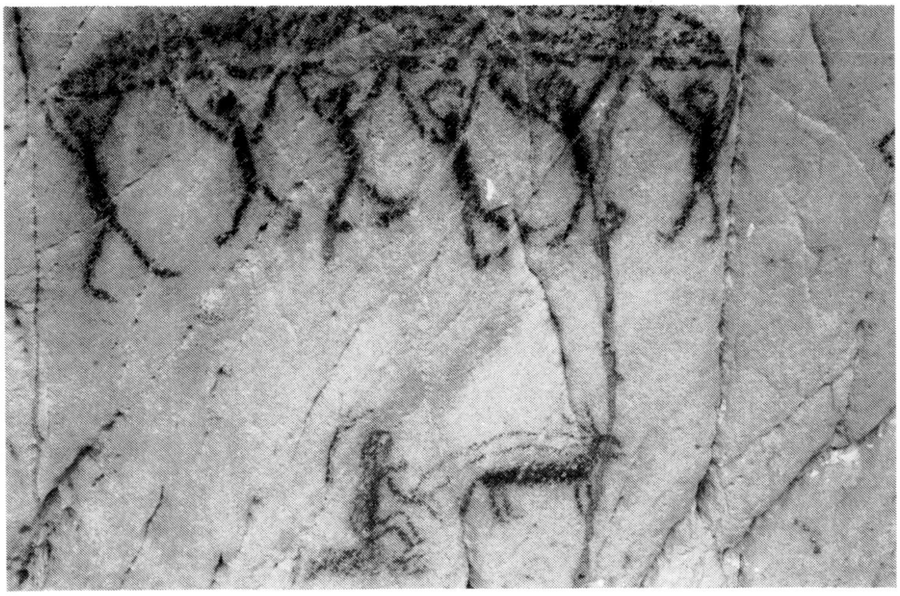

Plate 4. *These drawings include a group of people carrying a canoe, and a man guiding an animal with reins. The subject matter of this later drawing indicates that it was created after Europeans came to Malaysia.*

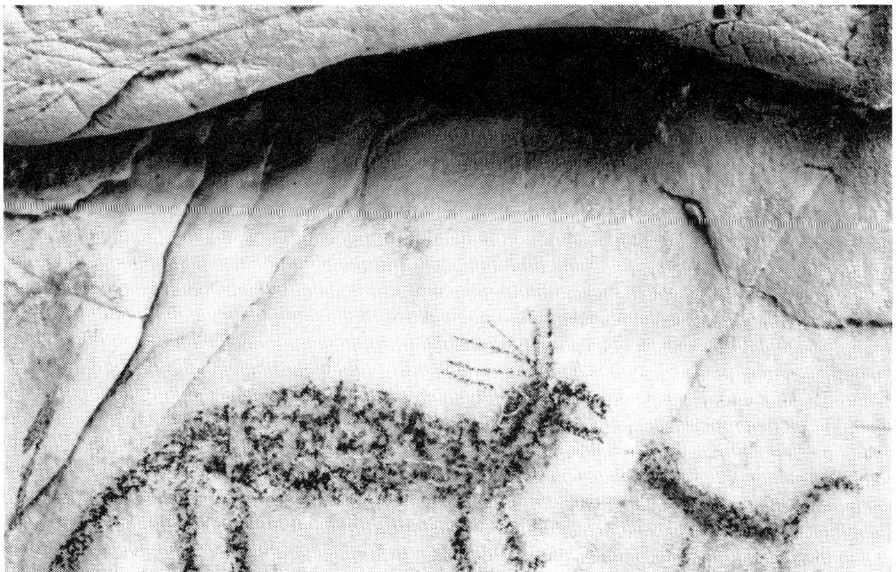

Plate 5. *This drawing portrays two unidentified animals. The large figure appears to have a cat-like body and antlers, and may be a fanciful creation.*

5

CONSERVATION OF INDIAN ROCK ART

Gajendra S. Tyagi

Abstract. The Author seeks to create an awareness of the infinite number and variety of rock art treasures in India and their intrinsic value, some of which date back to the Mesolithic period or earlier.

Introduction

In the last six years the author has visited 1300 rock shelters again and again in the state of Madhya Pradesh in central India, and in this short span of time has observed such deterioration, that in his opinion all these treasures will eventually be lost to mankind if immediate steps are not taken to counter this menace. Deterioration is caused by vagaries of nature, by mammals, insects and human agencies.

These treasures are in great danger of absolute destruction, due to total lack of conservation measures. In order to consider the question of conservation and preservation, it is necessary, in the first instance, to have an organisation in India. This idea of an Indian Rock Art Research Association was first conceived by the late Dr V. S. Wakankar, the pioneer of research in Indian rock art. His untimely demise early in 1988 prevented him from fulfilling this task.

The Indian Rock Art Research Association (IRARA) should consist of, amongst others, eminent scholars and have bureaucratic participation. The IRARA should also be a branch of or affiliated to similar international organisations.

It is hoped that the IRARA will also have a subsidiary organisation to deal with conservation, preservation and liaison with similar international organisations.

After over a century since the discovery of rock paintings in India (Smith 1906: 185-90) very little interest has been evinced in the conservation and preservation of rock art, even by the scholars of this country.

Due to the paucity of funds, India, a developing country has inadequate funds even for the upkeep, restoration and conservation of historical monuments, and so cannot have the resources to consider the preservation and conservation of rock art sites which span hundreds of square kilometres of area, although these sites house treasures of immeasurable importance and value to the country and the world at large.

The aim of this paper is to create an awareness of the need for the conservation and preservation of all rock art sites all over the world, but more specially in India.

The areas of the rock art sites that the author has been studying is in the state of Madhya Pradesh in Central India, which are sub-montane regions of the Vindhyachal Range in the Raisen district. He has attempted to categorise the causes of rock art damage and destruction and proposes to submit ways and means of preventing further deterioration.

Deterioration of rock art sites

The causes can be broadly categorised under three headings:

1. DUE TO VAGARIES OF NATURE
2. DUE TO ANIMALS AND INSECTS
3. DUE TO HUMAN AGENCIES

Firstly, the paintings of some of these shelters are exposed to direct sunlight which has a very damaging effect on them. The scorching heat of the Indian summer adversely affects the colours of the paintings. It could be argued that if they are still visible after so many years the sun should not affect them now. The submission of the author is that with the advent of 20th century civilisation, man has indiscriminately cut down trees and foliage rendering the shelters bare. Not only does the sunlight adversely affect the paintings in the shelters, but the high winds in the area also affect the paintings due to the lack of trees and foliage. Dust-laden dry winds not only erode the paintings but form an opaque layer over the patination, and it is difficult if not impossible to see the rock painting thereafter.

Fungus formation is also another damaging instrument to the rock paintings. Ways and means to counter this menace have to be devised.

Secondly, mammals and insects are also the cause of considerable damage to the rock shelters and the paintings inside. With the influx of civilisation in the surrounding areas, the monkey and langur tribes, which in earlier days were spread over vast areas have now congregated in these less inhabited hilly regions. They live in the rock shelters in large numbers and the acidity of their urine and faeces gravely damages the paintings. Another problem is termite damage. It is not known how many valuable paintings have been lost over the years due to this problem. Another intriguing damaging factor has been observed by the author. Due to the indiscriminate deforestation of the surrounding areas, hundreds of bee hives now exist in these shelters. The wax of the bee hives melts in the direct sunlight of summer and drips on the rock paintings, sometimes completely defacing them. The fact that deforestation has led to the number of beehives increasing in the rock shelters has been proved by the author who has observed rock paintings depicting trees with scores of beehives on them, and dwellers collecting honey. This supports the view that, due to deforestation, the bees have now shifted into shelters.

Thirdly, and most importantly, the greatest damage and destruction is caused by man. Local porcupine hunters create smoke fires in these shelters to drive out their game. This smoke totally blackens the rock shelters, completely obliterating the paintings. The rock shelters are also used as temporary abodes by mendicants (Sadhus) and shepherds who cook their meals on cow dung and wood fires, again damaging the paintings. Of late, there is an influx of visitors and tourists, and graffiti by irresponsible elements play havoc with the rock art. Strangely enough, even the so-called scholars and experts have indiscriminately and with abandon, been improperly marking and numbering, for their private convenience, paintings, and spoiling the superimpositions of older paintings. These so-called experts also seriously damage the patination by using water spraying techniques. Some of them even tried to chip away the colour pigments of the paintings for reasons best known to themselves.

Indian Rock Art Research Association

The author seeks advice and assistance from the members of the congress for ways and means of arresting this deterioration.

As a first step it is necessary to follow the advice of the late Dr V. S. Wakankar, the pioneer of Indian rock art research and the mentor of the author, who was not able to fulfil his ambition due to his sudden demise in Singapore in 1988, of firmly establishing the Indian Rock Art Research Association (IRARA).

The IRARA must also consist of eminent scholars of rock art, archaeology and anthropology, and have bureaucratic participation. The association must also be affiliated with ICOMOS and any other international organisations concerned with the conservation of rock art.

The author is of the firm view that the AURA Congress should table a motion to ICOMOS that the rock art sites in India, specially in the Raisen District of Madhya Pradesh, and the adjoining areas in Central India, merit full scale conservation efforts.

Conclusion

To sum up, therefore, the problems of conservation of Indian rock art are two-fold. One is the lack of awareness in the minds of the people and two, the paucity of resources and expertise to conserve and preserve the priceless wealth of rock art in India.

Acknowledgments

I am extremely grateful to my mentor, the late Dr V. S. Wakankar, for the encouragement and advice he has given me from time to time.

I am indebted to Smt. Pupul Jayakar for having taken time off from her onerous duties to give me a patient hearing and encouragement to attend the First AURA Congress in Darwin, Australia.

I am also very grateful to Mr Ashok Vajpaye, Secretary, Department of Culture, Government of Madhya Pradesh, India, for his interest, encouragement and assistance to attend the AURA Congress.

Special thanks are due to AURA, for the generous support enabling me to attend the Darwin congress.

Finally my profound thanks go to Prof. Jack Steinbring, University of Winnipeg, Canada.

6

ABORIGINAL ART SITES NEAR ADELAIDE, SOUTH AUSTRALIA

Elizabeth Caldicott

Abstract. Over 30 small but unique Aboriginal art sites exist in the Mount Lofty Ranges close to the city of Adelaide. The art is located exclusively on privately owned farming land. This regional series of painting sites is separated by several hundred kilometres from the Flinders Ranges, the nearest rock art region. Past isolation can no longer protect them from the increasing human impact of mobile, informed urban tourists and the demand for passive recreation. Minimal legislative protection, limited human and financial resources for management and the absence of archaeological research in this area currently offers little future heritage protection.

Introduction

The presence of Aboriginal art in the south of South Australia is not widely known as people do not expect to find evidence of Aboriginal occupation in this area.

A widespread belief exists that Aborigines were desert dwellers occupying only the arid interior and tropical north of the continent, or that only in these areas can current evidence be found.

The Mount Lofty Ranges form the eastern boundary of metropolitan Adelaide, the state capital and city of about one million people. The study area was delineated as that adopted for the Mount Lofty Strategy Plan currently being formulated by the South Australian Department of Environment and Planning, extending roughly 150 km from the Barossa Valley in the north to the southern tip of Fleurieu Peninsula (Fig. 1).

Within this area there are five known localities where rock paintings have been recorded. Some thirty-one sites have been registered with the Aboriginal Heritage Unit of South Australia, but local farmers indicated during interviews that further sites are known to exist in the area. These paintings are situated in small rock shelters close to seasonal water courses or near permanent springs along the escarpments of the ranges.

Description of the rock art sites

The semi-arid nature of this region, with precipitation averaging 250 mm per annum, falling mainly as winter rain, and followed by a hot dry summer, supports plant communities of grassland with sparse low shrubs and tall red gums (*Eucalyptus camaldulensis*) near streams.

The predominant land use today is low density pastoralism, wool sheep and beef cattle, with limited cultivation of cereal crops on the more undulating land.

The land tenure consists of private freehold property, apart from minor areas of government reserves. Recent subdivision of larger holdings into smaller allotments for hobby farms has resulted in fragmented land management, with absentee landlords, or at best, weekend urban commuters retreating to the rural fringe. Most of this area is little more than an hour's drive from the centre of the city of Adelaide.

This little known collection of rock art has attracted spasmodic interest over the years since the first official site recording by Stirling in 1902 at South Para.

Hossfeld (1926), a geologist, discovered two sites at Eden Valley. Mountford (1957, 1960) recorded several other sites in the 1950s near Macclesfield and Kanmantoo.

Further sites were recorded by Ross and Ellis (1974) during an archaeological survey preparatory to the proposed development of the new city of Monarto.

Gunn (1981) undertook detailed recording of other sites already registered.

The art consists of numerous small, simple figures variously reported by recorders to represent human, animal or abstract motifs (see Figures 2 and 3). A wide range of colours have been used, including red, yellow

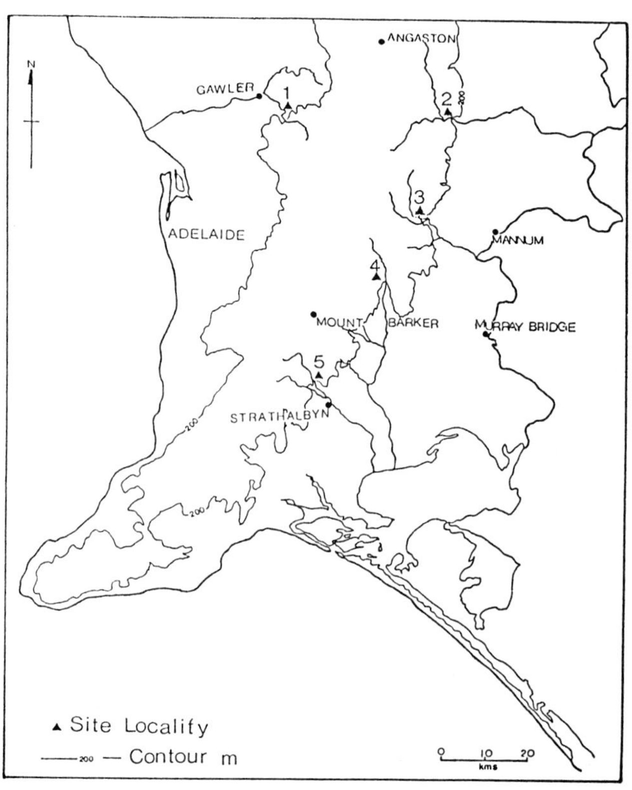

***Figure 1.** Map showing the relative location of Aboriginal Art Sites in the southern Mount Lofty Ranges, South Australia.*

and brown ochres, white pipe clay and a black substance, either charcoal or some mineral pigment.

The source of ochres has not been established but small local ochre mines have been referred to by Brown (1908) and Ludbrook (1980). Commercial extraction of ochre for industrial pigments was undertaken in the area earlier this century by Europeans.

With few exceptions the art has been executed by painting, with minor drawings. Use of stencilling has not been recorded in the study area. There have been several unconfirmed reports of engravings in the region (Gara 1986).

When compared with the larger and more elaborate galleries of rock art in the Northern Territory and Queensland, which attract thousands of tourists each year, these small and less spectacular paintings have been generally regarded as insignificant by the landholders on whose properties they occur. They have been neglected, to date, in literature pertaining to Australian rock art.

I believe that this collection of art sites is important as a specific regional example of prehistoric rock art, for the distinct style employed by the artists and the proliferation of paintings within a relatively small and discrete area.

The next nearest known occurrences of Aboriginal paintings are found in the Flinders Ranges near Hawker and in the Olary province near Yunta, both over 400 km to the north. These are stylistically quite different from those in the southern Mount Lofty Ranges.

Condition of sites

To date the Mount Lofty sites have been relatively undisturbed, other than by the natural processes of physical and chemical weathering. Damage from biological agents such as mud wasps and nesting martins has caused encroachment on painted surfaces, affecting them in various degrees. Grazing animals rubbing against the rock walls have left grease deposits on the lower surfaces.

The semi-arid climate has limited algal and fungal damage although lichen development on rocks outside the shelters is considerable. Rain wash and seepage do not appear to present a major concern.

Most sites are remarkably free of graffiti and other deliberate human damage, with the exception of those sites well exposed to the public over many years. Such sites include the shelters at Eden Valley which are close to a favoured local picnic area where inscriptions dating back to 1908 have acquired somewhat of an historical significance. The Palmer West paintings have been completely obliterated over recent years by extensive signwriting on the large boulder, a prominent landmark beside a major road (see Figure 4).

Considerable damage occurred at the South Para sites several years ago necessitating the installation of mesh screens by the Heritage Branch as a protective measure.

Probably the major contributing factor to the well-preserved state of this art is the relative isolation of the sites in inaccessible areas of privately owned land. Access involves traversing rough terrain, limited to 4WD, foot or horse-back movement only.

Cultivation of the land, although limited, has during ploughing revealed a number of stone hearths indicating large-scale Aboriginal occupation of the district at some time in the past. To date there has been no significant archaeological research undertaken in this area, without which evidence it is not possible to attempt the dating or reconstruction of past human activities.

Site ownership

The last published reference to Aboriginal presence in this region was in the 1860s, soon after the land was released to European settlers. Any natives still in the district at that time were relocated on the missions along the Murray River where traditional culture and land associations were actively discouraged, along with the fragmentation of family and tribal groups.

Following the Harvard-Adelaide Universities' expedition in 1939, Tindale produced maps outlining tribal boundaries based on the oral records of descendants of Aboriginal groups known to have occupied the land before European settlement, some eighty years previously in the case of the study area. Tindale's maps (1974) have been used as a basis for arbitrarily determining land rights claims, but the validity of these boundaries may be open to future debate. Later maps produced by linguists Walsh and Wurm after 1981 (n.d.) indicate that tribal language divisions may vary considerably from those established by Tindale. Today descendants of the three major tribal groups to have occupied the study area have largely intermarried under the mission system to form a new group, the Adelaide tribe, which has already initiated several claims to ownership of particular sites.

Site protection

These art sites are currently [1988] officially protected under the South Australian Aboriginal and Historic Relics Act (1965), but limited resources for staff to protect and research the sites affords minimal effective management.

Grants to date have been directed to the better known collections of art in the Flinders Ranges, already established as a major tourist area with large scale future promotion proposed. A new Bill (the Aboriginal Heritage Act 1988) is currently under review by the State government and expected to be declared law by the end of 1988. However, unless increased financial support is offered, adequate management is unlikely to be implemented to conserve those art sites on Crown Land, in the National Parks or on pastoral leases, much less those on private land which offer no access to the public.

An increasing interest in Aboriginal heritage shown by a more informed and mobile urban population presents a real threat to the protection of these sites and conservation of the adjacent environment. Damage includes both wilful destruction and that caused by the unconsidered actions of people resulting from the sheer pressure of visitor numbers.

Establishment of a number of walking trails, such as the Heysen Trail, through the Ranges has encouraged greater access to private land with various consequences. Indeed one landowner requested that the proposed route be re-aligned to avoid passing within metres of the South Para sites.

Repeal of the Trespass Act in 1987 has virtually eliminated any legal protection for private property and has antagonised landowners towards those people seeking permission to visit the sites. Despite this known hostility, groups from both private and government bodies are proposing greater promotion and development of the art sites in this area. A number of the sites are on already established tourist routes, and some of them are being regularly included on educational excursion itineraries.

The landholders have no indemnity against public risk liability, even when the entry to their properties is unauthorised.

An additional hazard emerging to threaten the continued survival of these heritage areas is the rapid subdivision of large land holdings into smaller, fragmented hobby farms.

Without the inclusion of some form of heritage agreement, or covenant on the new titles created, there is at present no guarantee that the art will remain undisturbed in the future. The growing number of absentee landlords has led to reduced personal involvement in management of the land, many of the properties becoming purely speculative investments, largely leased to local farmers, with little or no interest in heritage conservation being exhibited by the new owners.

As an income supplement, landholders in the area have encouraged surface disturbance on a significant scale by permitting commercial operators to collect, somewhat indiscriminately, large lichen covered boulders popularly known as 'moss rocks', and receive royalty payments in return.

These rocks are much in demand for use by the landscape industry in the currently fashionable enhancement of urban gardens. As neither quarrying nor mining licences are required there is total lack of regulation of this industry by the government, and a general ignorance among operators of the potential impact of their activities on the environment. Removal of boulders up to 5 tonnes from the immediate vicinity of paintings is a cause for concern.

The problems of management and continued conservation of these sites is multifaceted, with conflicting interests of landholders, tourism promoters and conservationists. Lack of resources and divided opinions within the relevant government departments responsible for the protection of the environment combine to present a bleak future for the survival of these paintings, part of both the Aboriginal and national heritage.

Elsewhere most management strategies have been applied, with varying degrees of success, in national parks where the broader principles of visitor management can be applied to the specific needs of the Aboriginal art sites.

The dilemma now facing the future of the sites on private land is how to determine who should have the responsibility of custodianship, and consequently bear the financial burden of undertaking conservation measures. Should these sites be promoted and thus further exposed to visitor pressure without planned management, or should they be kept remote, as museum pieces to quietly weather away, until such times as the resources for wise management become available?

REFERENCES

BROWN, H. 1908. *Record of the mines of South Australia*, fourth edition. Government Printer, Adelaide.

GARA, T. 1986. Two Aboriginal engraving sites in the Mount Lofty Ranges. *Journal of Anthropology* 24(2): 3-9.

GUNN, R. G. 1981. Aboriginal rock art of the Mount Lofty Ranges, S.A. Unpublished report to the Aboriginal Heritage Unit of S.A.

HOSSFELD, P. 1926. The Aborigines of South Australia: Native occupation of the Eden Valley and Angaston district. *Transactions of the Royal Society of South Australia* 50: 287-297.

LUDBROOK, N. 1980. *A guide to the geology and mineral resources of South Australia*. South Australian Government Printer, Adelaide.

MOUNTFORD, D. 1957. Aboriginal cave paintings in South Australia. *Records of the South Australian Museum* 13, No. 1.

MOUNTFORD, D. 1960. Cave paintings in the Mount Lofty Ranges, South Australia. *Records of the South Australian Museum* 13, No. 4.

ROSS and ELLIS 1974. Aboriginal relics in the Lower Mount Lofty Ranges, Murray River and Monarto area Unpublished report for the South Australian Museum.

STIRLING, E. 1902. Aboriginal rock paintings on the South Para, Barossa Ranges. *Transactions of the Royal Society* 26.

TINDALE, N. 1974. *Aboriginal tribes of Australia: Their terrain, environmental controls, distribution, limits and proper names* Australian National University Press, Canberra.

WALSH and WURM (n.d., using 1981 data). *Australia and Tasmania*, Linguistic map of south eastern Australia in the atlas Part I, Australian Academy of the Humanities.

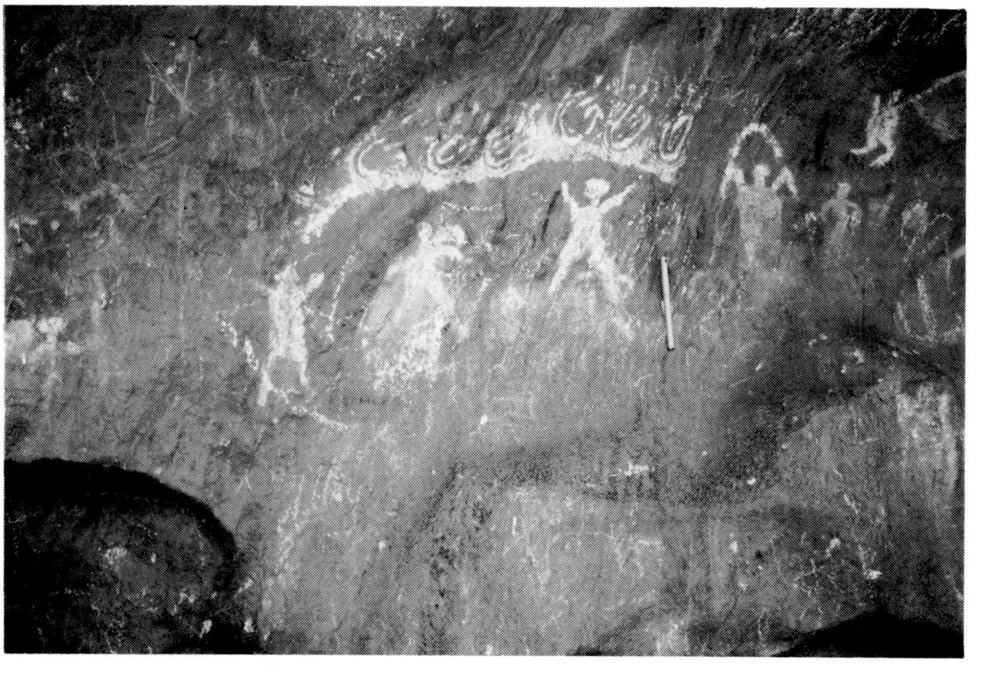

Figure 2*. Eden Valley paintings, Cave A, near Springton, South Australia, December 1987.*

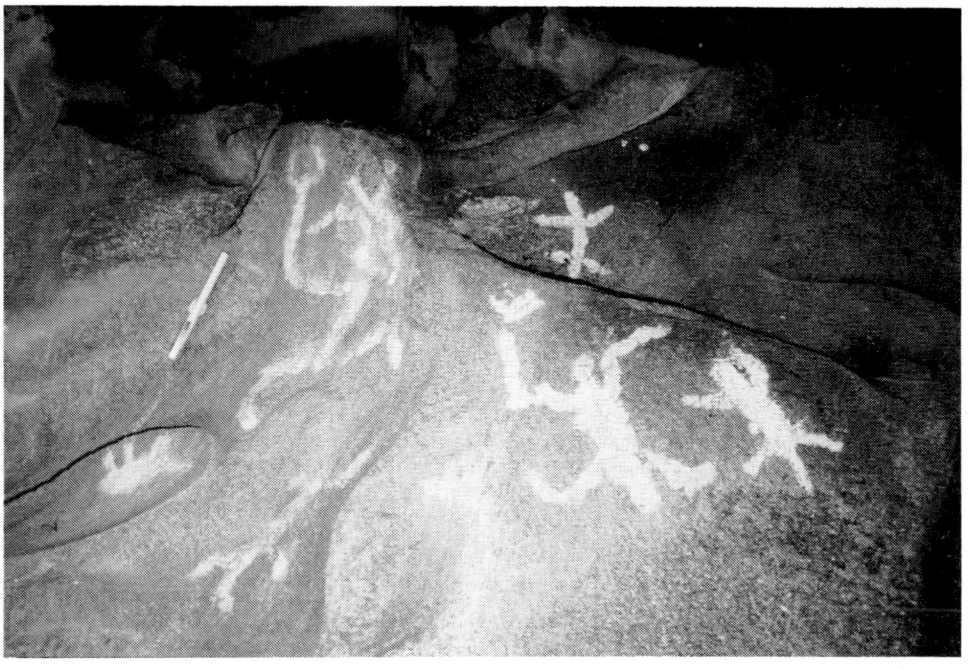

Figure 3. *Eden Valley paintings, Cave B, near Springton, South Australia, December 1987.*

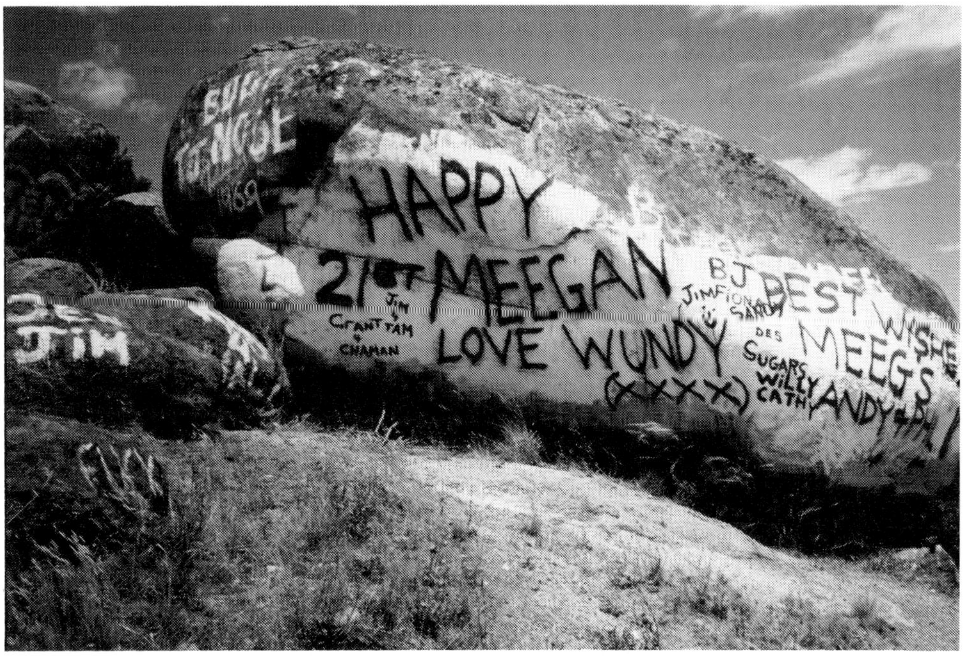

Figure 4. *Graffiti has totally obliterated the Aboriginal paintings on the boulder at Palmer West in South Australia.*

7

FLINDERS GROUP NATIONAL PARK SITE MANAGEMENT:

Regional cultural resource evaluation, preservation and selective presentation in far north Queensland.

GRAHAME L. WALSH

Abstract. Tourism and its unavoidable accompanying development pose an on-going and ever increasing threat to rock art sites, particularly in publicly accessible areas such as national parks. Tourist pressures are most evident in the northern states of Queensland and the Northern Territory. Over the past decade the Q.NPWS Historic Sites Section has developed a multi-phase approach to service these cultural resource management needs. Project Flinders involves the introduction of effective management to some of the most spectacular island art galleries on Australia's east cost; in the Flinders Group NP. A concerted site recording survey established an essential cultural resource inventory for the little known islands. Data analysis identified sites suited for development and permitted accurate drafting of management plans. Boardwalk materials transported by barge from Cairns were heli-copter-airlifted to the island sites, where Army volunteers assisted with assembly. A summary of operations may assist other state authorities develop conservation strategies for endangered rock art in logistically difficult settings.

Introduction

In a time when economic levelling off is being experienced in many sectors of primary and secondary industry, Tourism currently represents one of Australia's major growth industries, and is heralded in many quarters as the future hope for the nation. Major promotion campaigns launched internally and abroad by developers and government agencies have ensured this trend will continue at least into the foreseeable future.

Greater interest is being shown in Australia's cultural resources, as has been emphasised in an article in *The Weekend Australian*, 16.7.1988. 'A survey by the Queensland Tourist and Travel Corporation (QTTC) shows that many overseas tourists want to see more of the "real" Australia and experience Aboriginal and Islander cultures first-hand.'

Queensland and the Northern Territory are becoming increasingly favoured tourist destinations, and in both areas a selection of rock art sites have been protected, made accessible to the public and actively promoted as a significant component of regional attractions. In almost all cases the presented and interpreted galleries are on areas of national park estate.

Anticipating the demands which will be made on rock art sites, and ensuring protective development precedes pressures associated with tourist expansion, represents a major obligation of cultural resource managers. Queensland is particularly subject to these needs, as it is claimed that within the next two years it will have almost doubled its present stock of tourist accommodation. At present more than 50 per cent of all Australian projects firmly committed or under construction are in Queensland. This involves 14 600 rooms, at the cost of close to $5 billion.

Aboriginal rock art has been a major attraction in some Queensland national parks for more than half a century, with some of the most widely known examples located in Carnarvon Gorge. Carnarvon National Park is now Queensland's most popular inland park, and the initial 86 square miles of its area were gazetted on 30 April 1932 (Walsh 1984: 12).

A progressive increase in public visitation and interest in the Carnarvon art sites was largely responsible for the development of management policies by Queensland National Parks and Wildlife Service for the conservation, presentation and interpretation of publicly accessible rock art sites. Expertise was gained largely through trial and error, as the lack of precedents in this specialised area of cultural resource management has been emphasised by Gale (1983: 40). 'The whole field is in a very early developmental phase in Australia...'

Regular revision of management techniques and policies has been necessary to cope with specialist needs of the diverse range of rock art sites and associated environments continually being inherited with the expanding Q.NPWS estate. Mainland management responsibilities of Q.NPWS currently include 314 national parks and 150 environmental parks, totalling over 3 500 000 hectares, or more than 2 per cent of the state's area. Additional responsibilities include day to day management of the Great Barrier Reef Marine Park (the largest marine reserve in the world) which encompasses many islands in its 385 000 square kilometre area.

Logistics associated with the management of cultural resources scattered over this estate are enormous, with the distance from Sundown NP (near the southern border) to Jardine River NP on Cape York Peninsula being more than 2200 kilometres.

Particular consideration has been given to developing management strategies which fully capitalise on the often absent or ill-defined visitor use patterns associated with rock art sites on newly acquired national park areas.

A process of systematic site survey, recording and evaluation was developed, based on specially designed computerised recording proformas (Walsh 1984a: 47), to permit a comprehensive cultural resource register to be established for newly acquired national park areas.

Project Duwadari's contribution to the management of Lawn Hill NP verifies the values of these strategies (Walsh 1985a: 80-88). The implementation of site management coinciding with tourist visitation has also been mentioned in the NT case study by Gale (1983: 32). 'Unlike southern areas of Australia it was possible to develop ways of protecting and interpreting the art (at Kakadu NP) virtually soon after the tourist boom began'.

Flinders Group NP

The need for a similar management oriented survey of the Flinders Group NP, located in the remote eastern Cape York setting of Princess Charlotte Bay (PCB), first became apparent in early 1982 when unconfirmed reports of interference with cultural sites began filtering in (Walsh 1984b: 3).

The Flinders Group NP consists of a 2962 hectare area encompassing the continental islands of Blackwood, Denham, Flinders, Macleay and Stanley, and was gazetted on 2.12.1939. This park's wealth of cultural resources were largely protected by remoteness and inaccessibility over the ensuing years, in a setting devoid of development and receiving only infrequent patrols.

As commercial fishing operations began to focus their attention on the rich resources of adjacent Princess Charlotte Bay, the effectiveness of isolation as a form of management for this park began to diminish rapidly in the last 1970s. Logistics associated with trawlers operating at such distances from the main Cooktown base were overcome when a fuel barge with trading post and freezer facilities was moored seasonally in Owen Channel. Long recognised as one of the safest moorings in far North Queensland, this 'development' served also to attract the dramatically increasing number of passing tour boat operators and pleasure craft to make the islands a stopover.

By 1982 as many as 50 boats were using Owen Channel for nightly mooring, with numbers continuing to increase. 'The barge operator advised that approximately 75 trawlers had been operating out of the Flinders Group in 1985 and indications are that this figure will increase to in excess of 150 boats during the 1986 season' (Walsh 1985d: 7). Recreation time for crews of moored boats frequently involved forays on the Flinders Group islands, searching for cultural sites. Details of discoveries were soon common knowledge to private craft owners and tour operators, who incorporated visits to the sites in their itineraries. Directions (together with a map) to one particularly sensitive site were published as well-intended tourist locational detail in a popular yachting book dealing with the Queensland coast.

The situation which developed is aptly summarised in a general statement of Rosenfeld (1985: 1), 'In Australia, Aboriginal political pressure has created a wider awareness of the significance of their culture, which together with development in formerly remote areas has greatly increased tourist pressures on Aboriginal rock art sites'.

Additional control over the waters surrounding the national park islands became effective on 25.8.1983, with the declaration by the Commonwealth Government of the Far Northern Sector of the Great Barrier Reef Marine Park.

History of Flinders Group sites

The Flinders Group will always hold particular significance in Australian rock art studies, as the discovery of paintings on Clack Island (the group's most remote island) is claimed to be second only to the Chasm Island discoveries by Matthew Flinders on 14 January 1803, (Flinders 1814: 188).

On 22 June 1821 Allan Cunningham discovered the galleries when visiting Clack Island to obtain a shell collection (King 1837: 27). Descriptions published in King's journals have resulted in reference to the Clack galleries being second only to George Grey's 1838 Wandjina discoveries in documentation of Australia's Aboriginal history (Walsh 1988: 2). The Clack galleries have since been referenced in a diverse range of books, from Meston (1895: 134) to McMinn (1970: 45). The interest these discoveries aroused among mariners in this early exploration period is evident in comments by Stokes (1846: 352) and Coppinger (1883: 191). Coppinger personally visited the galleries and published the first sketches of the motifs in 1883.

Rock art galleries on Stanley Island and the adjacent mainland appear to have remained unknown until 1927, with their discovery and descriptions being published by Hale and Tindale (1934: 148-49). The next scientific investigation of the art was 47 years later, when the AIAS expedition visited sites on Clack and Stanley Islands between 30 September and 4 November 1974. This was an important visit, as they were accompanied by three Aboriginal elders who originated from the area, Johnny Flinders, Bill McGreen and Bob Flinders. 'Bob (Flinders) himself (*Othelarrni, Ngumartaydini*) was born at Flinders Island in 1913' (Sutton, Lorblanchet and Layton 1975: 34). The Flinders Group art also received attention during the preliminary archaeological survey by Dr. John Beaton and Tia Negerevich in 1979, detailed in a later report by Negerevich (1979).

Q.NPWS cultural resource surveys

In 1982 an unsuccessful attempt was made by the Q.NPWS Historic Sites Officer to make a crossing from Lakefield NP to the Flinders Group to investigate unconfirmed reports of site interference. It was July 1983 before a combination of circumstances and favourable weather conditions permitted the author to make a crossing of Rattlesnake Channel from Bathurst Heads in order to pursue these investigations. A rapid deterioration in the weather reverting to the normal high seas and south easterly winds confined mobility in the 5 m aluminium dinghy (the only transport available) to areas of Stanley and Flinders islands. An attempt to reach the important Clack Island galleries with the assistance of a local 11.5 m trawler was also thwarted by the weather, and eventually the services of this craft had to be sought to return the survey team from Flinders Island to the mainland.

A primary recommendation made in the report on this preliminary survey was to 'organise at the first available opportunity a site location and recording program for the area to be able to more adequately assess the area's cultural resources and advise on conservation and management needs' (Walsh 1984b: 83).

In July 1984 another attempt was made to undertake a preliminary investigation of the Clack Island galleries, using the flexibility of a Grumman Mallard Seaplane. Although a landing was made off Clack Reef, high seas again made landing impossible. The two person survey team was alternatively transferred to an Apa sandspit base camp on Flinders Island to further investigations in that area (see Plate 1).

In an attempt to overcome the inherent weather problems and logistics which persistently plagued attempts to obtain the most basic overview of the island's cultural resources, another attempt with different transport was undertaken between 22-27 April 1985. Working in conjunction with a Q.NPWS patrol of the Great Barrier Reef Marine Park, the survey was undertaken on the charter vessel *Sunbird*, a 14 m engine/sail powered catamaran research vessel, operating out of Lizard Island Research Station.

A landing was made on Clack Island, using a 5 m inflatable boat to make an eventful crossing of Clack Reef. The extent and significance of the galleries was immediately evident, as was the apparent marked deterioration of the artwork compared to the early discoverers' reports. Landings and preliminary investigations of art sites were also made on Stanley, Flinders, Denham, the three Cliff islands and the mainland Cape Melville National Park.

Observations on the rock art sites were made in a subsequent interim report, with the needs summarised in a brief statement. 'My earlier optimism regarding archaeological potential of this important area, and concern for rapidly advancing developmental pressures have been further compounded as a result of this expedition. Hopefully the 1985/86 budget will permit further detailed follow up work which will ensure more comprehensive documentation of Service estate cultural resources, and a move towards public management of same' (Walsh 1985b: 7).

Project Flinders

In spite of the unavoidable constraints enforced on the three preliminary Flinders Group site surveys by unfavourable weather conditions, sufficient observations had been made to verify the extent and importance of local cultural resources. The rapidly increasing presence of uncontrolled visitors and the resultant interference and degrading of important sites was also apparent.

The situation at that time has since been aptly described in a general statement by Gale and Jacobs (1987: 3). 'Those concerned with protecting cultural and natural heritage sites are now faced with a frightening dilemma—how do they protect sites under constant pressure from increasing numbers of visitors when the natural protection of isolation can no longer be relied on?'

In order to develop the urgent management strategies required for the now endangered cultural resources of Flinders Group NP, a major recording program (Project Flinders) was planned for the latter part of 1985. This was planned in conjunction with Q.NPWS marine park and island management activities in the area, with support from the Great Barrier Reef Marine Park Authority. Optimum weather conditions for the all important survey component of inter-island mobility were considered by experts to be November. At this time the lull which precedes the onslaught of the cyclone season often produce calm seas.

The four person survey crew and necessary equipment were transported to a base camp at Frederick Point on Flinders Island by the *Aurora*, a 13 m charter boat. The half-cabin Marine Park patrol boat *Remora* was also towed to this base by the *Aurora* for use during the survey. From 8 to 28 November, two Historic Sites Officers and two assisting Marine Parks' rangers systematically surveyed Flinders, Stanley, Denham, Blackwood, King, Clack, the two Rocky Isles, the two Macleay Islands, the three Cliff Islands and part of Cape Melville NP foreshore. Brief periods of calmer weather were used for the most dangerous forays across unprotected expanses of ocean, to such areas as the Cliff Islands, a distance of approximately 50 km.

Cultural resource recording

A special set of recording forms were designed to suit the art, environment and management needs observed during the preliminary surveys. These involved modification of the original Q.NPWS computer-based proformas developed for the Central Highlands Management Project, shown in Walsh (1984a: appendix C). A special 43 question recording proforma, accompanied by an illustrated shell identification chart, was devised to deal with the extensive middens associated with art sites. Additional specialist recording proformas were devised, with up to 33 questions to permit analysis of the obvious motifs dominating the art body. These included turtle, moth, CHZ (crescent head zoomorph), and boat motifs.

In the island segment of the Princess Charlotte Bay (PCB) survey, 35 art sites, three burial sites, 72 middens and one stone arrangement were recorded in detail by this system. The infinite options provided by these recording proformas' sub-questions has permitted 425 individual primary sets of histograms to be drawn up from the computer analysis. These provide a working base for reference, research, comparison and management planning of Flinders Group NP cultural resources which can be taken to infinitely finer degrees where highly specialised detail is required.

Sites and art detail were recorded on 35 mm colour transparency with 6x6 cm B&W negative, with 6x6 cm transparency reserved for specialist detail. Over 400 motifs were sketched by eye on individual A4 sheets, accompanied by pertinent site, orientation, dimension, technique and colour documentation.

In spite of its geographical proximity to the Laura art body, the Flinders Group NP art represents the core of a discrete regional body of art, referred to as the PCB Style by Walsh (in press). The art was found to be predominantly freehand, ranging from early monochrome red styles to bichrome recent styles featuring complex decoration, and in cases distinctive white dot outlining. Hand stencils were rare and restricted to a small number of sites, while evidence of any form of engraving was totally absent.

Many motifs were of a marine theme, with a number suggestive of being associated with local mythology (of which remnants have been documented). The varying degree of schematisation present in the art does not preclude the possibility of less readily identifiable motifs having similar associations. The almost invariable association of this art with rock shelters facing the ocean is consistent with observations made on the lifestyle of marine oriented north Queensland Aborigines made by Hynes and Chase (1982: 43). 'Looking out over the changing face of the sea appeals to many people - most particularly those Aboriginal groups whose myths largely involve actions of marine creatures'.

A distinctive feature of the PCB style art is a frequently depicted motif, shown with many variations, which was interpreted by local Aborigines in 1927 as '*Motjala*', meaning moth or butterfly (Hale and Tindale

1934: 155).

An equally interesting theme is evident in a wide range of boat paintings, first recorded on Clack Island and illustrated by Coppinger (1883: Fig. 3, opp. p. 192). Another example in the Stanley Island gallery was interpreted by local Aborigines in 1927 as a trepang cutter (Hale and Tindale 1934: 148). Other Flinders Group examples include a steam boat and sailing boats depicting fine detail, sometimes to the degree of letters of the boat's name and having a dinghy in tow. Mast, rigging and cabin details is in instances so accurate that one painting has been identified as representing the lugger, *The Mildred*. Schematisation is evident in earlier boat paintings, which range through five mast craft to a distinctive 'prau' shaped craft, an example of which is shown in Plate 2. A detailed discussion and analysis of this rock art body will be covered in a separate publication.

Evidence of the long period of contact with alien cultures was not only evident in the art but also in the proliferation of middens. Worked glass, fragments of pottery and ceramics, a silver teaspoon, an Irish clay pipe bowl, copper sheeting, cast bronze fittings and squared bronze nails were among the introduced materials observed during the survey. Some of this material undoubtedly represented trade and theft from visiting boats, while another source would have been the wreck of the *Frederick* on Cape Flinders, Stanley Island (King 1837: 18).

An interesting array of traditional material still survived in areas distanced from the access point of visitors. These included a baler shell dish, similar to Hale and Tindale (1934: 132, Figs 170, 171); a range of nautilus shell pendants, similar to Hale and Tindale (1934: 142, Fig. 207); and numerous blanked out melo shell ovals, similar to Hale and Tindale (1934: 101, Figs 90, 93).

The blanked out melo shell ovals are of particular interest, as they were traditionally manufactured as decorative objects for mounting on the hand grip extremity of woomeras, as shown in Hale and Tindale (1934: 161, Figs 86, 87, 88). Finished but unmounted specimens were much sought trade items with inland Aborigines regarding them as highly prized decorative pendants. Some indication of the vast distances these objects were traded over is indicated in McCarthy (1939: 93, map 14) and Akerman (1973: 124, Fig. 1). Detail of stencil outlines of these shell objects occurring in galleries on the Central Queensland Sandstone Belt has been described by Beaton and Walsh (1977: 46).

Survey findings

Findings of the survey were covered by Walsh (1985d: 1-14) emphasising the obvious damage resulting from an escalation of uncontrolled visitation in the seven month period between the two 1985 surveys. Well defined tracks had now been formed to several of the main sites, including particularly sensitive cultural sites. Beer cans now scatter approaches to these once remote scrub shrouded settings. Crisscrossing pads pounded up to 10 cm deep through loose shelter floor deposits where sizeable excavation holes evidenced vandals scavenging for relics and human remains. The subsequent dust damage to sloping art surfaces in the main Ship Shelter was comparable to this increased activity. Feral pigs were causing considerable damage to Stanley Island middens, while on Flinders Island feral goats were rubbing art surfaces and reducing shelter floors to the appearance of stockyards.

Two major Stanley Island art sites, Ship Shelter and Endaen Shelter, were considered suitable for development, and boardwalk plans were illustrated in Walsh (1985d: Figs 1-3).

Aboriginal involvement

The problems and urgency surrounding rapidly increasing risks of interference with Flinders Group NP cultural sites were discussed with members of the Hopevale Community during visits by Q.NPWS staff involved in the marine park. Bob Flinders was then the last surviving member of the group identified by the 1975 AIAS expedition as directly associated with the area, and his advice was sought in developing management strategies, particularly for the more sensitive sites. Mr Flinders' age did not permit him to participate in the subsequent arduous development phase of the project, but two other Aborigines were employed to participate and represent the Hopevale community.

Project Flinders Development

A plan was developed to undertake all segments of the preliminary phase of Flinders Group NP management developments in a single major operation. May 1986 was selected as most suitable, and all resources were to be pooled in order to overcome the daunting logistics problems.

Approximately 80 tonnes of materials and equipment were shipped over 400 km from Cairns to an Owen Channel mooring on the 110 tonne barge *MV Tasma*. Site development materials alone included 12 tonnes of construction timber, 1 tonne of sign timber and posts, 4 tonnes of cement powder, 6 tonnes of aggregate and premix, three-quarters of a tonne of cement blocks and a similar weight of extruded mesh. A small 4WD vehicle was brought to transport gravel and water for track consolidation. In addition to this two demountable buildings and prefabricated canopy represented the nucleus of a permanent base, and a small D3 dozer was brought to haul them to the site.

The project's work force included eight Q.NPWS rangers, two Historic Sites Section officers, two Aborigines from Hopevale, four contract carpenters and twelve Army volunteers from Townsville's Laverack Barracks. A chartered vessel, the *Nimrod* transported the group from Cairns and Bathurst Head rendezvous points to reach the Flinders Group NP on 20 May.

With the assistance of the Queensland Government Squirrel helicopter and its three man crew the entire operation was completed in 16 days. A Bell Jet Ranger helicopter had originally been used by Q.NPWS to pioneer the technique of airlifting pre-cut boardwalk materials into remote art sites during the May 1980 Carnarvon NP project. This technique is shown in Walsh (1984a: Pl. 16A) and Gale and Jacobs (1987: 68). The additional power of the Squirrel helicopter used at Stanley Island proved advantageous in coping with gusty wind conditions while hovering for payload coupling above the moored barge's deck, see Plate 3. Materials had been precut and packaged into suitably weighted loads to increase trans-shipping efficiency. Forty airlifts were required to move all materials to the respective construction dropsites approximately one km distant. The helicopter was also used for aerial survey work of Clack Island sites and the Flinders Island stone arrangement. The values of helicopters in site survey work have been dealt with in Walsh (1981: 78; 1983b: 53; 1984a: Pl. 13A).

Boardwalk design was based on the modular system which had been developed in the original 1980 Carnarvon design, the values of which have been covered by Gale and Jacobs (1987: 69) and Devine (1986: 9). This facilitated ease of removing modules from the construction to permit future archaeological investigations. Particular attention was paid to selecting construction materials which would cope with the variations of extremes in the salt-laden atmosphere of this tropical setting, where a long rain-less winter changes to the cyclonic downpours of the wet season. The twelve tonnes of hardwood used in the boardwalks consisted of selected rosegum, yellow stringybark and red stringybark, with the latter used mostly for boardwalk decking. All upright material was painted, decking treated with linseed oil, and galvanised bolts and fittings used throughout the project.

The basic construction was a variation of the original 1980 Carnarvon NP design which has proven so successful (Gale and Jacobs 1987: 69). Various segments of this style are shown in Gale and Jacobs (1987: Pls 25, 30), Walsh (1984c: Pls 10A, 10B, 11A, 11B), Walsh (1985a: Pls 42A, 42B, 43A).

Extreme floor level variations present in the Stanley Island shelters necessitated the use of a modified mounting system. Broad cement pads were laid (without disturbing the archaeological deposits) to distribute structured weight and minimise compression. Square cement blocks were used to form reinforced bearer support piers, varying more than a meter in height, as shown in Plate 4. When painted Service brown (together with the supported structure) this represented a strong moisture, fire and termite resistant foundation which blended in well with the robust squared timber theme.

As with all Q.NPWS site developments, the boardwalk alignment was devised by exhaustive on-site tests with pegged profiles. To be effective, boardwalks must permit optimum viewing and photographing angles of all main art panels while sufficiently distanced from the art to avoid touch by the curious. Cameras and lenses of different format and focal length were used to determine the final alignment. The value of catering for visitor photographic needs has also been stressed by Gale and Jacobs (1987: 69). Provision for seating facilities were incorporated into the designs, positioned where visitors can pause to appreciate the most striking panels of art, impressive views of the gallery and the associated surrounding environment.

Artwork on the sloping shelter walls was covered with plastic sheeting to minimise dust damage during the construction, as shown in the Ship Shelter (see Plate 5). Materials and equipment were carried from the drop points up steep slopes to the teams which had been allotted to each site. Alternating labour intensive tasks between the sites permitted construction to proceed in tandem with the most efficient use of the available work force.

With the assistance of Conrad Rosendale, one of the Hopevale Aborigines on the project, tree species providing food, medicine and manufacturing needs of the former inhabitants were identified in the immediate site's area. A circuitous track was then planned to provide access to the sites and pass close to a selection of 16 different species of these resource trees.

Lowland sectors of the track crossed sand flats, but the steep slope areas consisted of powdery leached soils which posed major dust and erosion problems. Consolidation of some steep surfaces was unavoidable if art surface dust damage was to be minimised and track survival was to be attained under alternating seasonal stresses of heavy pedestrian traffic and cyclonic downpours. A distance of 115 m between the two art sites, and an additional 67 m segment exiting to a look-out was cleared for the track and covered with plastic sheeting to protect the deposits. More than nine cubic metres (18 tonnes) of concrete was mixed to form the cement path, with combinations of red and black oxide used to create a surface colour compatible to adjacent soils. The steep inclines required cement barrows to be pushed by one individual and pulled by a rope by two others. In sensitive areas all cement was carried by bucket brigades, as shown in the Endaen site, Plate 6.

The return leg of the track is a steep 40 m descent from the look-out to the beach, crossing sloping dunes which form the scattered extremity of a major midden. To avoid dune erosion and disturbance of midden material, a track surface formed by 20 sheets (each 2000 x 450 mm) of G.I. extruded steel mesh has been laid down. Linked end to end and mounted on one side by loops around angled galvanised steel star pickets, this design's flexibility facilitates easy adaptation of the surfacing to natural variation of dune contours.

Stanley Island development centred around a focal point of the visually attractive rock art galleries, but strived through interpretation to foster a far greater overview. A prime objective was to encourage public appreciation of the complex Aboriginal physical and spiritual relationship with the environment, resources and creation of the landscape. Detailed plans for text, illustrations and signage for the Interpretive Trail theme were provided in Walsh (1986a: 1-32), with aluminium signs considered most suited to extreme local conditions.

Standard Q.NPWS routed hardwood signs guide visitors along the east bank to the trail's first interpretive sign, which advises:

> This point marks the commencement of a self-guiding trail interpreting Aboriginal use of the area, centred around two major art/habitation shelters. Sixteen tree species used by the Aboriginals have been identified along the trail. Text identifies Aboriginal name for species in Kokoyimidir language, common name, botanical name, and food, manufacturing or medicinal values. Location indicates distance of tree behind each identification sign.

Wooden stairs guide visitors to the boardwalk system of the 58 m long Endaen Shelter (see Plate 7), where 12 hand rail interpretive signs (with motif image and directional arrow) similar to the example shown in Plate 8, provide detail on a range of the 168 freehand motifs. Information also deals with the boardwalk's purpose, art mythology, a fresh water source, adjacent food resources, the shelter's extensive midden, history of Aboriginal use and excavation details. Endaen Shelter was first recorded by Europeans on 14.1.1927 (Tindale n.d.: 32), and excavated by Dr John Beaton during his 1979-80 archaeological survey, revealing an occupation date of 2370 ± 100 years (Beaton 1985: 6).

A feature has been made of a wooden sleeping platform in the site (see Plate 7), originally referred to by Hale and Tindale (1934: 127). Mention was made of the structure still standing in 1979 by Negerevich (1979: 21), but it had collapsed by the time of the author's 1983 visit. Plate 9 shows restoration of the structure in 1986, based on detail obtained from historic photos from Hale and Tindale (1934: Figs 158, 160).

A concrete-surfaced path guides visitors past seating facilities, which provide views of the ocean and into Endaen Shelter (as shown in Page 7), meandering through scrub up the slope to the Ship Shelter. A more complex boardwalk and platform system (see Plate 4) is required to permit public appreciation of the 270 freehand motifs decorating the walls of this 12 m-wide cavernous shelter. Fourteen handrail signs interpret the extensive entrance midden, adjacent food resources and a range of the art, such as Plate 3, which is one of the 12 boat motifs present. Plate 10 shows a panel of bichrome double bar decorated birds (erroneously referred to in Trezise [1969: Pl. 24] as being in 'the Clack Island Gallery') which in this form of presentation is unique to the PCB style.

A concrete-surfaced path leads 65 m down to a developed look-out area, where visitors may appreciate a panoramic view of the site setting, surrounding environment and adjacent islands. Five hand rail signs provide historic information, place names and detail of features with mythological association with the whale and long-tailed stingray (which are also depicted in the rock art).

Summary

Phase one of the Flinders Group NP management development program, including construction of a field base for management of the island and the surrounding marine park, was completed at Frederick Point by 5 June. Interpretive signs were installed by November 1986. Some indication of this project's effectiveness may be gained by evaluating the 538 visitor book comments added over the first ten months of operation. Favourable general comments - 174, favourable boardwalk comments - 71, negative boardwalk comments - 16 before installation of explanatory signs, but only four after. Origin of visitors was: Queensland - 128, far north Queensland - 109, interstate - 112 and overseas - 47.

Benefits resulting from planned presentation, interpretation and increase official presence in the Flinders Group NP are the better servicing of public educational requirements, increased protection for cultural resources, and benefits to Aboriginal culture through the fostering of greater public understanding, appreciation and respect.

The Q.NPWS policy of presentation and preservation of rock art sites has extended a stage further since Project Flinders. Boardwalk developments in the Nara Inlet Gallery on Hook Island NP (in the Whitsunday Group of central coastal Queensland) were completed in March 1988. The Stanley Island and Hook Island developments now make the two best known examples of island rock art on the east coast of Australia accessible in protected form for public appreciation.

REFERENCES

AKERMAN, K. 1987. Aboriginal baler shell objects in Western Australia. *Mankind* 9(2): 124-25.

BEATON, J. M. 1985. Evidence for a coastal occupation time-lag at Princess Charlotte Bay (North Queensland) and implications for coastal colonisation and population growth theories for Aboriginal Australia. *Archaeology in Oceania* 20: 1-20.

BEATON, J. M. and WALSH, G. L. 1977. Che-ka-ra. *Mankind* 2: 46-48.

COPPINGER, R. W. 1883. *Cruise of the alert.* W. Swan Sonnschein and Co., London.

DEVINE, D. 1986. Cultural site protection - the Stanley Island experience. *Australian Ranger Bulletin* 4(1): 8-9.

FLINDERS, M. 1814. *A voyage to Terra Australia* (2 vols). G. and W. Nicol, London.

GALE, F. 1983. The protection of Aboriginal rock art from tourists at Ubirr. In H. Sullivan (ed.), *Visitors to Aboriginal sites: Access, control and management*, pp. 32-40. A.NPWS. Canberra.

GALE, F. and JACOBS, J. M. 1987. *Tourists and the National Estate.* Australian Heritage Commission Special Australian Heritage Publication Series No. 6., Canberra.

GILLESPIE, D. A. 1983. The practice of rock art conservation and site management in Kakadu National Park. *The rock art sites of Kakadu National Park.* Special Publication 10. A.NPWS, Canberra.

GILLESPIE, D. and SULLIVAN, H. 1983. Report on the Plenary Session, In H. Sullivan (ed.) *Visitors to Aboriginal sites: Access, control and management*, pp. 114-23. A.NPWS. Canberra.

HALE, H. M. and TINDALE, N. B. 1934. Aborigines of Princess Charlotte Bay, north Queensland, Part II. *Records of the South Australian Museum* 5(1): 63-116.

HYNES, R. A. and CHASE, A. K. 1982. Plants, sites and domiculture: Aboriginal influence upon plant communities in Cape York Peninsula. *Archaeology in Oceania* 17(1): 38-50.

KING, P. 1937. *Narrative of a survey of the intertropical and western coasts of Australia between the years 1818 and 1822.* London.

LAMBERT, D. 1988. Ward G. K. (ed.). *Conserving Australian rock art. A manual for site managers.* Aboriginal Studies Press, Canberra.

McCARTHY, F. D. 1939. 'Trade' in Aboriginal Australia and 'trade' relationships with Torres Strait, New Guinea and Malaya. *Oceania* 9(4): 80-104.

McMINN, A. 1970. *Allan Cunningham, botanist and explorer.* Melbourne University Press, Melbourne.

MESTON, A. 1895. *Geographic history of Queensland.* Government Printer, Brisbane.

NEGEREVICH, T. 1979. *A report on the Bathurst Bay/Flinders Island Group area.*

ROSENFELD, A. 1985. *Rock art conservation in Australia.* Australian Heritage Commission Special Heritage Publication Series No. 2. Australian Government Publishing Service, Canberra.

ROTH, W. 1899. Distribution of gifts to coastal Aboriginals. Unpublished report to the Commissioner of Police.

ROTH, W. 1902. Games, sports and amusements. *North Queensland Ethnography.* Bulletin No. 4. Government Printer, Brisbane.

STOKES, J. L. 1846. *Discoveries in Australia* (2 vols). T. and W. Boone, London.

SULLIVAN, K. M. 1983. Monitoring visitor use and site management at three art sites, an experiment with visitor's books. In H. Sullivan (ed.), *Visitors to Aboriginal sites: Access, control and management*, pp. 43-53 A.NPWS, Canberra.

SUTTON, P. 1976. Proper names of patrilineal descent groups in the Cooktown - Princess Charlotte Bay area, north Queensland. Unpublished typescript held in AIAS (A3 141).

SUTTON, P., LORBLANCHET, M. and LAYTON, R. 1975. Flinders Island Expedition. *AIAS Newsletter* No. 3: 34-37.

TINDALE, N. B. n.d. Journal of a visit to Cape York Peninsula, north Queensland, November 1926 - March 1927. Unpublished typescript document.

TREZISE, P. J. 1969. *Quinkan Country.* A. H. and A. W. Reed, Sydney.

WALSH, G. L. 1981. Site locating by helicopter in Queensland's Central Highlands. *AIAS Newsletter.* No. 15: 73-78.

WALSH, G. L. 1983a. Archaeological site management in Carnarvon National Park. In H. Sullivan (ed.), *Visitors to Aboriginal sites: Access, control and management*, pp. 1-14. A.NPWS, Canberra.

WALSH, G. L. 1983b. *The roof of Queensland.* University of Queensland Press, Brisbane.

WALSH, G. L. 1984a. *Managing the archaeological sites of the Sandstone Belt.* Q.NPWS, Brisbane.

WALSH, G. L. 1984b. Archaeological significance of Flinders Island/Princess Charlotte Bay. Unpublished illustrated report to Q.NPWS, Brisbane.

WALSH, G. L. 1984c. Archaeological significance of Hook Island. Unpublished illustrated report to Q.NPWS, Brisbane.

WALSH, G. L. 1985a. Archaeological significance of Lawn Hill Gorge. Unpublished illustrated report to Q.NPWS, Brisbane.

WALSH, G. L. 1985b. Visit to Cape Melville and Princess Charlotte Bay. Unpublished illustrated report to Q.NPWS, Brisbane.

WALSH, G. L. 1985c. Flinders Island Group Conservation Project. Unpublished report to Q.NPWS, Brisbane.
WALSH, G. L. 1985d. Site recording report and site development suggestions, Flinders Group NP. Unpublished illustrated report to Q.NPWS, Brisbane.
WALSH, G. L. 1986a. Stanley Island Historic Sites Project. Unpublished report to Q.NPWS, Brisbane.
WALSH, G. L. 1986b. Cultural resource management in Queensland National parks. *Australian Ranger Bulletin* 4(1): 5-8.
WALSH, G. L. 1988. Clack Island report. Unpublished illustrated report to AIAS, Canberra.
WALSH, G. L. 1988. *Australia's greatest rock art.* E. J. Brill/Robert Brown and Associates (Aust.) Pty. Ltd., Bathurst.
Newspapers:
September 1986. Stanley assault. *Newspaws.*
30.5.1986. New Ranger base for Flinders Island. *Cairns Post.*

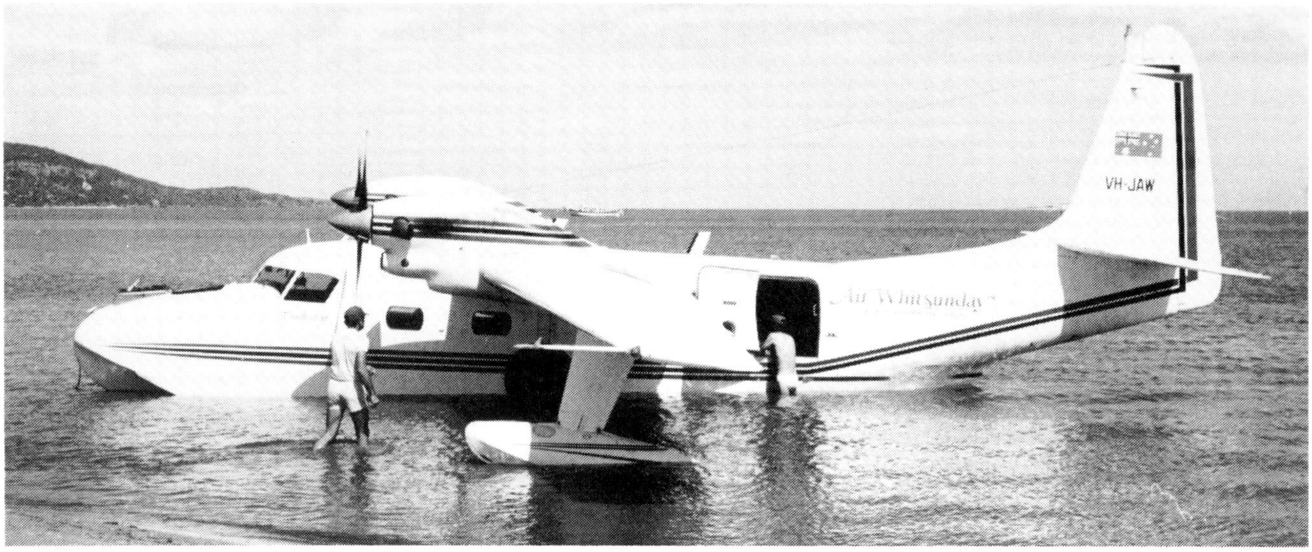

Plate 1. *The Grumman Mallard seaplane used during the preliminary site assessment stage in an attempt to overcome logistics problems associated with gaining access to the remote Flinders Group NP islands.*

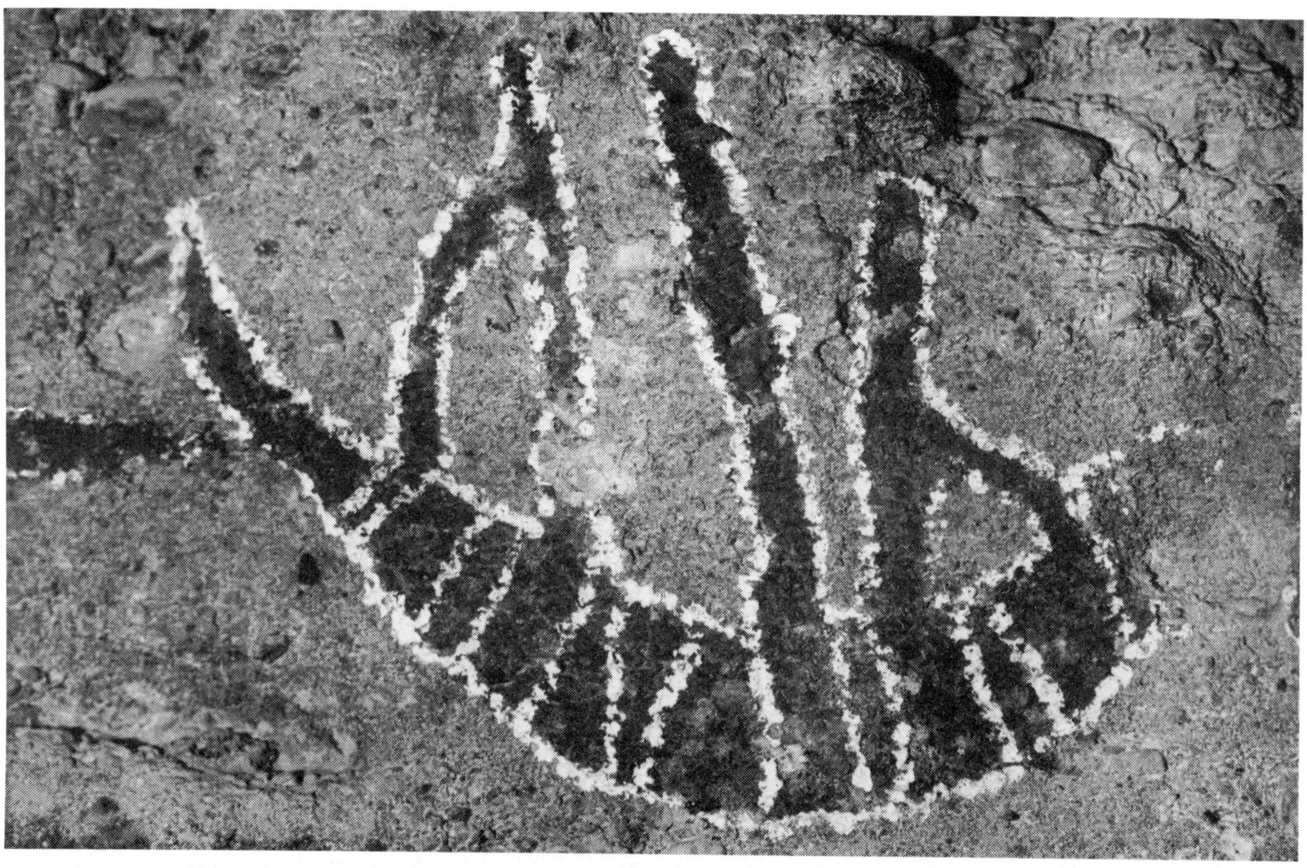

Plate 2. *One of the Ship Shelter boat paintings depicted in the PCB art style. This particular type of craft recurs frequently in this area's art, appearing to be among the earliest depictions in a range of often very detailed boat paintings.*

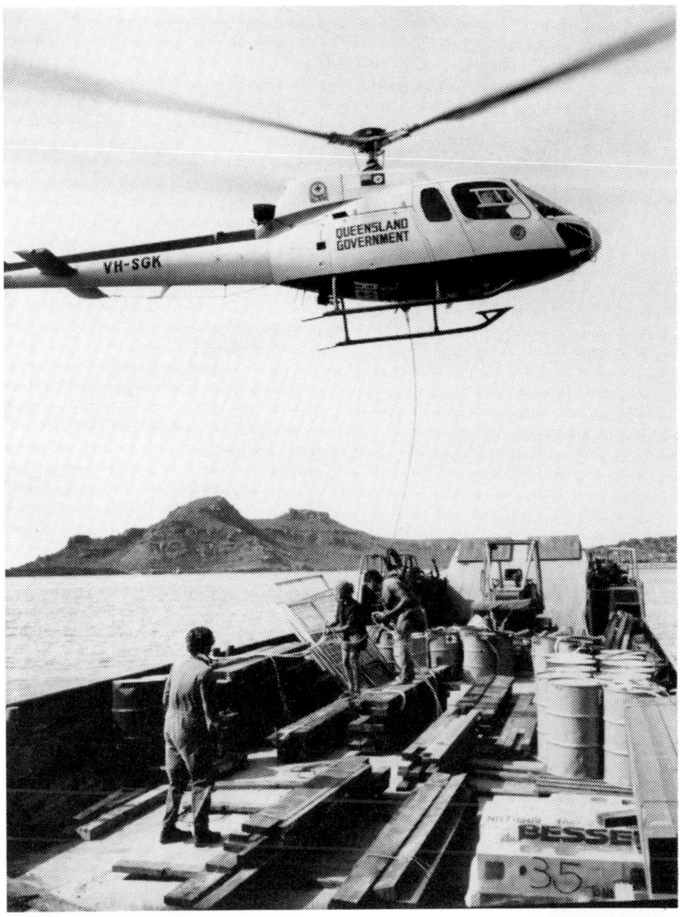

***Plate 3.** The Queensland Government helicopter airlifting pre-packaged bundles of boardwalk materials from the deck of the barge to construction crews waiting at the respective island drop sites.*

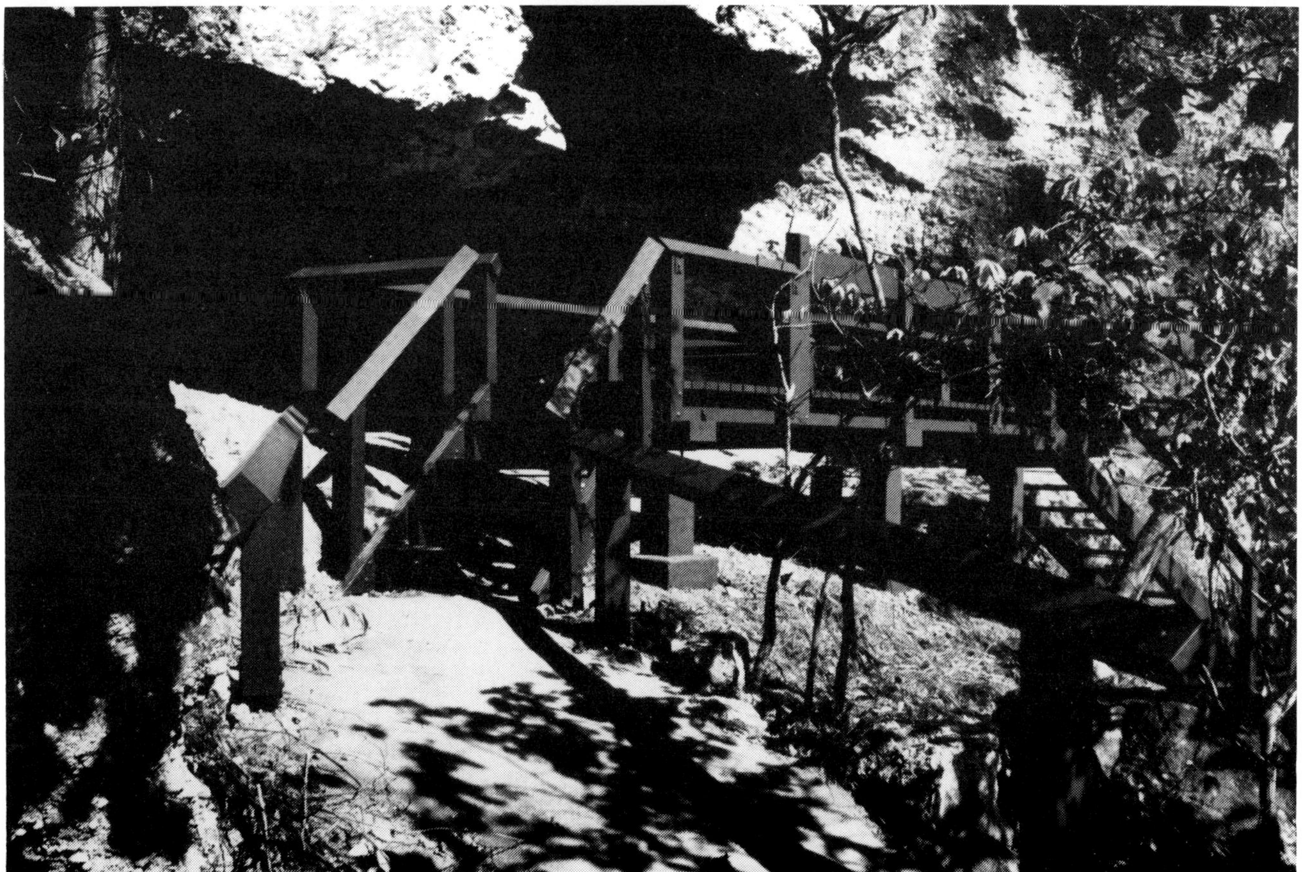

***Plate 4.** Approach to the completed Ship Shelter development, with the seating equipped outer observation deck and access staircases spanning the site's extensive entrance area midden.*

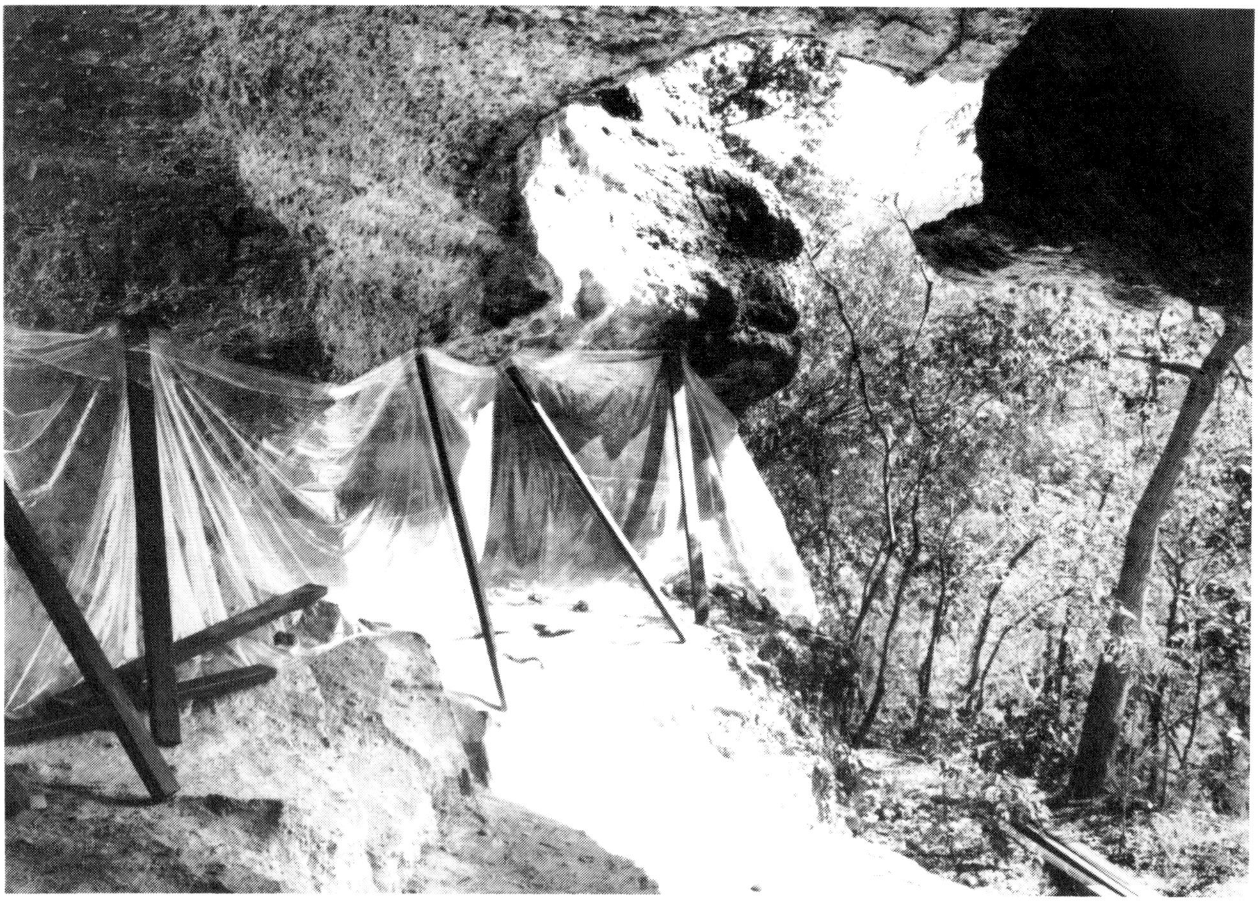

Plate 5. *Protective transparent plastic sheeting draped over the extensive art panels of Ship Shelter as a means of minimising dust damage during boardwalk construction.*

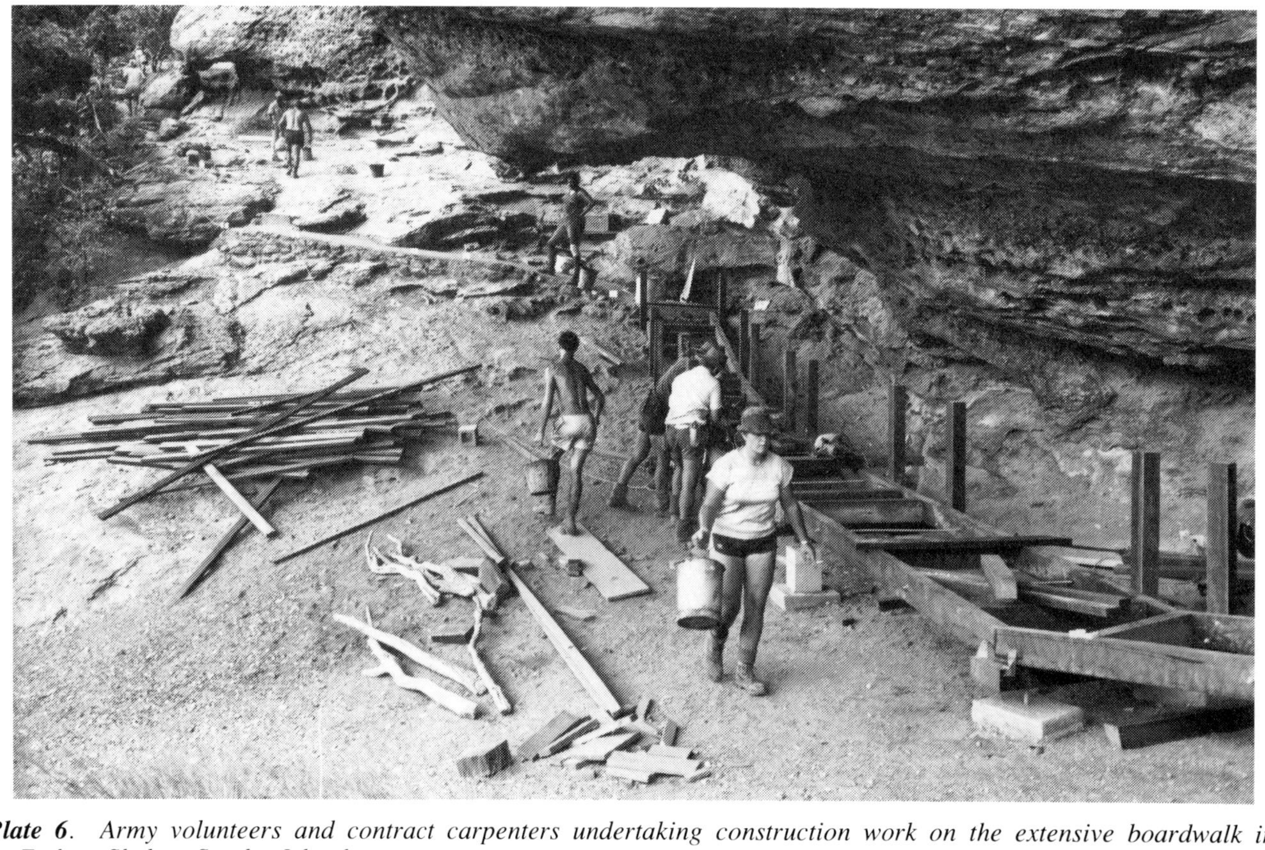

Plate 6. *Army volunteers and contract carpenters undertaking construction work on the extensive boardwalk in Endaen Shelter, Stanley Island*

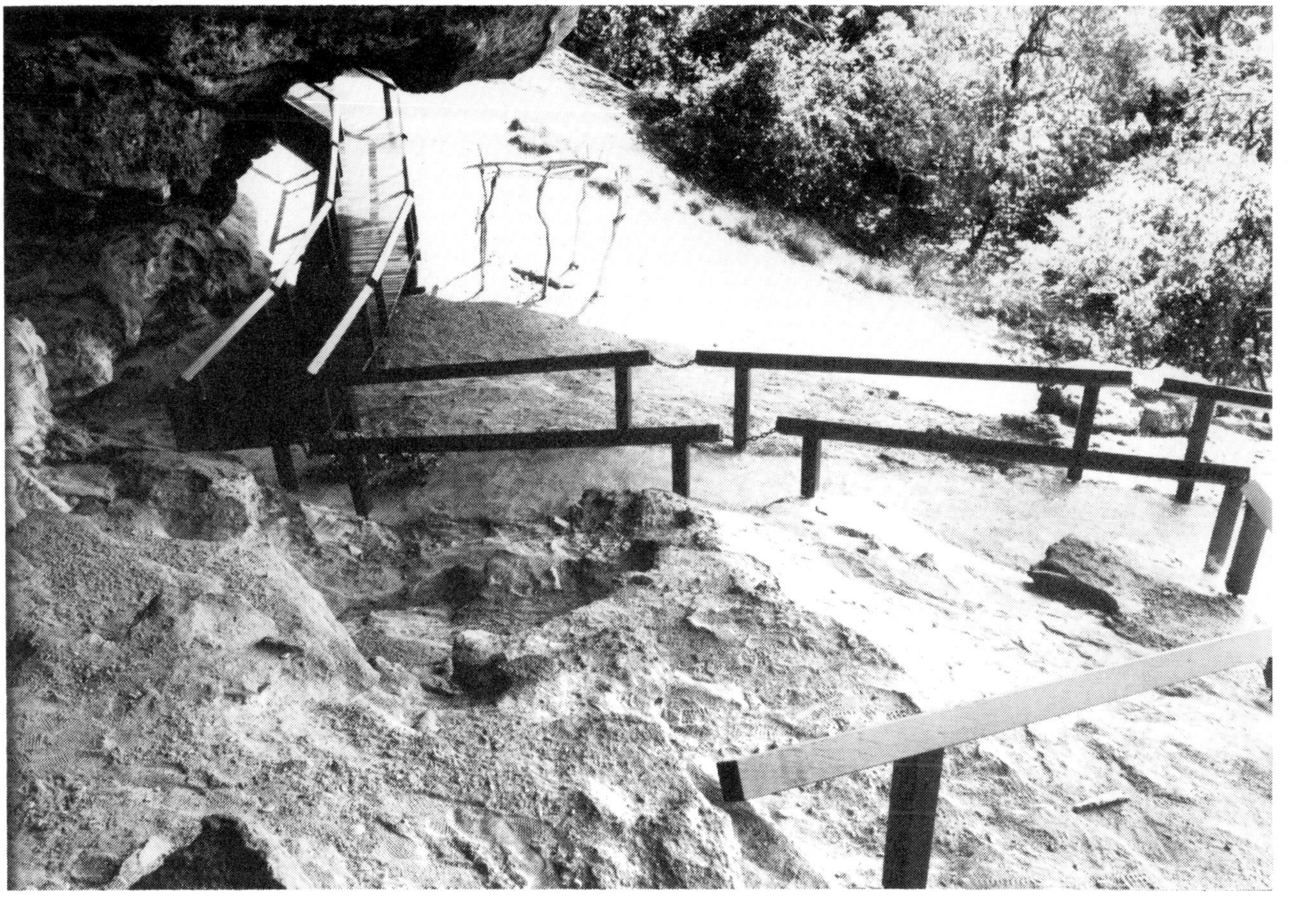

Plate 7. The eastern approach to the Endaen Shelter's completed developments. Wooden handrails line dust reducing concrete surfaced access tracks which lead to the wooden boardwalks. The restored Aboriginal 'sleeping platform' can be seen standing in its original location.

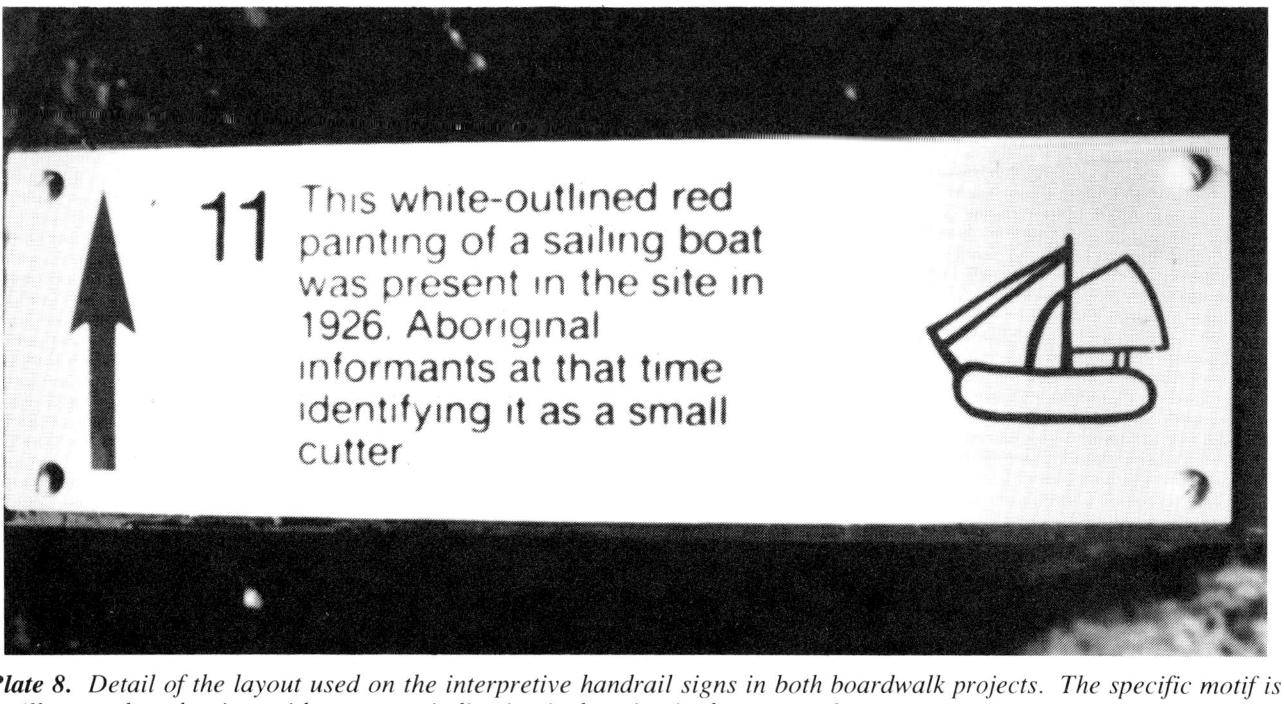

Plate 8. Detail of the layout used on the interpretive handrail signs in both boardwalk projects. The specific motif is illustrated on the sign, with an arrow indicating its location in the art panel.

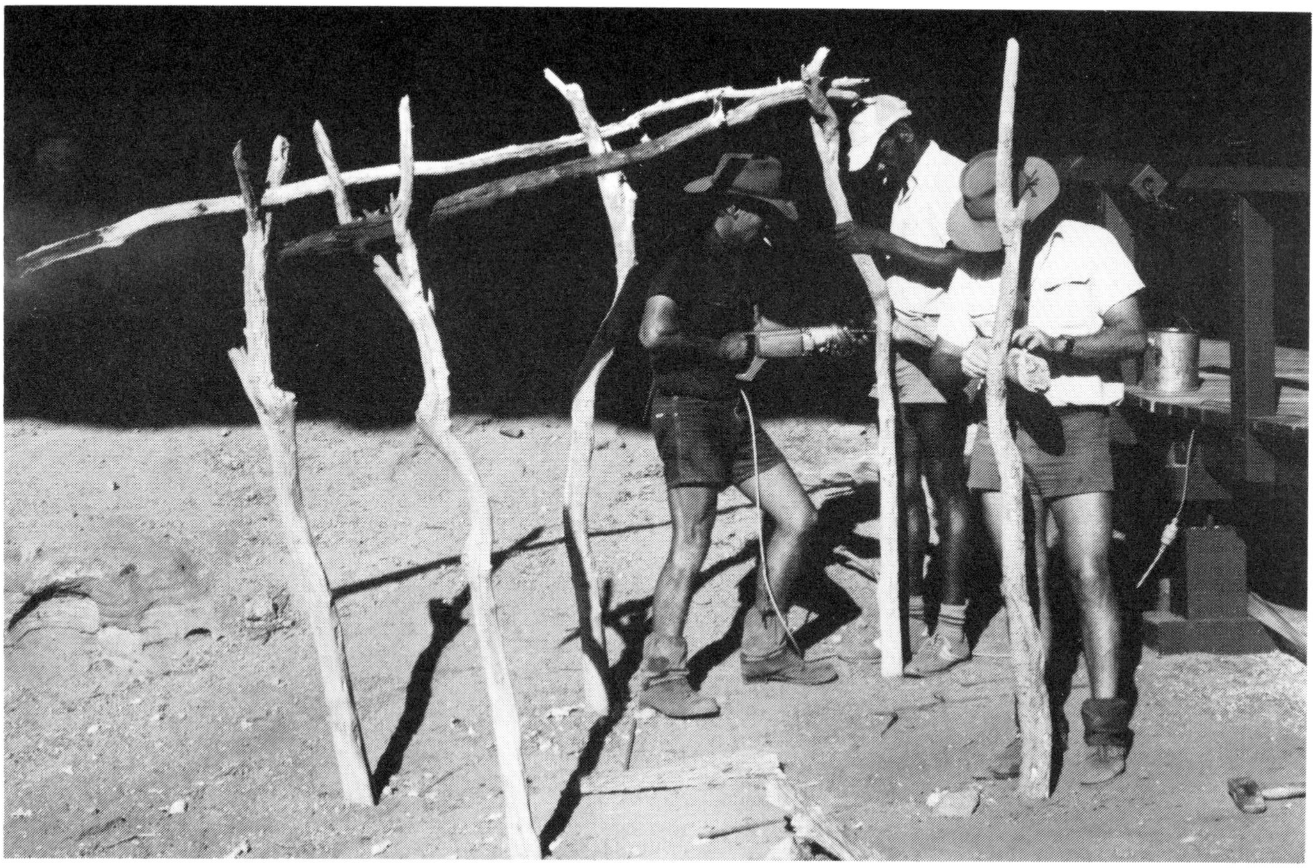

Plate 9. *Reconstruction work under way on the collapsed Aboriginal 'sleeping platform' in Endaen Shelter. Photographs taken by the 1927 Hale and Tindale expedition were used to accurately reassemble the deteriorated woodwork.*

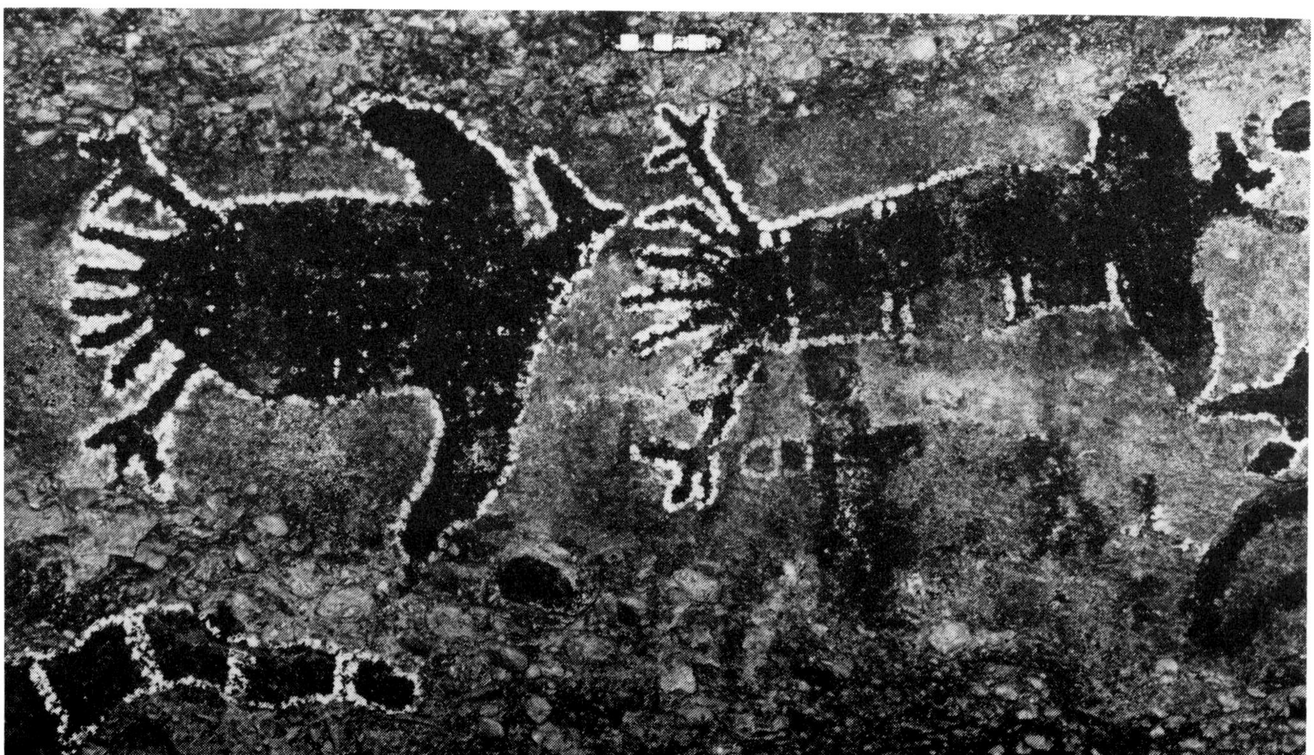

Plate 10. *An example of bichrome rock art from the Ship Shelter. The two birds in flight are depicted in the white dot outline and double bar decorative infill common to the recent PCB art style.*

8

MT CAMERON WEST ROCK ART PROTECTION PROJECT:

An assessment of the conservation program undertaken to June 1988

Tony Blanks and Steve Brown

Abstract. This paper briefly describes the rock art at Mt Cameron West and its setting; outlines the problems involved in short and long-term site conservation, and describes the protection strategy adopted. A preliminary assessment of the success of the strategy is included. The short-term protection strategy adopted consists of repairing the breach in the foredune and stabilising the adjoining extensive sand blow with marram grass. A boundary fence will then be extended across the sand thereby excluding cattle. This strategy appears to be succeeding. Long-term protection measures will necessitate ameliorating the likely impact of possible rises in sea level. This problem has not yet been addressed.

Background and Previous Investigations

It is generally agreed that in the early 1830s when George Augustus Robinson passed close by the Mt Cameron West engraving site (Figure 1) it was covered with sand which supported a thick ti-tree and other scrub cover (Plomley 1966). It is likely that this sand cover remained until shortly before the site was reported to Meston in the early 1930s (Meston 1932, 1934). Since that time many people have commented on the state of preservation, and problems of conservation, of the site. Some slabs from the site were removed to Tasmanian museums in the 1950s and 1960s.

In 1969 Rhys Jones and Harry Lourandos undertook a major archaeological study of the site (Jones 1981), and McCarthy recorded the engravings in detail (McCarthy 1979). The results of the archaeological study indicate that the engravings were completed prior to 1,350 years BP, and that from that time until about 800 years BP the site was used as a campsite. After 800 BP the site was abandoned and became covered with sand and scrub.

The 1969 excavations may have exposed only a tiny fraction of the site, and the entire blocky talus slope may be covered with engravings (Plate 1). The site is widely regarded as being of international cultural significance (see Brown, in press). The site is of considerable importance to present day Tasmanian Aboriginal people, some of whom are seeking freehold title vested in the Tasmanian Aboriginal community ('land rights') for the area.

The Mt Cameron West Aboriginal Site (560 ha) was proclaimed a State Reserve under the National Parks and Wildlife Act on 20 June 1970, having previously been compulsorily purchased by the Tasmanian Government from the adjoining Woolnorth property of the Van Diemens Land Company.

In 1979 a preliminary geological and geomorphological investigation in relation to the deterioration, conservation and management of the site was undertaken (Hughes, Watchman and Sullivan 1981). This report recommended that comprehensive erosion control measures be implemented as a first step.

Setting of the Site

The Mt Cameron West rock engravings are located on calcarenite outcrops in an area of unstable blown out sand dunes (Plate 2). The high tide mark is approximately 20 m to seaward of the main exposed calcarenite bluff. The lowest engraved slabs are virtually at HWM. The adjacent blow now covers an area of approximately 170 ha extending inland 2.2 km from a breach in the foredunes 850 m wide. An examination of aerial photographs taken in 1979 and 1983 shows little advance of the slip face of the sand blow.

Potential Threats

The items most at risk at Mt Cameron West Aboriginal Site are the engravings situated on both sides of the creek at the northern side of the blowout. To attempt to protect the engravings without controlling the sand blow is to waste money. The sand blow is the key to any plan purporting to provide more than immediate short term protection for the Mt Cameron West engravings (Foster 1981).

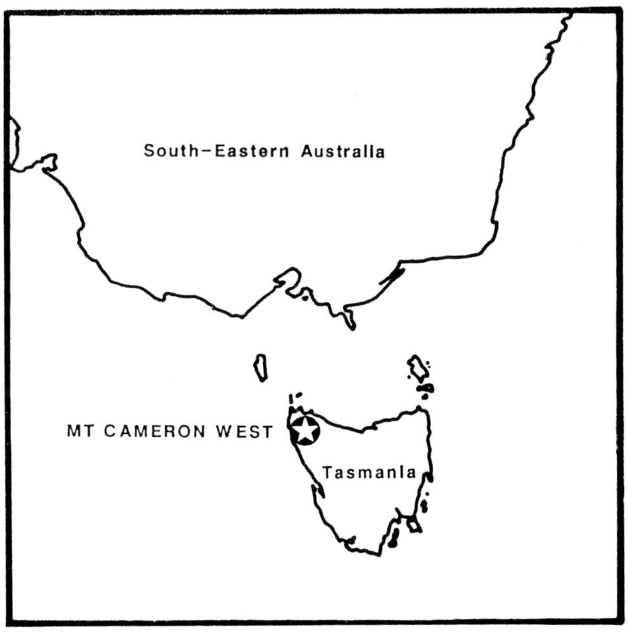

Figure 1. *Mt Cameron West engraving site, location map.*

A large part of the reserve including the area in the vicinity of the engravings is vegetated with marram grass (*Ammophila arenaria*). This is selectively and preferentially grazed by cattle.

The reserve has not yet been completely fenced because a stock proof fence cannot be maintained across the sand blow which extends over the boundary into Woolnorth.

Cattle have frequently been found in and removed from the reserve. The marram on the outcrop carrying the engravings is grazed by cattle from time to time. The engravings are therefore at risk of mechanical damage resulting from trampling. In an attempt to minimise this problem a temporary fence has been erected around the major engraving site.

Longer term threats are posed by beach recession and the possible rise in mean sea level predicted by the greenhouse hypothesis.

Beach recession is believed to be occurring at present although no quantitative studies have been made. Local residents have expressed the opinion that the high water mark has moved closer to the bluff in recent years. Storm waves now wash around the base of the bluff.

In an ideal cycle of sand removal and redeposition, storms remove sand from the beach to an off-shore bar. Sand is returned to the beach from the bar by the normal onshore swell. This raises the beach berm. Dry sand on the berm is carried landwards by onshore winds and a foredune forms at the back of the beach providing a barrier to further landward transport.

More intense storms may erode the foredune as well as the berm. This is a normal process and the foredune quickly reforms.

Any loss of sand from the system, either by movement along the coast or through a sand blow to the rear of the dune system will, if not compensated by a supply of sand, result in long term beach recession.

Foster's 1981 study of the Mt Cameron West site suggested that sand is still being supplied from off-shore and that the main loss of sediment from the beach is to the sand blow.

Remedial Options

The removal of the engravings to a safe site has been suggested in the past and rejected. The remaining courses of action available were limited to doing nothing or attempting to provide protection for the engravings on their present site.

On-site hard protection by means of a sea-wall or rock-filled mattress with or without dune stabilisation work has been suggested (Foster 1981; Hunching and Spur 1982). The cost of this protection far exceeded the available funds. The difficulty of obtaining suitable rock to fill the mattress or construct the sea wall was a major factor in the high cost. The cost, the difficulty of predicting a life-span for the structure and the impossibility of guaranteeing that it will survive for any period combined with concern over the risk of damage to the site ruled out the hard protection option.

Soft protection by reconstruction of a foredune across the adjoining sand blow, thus trapping sand in the immediate vicinity of the beach berm and foredune and ultimately broadening the berm to provide protection from storm attack, provided the last option.

The projected cost of initial works and subsequent maintenance of the soft protection option was lower than that of the hard protection option. The risk of mechanical damage to the site was considered low. This low option was therefore pursued on two fronts.

It was decided to attempt to reform the fore-dune in the vicinity of the bluff to prevent further loss of sand from the beach system, and to stabilise enough of the main blow to permit construction of a stock proof boundary.

Techniques Applied

1. *Dune Formation*

The accumulation of sufficient sand to reform a stable accreting foredune can be achieved by either the use of heavy machinery or sand trapping wind breaks or drift fences.

Heavy machinery (such as bulldozers, scrapers), could have been used to rapidly move sand from either the beach berm or the hind-dune area to the location of the new fore-dune. In the Mt Cameron West situation this involved unacceptable risk of damage to the engravings and nearby shell middens, and therefore could not be used.

Dune formation by drift fencing involves the construction of single or multiple fences along the intended line of the fore dune and as far as possible at right angles to the predominant sand carrying wind. Additional fences are constructed as required. When the desired dune profile is achieved the newly constructed dune is planted with marram grass.

Drift fencing uses a material of 40-60% porosity to reduce wind speed through a breach in a dune and so causes sand deposition in and immediately behind the beach. The fences must be tied into a wind breaking structure at each end—rock, dune or scrub, and sited in such a way that the sand trapping effect is maximised while the risk of being washed away by storm waves is minimised.

In the literature, recommendations on the siting of the initial fences vary from a position generally in line with or slightly behind the existing fore-dune to up to 200 m back from the high tide line (Anon. 1; Adriani and Terwindt 1974).

It appears that dual fences provide a significant increase in the amount of sand trapped when compared to a single fence (Manohar and Brun 1970; Barr and McKenzie 1977). The spacing between fences can vary but a distance of four times the fence height appears satisfactory under most circumstances (Manohar and Brun 1970).

Materials which have been used for drift fences include scrub (in Tasmania principally *Leucopogon* and *Melaleuca* spp.), pickets, hurdles and natural and synthetic meshes.

Scrub can be used in the form of cut lengths of head and branch laid horizontally against posts and wired in place to form a fence up to one metre high; by simply sticking cut stems into the sand to form a 'hedge' on the desired line, or as matting. This is highly labour intensive and dependent on supplies of suitable scrub being available close at hand (Barr 1966).

Pickets can be fastened together with fencing wire using a specially developed machine (Anon. 2; Alsop 1973; Manohar and Brun 1970). Hurdles are used in Europe for short sections of fence but are not available in Australia.

Natural meshes, principally jute mesh ('Soil Saver'), have been used (Barr 1966) but the porosity of the material is greater than desirable and supply from Bangladesh has been intermittent.

There are numerous synthetic meshes available either knitted or extruded. Those available in Tasmania include Humes Weatherscreen, Netlon Sandfencing, Cyclone Soil Stabilisation Mesh and Nylex Universal Mesh. In general the performance of synthetic meshes is inferior to that of scrub or pickets (Barr and McKenzie 1977) but this must be balanced against the relative costs involved.

In the case of Mount Cameron West where labour costs are high and suitable scrub hard to obtain, synthetic mesh provided the most economical drift fence material.

At Mt Cameron West drift fences were erected in 1986 (Plate 3). They initially consisted of a dual fence 0.9 m high separated by a space of 3.6 m running from a point adjacent to the bluff on which the engravings occur, southwards to the first stable remnant fore-dune hummock.

The fences, as far as possible, followed a line approximating the estimated position of the toe of the blown out foredune but with the seaward fence set back inland from the line a distance of five times the height of the existing remnant foredune hummocks. This involved fencing across the creek and ultimately a lagoon or soak may form in this area. To date sand has not accumulated in the creek to any extent and this remains a weak point in the reconstruction. Eventually, however, the reconstructed dune on the south of the creek will form a slip face which will probably slump into the creek. This may wash away or stabilise and in the latter case the creek will become a soak through the dune.

The fences are monitored for undermining and sand accumulation. Any holes are filled with scrub or mesh. Except for along the creek course, the first drift fences have successfully trapped sand enabling a second series of fences to be erected.

The second and subsequent fences were erected both to the seaward and inland side of the preceding fences, the distance between the new and buried fences being equal to the fence height (Plate 4). This is intended to have the effect of increasing the height of the dune and moving the windward toe of the dune to seaward. The second and subsequent pairs of fences are likely to be buried at a slower rate than the first fences but the quantity of sand trapped will be greater. Finally the new dune will be stabilised with marram grass at some time in the future after the third fence begins to fill. The grass will be planted by hand in this area.

To date the inland fences have been covered very rapidly but the seaward fence has trapped little sand. This appears to indicate that the relatively dry northeasterly winds which occur at the site in summer are the winds most active in moving sand into the fence zone and may in fact be more important than the prevailing onshore wind.

2. *Marram Grass Planting*

The second part of the protection strategy involves planting marram grass on the sand blows to enable fencing and exclusion of cattle.

Mechanised planting of marram grass has been used in Tasmania to stabilise coastal sand masses since the mid 1950s. In Tasmania marram grass is propagated vegetatively by transplanting divisions from established tussocks. If a large area is to be planted the initial step is the preparation of a nursery area to provide a marram supply. Any reasonably level site on which marram is already established will suffice.

The nursery is burnt in late winter (July to September) to remove top growth and dead material. It is then cultivated using a disc plough and fertilised using either a high nitrogen fertiliser such as ammonium sulphate, or an N.P.K. mix. (NPK mix is a general term for remixed proprietary fertilisers containing a source of nitrogen [N], phosphorus [P] and potassium [K], hence NPK. It is usually sold as, for example, 'NPK 8:4:10' which is 8% nitrogen, 4% phosphorus and 10% potassium. This mixture is marketed in Tasmania as a general garden fertiliser and has been effective as a marram fertiliser.) This promotes a flush of new vigorous growth which provides reliable planting material in the next season.

Planting material is obtained by 'skimming' the nursery. An angled steel wing attached to the three point linkage on a tractor is drawn through the ground in such a way that the surface of the wing is parallel to and approximately 30 cm below the sand surface. This disrupts the sand and breaks the marram rhizomes permitting easy collection. The divisions are collected by hand, bundled up and kept moist with an occasional hosing. Marram will keep in this way for up to 10 days.

The marram planting season in Tasmania runs from May to late August. The divisions are planted at approximately 30 cm spacings, and 30 cm deep in furrows one metre apart using a planter towed behind a bulldozer or tractor. The planter opens a furrow into which the operator places a division and the furrow is closed. In confined areas a smaller two row planter mounted on a large 4WD wheel tractor is used.

After planting, ammonium sulphate fertiliser is applied at a rate of 100 kg/ha. The possibility of applying the fertiliser at planting is currently being investigated.

In locations where mechanised planting cannot be used, marram divisions are placed in slots opened in the sand using a narrow spade or steel shod spike. This technique is surprisingly fast when used by an experienced planting team of two persons.

Three seasons of planting have now been undertaken at Mount Cameron West and 140 ha. of the sand blow have been planted with marram grass. The major planting program is now essentially complete and the grass seems to be taking well. Marram grass has been planted by hand near the engravings, and using a tractor towed planter over the rest of the sand blow.

Conclusion

In summary, the protection strategy which has been developed and is being implemented at the Mt Cameron West engraving site involved:

1. Reforming and restabilising the foredune using drift fences and marram grass, thereby stopping sand loss and raising the beach profile.
2. Planting marram grass on the sand blows to enable fencing and exclusion of cattle.

This strategy is a relatively long term one which commenced over three years ago. The results to date have been sufficiently good for us to believe that the protection strategy will be successfully brought to fruition. However, the possible rise in sea level predicted by the greenhouse hypothesis will pose a considerable threat to the long term survival of the engraving site. This outstanding artefact of past human coastal use in Australia is as much threatened in the long term as parts of Venice, London and numerous other world-wide coastal sites of present and past human occupation.

Acknowledgments

The protection strategy initiated and undertaken by the Tasmanian National Parks and Wildlife Service (subsequently Department of Lands, Parks and Wildlife) has been made possible by grants made through the National Estate Grants Program and the Australian Institute of Aboriginal Studies, Canberra.

REFERENCES

ADRIANI, M. J, and TERWINDT, J. H. J. 1974. Sand stabilisation and dune building. *Rijkswaterstaat Communications*, 68 pp. From Rijkswaterstaat, Directie Waterhuishouding en Waterbeweging, The Hague, Netherlands.

ALSOP, P. F. B. 1973. The stabilisation of coastal sand dunes. *First Australian Conference on Coastal and Ocean Engineering*. The National Committee on Coastal Ocean Engineering of the Institution of Engineers, Australia, Sydney.

ANON. 1. *Coastal sand dunes, their vegetation and management.* Beach Protection Authority of Queensland, Leaflet No. V-03.3

ANON. 2. *Coastal sand dunes, their vegetation and management.* Beach Protection Authority of Queensland, Leaflet No.V-03.5.

BARR, D. A. 1966. Jute mesh dune forming fences. *Journal of Soil Conservation NSW* 22: 123-29.

BARR, D. A. and McKENZIE, J. B. 1977. Progress in coastal and sand dune stabilisation and management experiments on South Stradbroke Island, Queensland. *Proceedings of the 3rd Australian Conference on Coastal and Ocean Engineering*, Melbourne 18-21 April 1977, pp. 207-13.

BROWN, S. in press. Rock art and Tasmanian prehistory: Evidence for changing cultural traditions in a changing environment. Paper presented at First AURA Congress, (1988) Darwin.

FOSTER, D. N. 1981. Coastal erosion of Mount Cameron West Aboriginal Site, north west Tasmania. *Technical Report No. 81/18*. The University of NSW Water Research Laboratory.

HUGHES, P. J., WATCHMAN, A. L. and SULLIVAN, M. E. 1981. A geological and geomorphological investigation of the Mt Cameron West Aboriginal Site, north west Tasmania. *Papers and Proceedings of the Royal Society of Tasmania* 115: 211-22.

HUTCHINGS, B. J. and SPURR, R. G. 1982. Preliminary report on proposed seawall to protect petroglyphs at Mount Cameron West Aboriginal Site, for National Parks and Wildlife Service, Tasmania. Reference 2033, September 1982. Hutchings, Spurr and Associates, Consulting Chartered Engineers, Hobart.

JONES, R. 1981. Rocky Cape, West Point and Mt Cameron West, north-west Tasmania. In *The Heritage of Australia*. Macmillan.

McCARTHY, F. D. (1979). Report on Mt Cameron West Engravings. Unpublished report to Australian Institute of Aboriginal Studies and the Tasmanian Museum and Art Gallery.

MANOHAR, M. and BRUN, P. 1970. Mechanics of dune growth by sand fences. *Dock and Harbour Authority 1970*, pp. 243-52.

MESTON, A. L. 1932/33. Aboriginal rock cavings in Tasmania. *Papers and Proceedings of the Royal Society of Tasmania* 1-6.

MESTON, A. L. 1934. Aboriginal rock carvings in Tasmania. *Antiquity* 8: 179-85.

PLOMLEY, N. J. B. (ed.) 1966. *Friendly mission : The Tasmanian journals and papers of George Augustus Robinson 1829-1934*. Tasmanian Historical Research Association, Hobart.

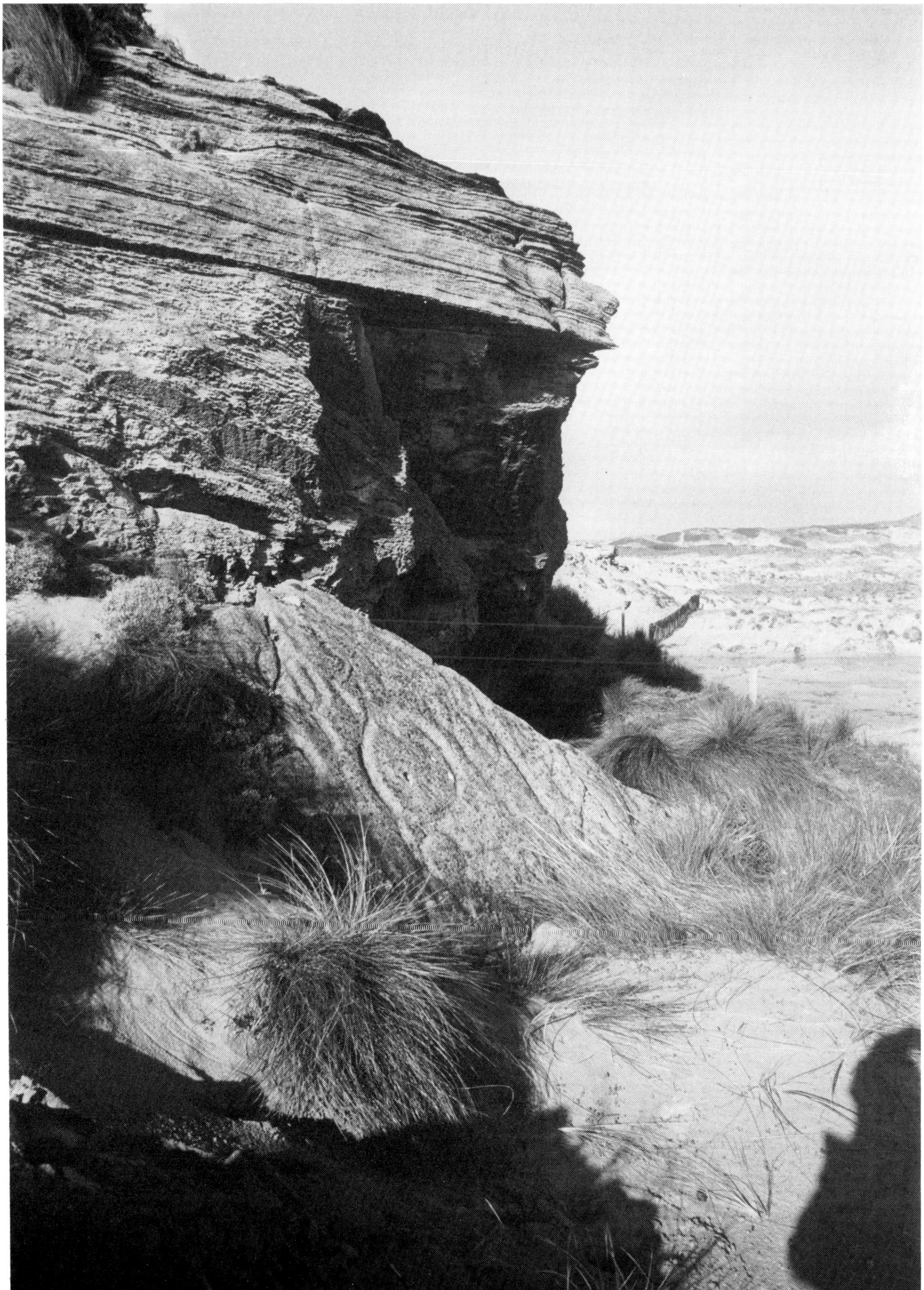

Plate 1. *View south of main concentration of engravings. Most of the engraved panels, which have been reburied with sand, lie under the slope beyond the lower right of the photograph.*

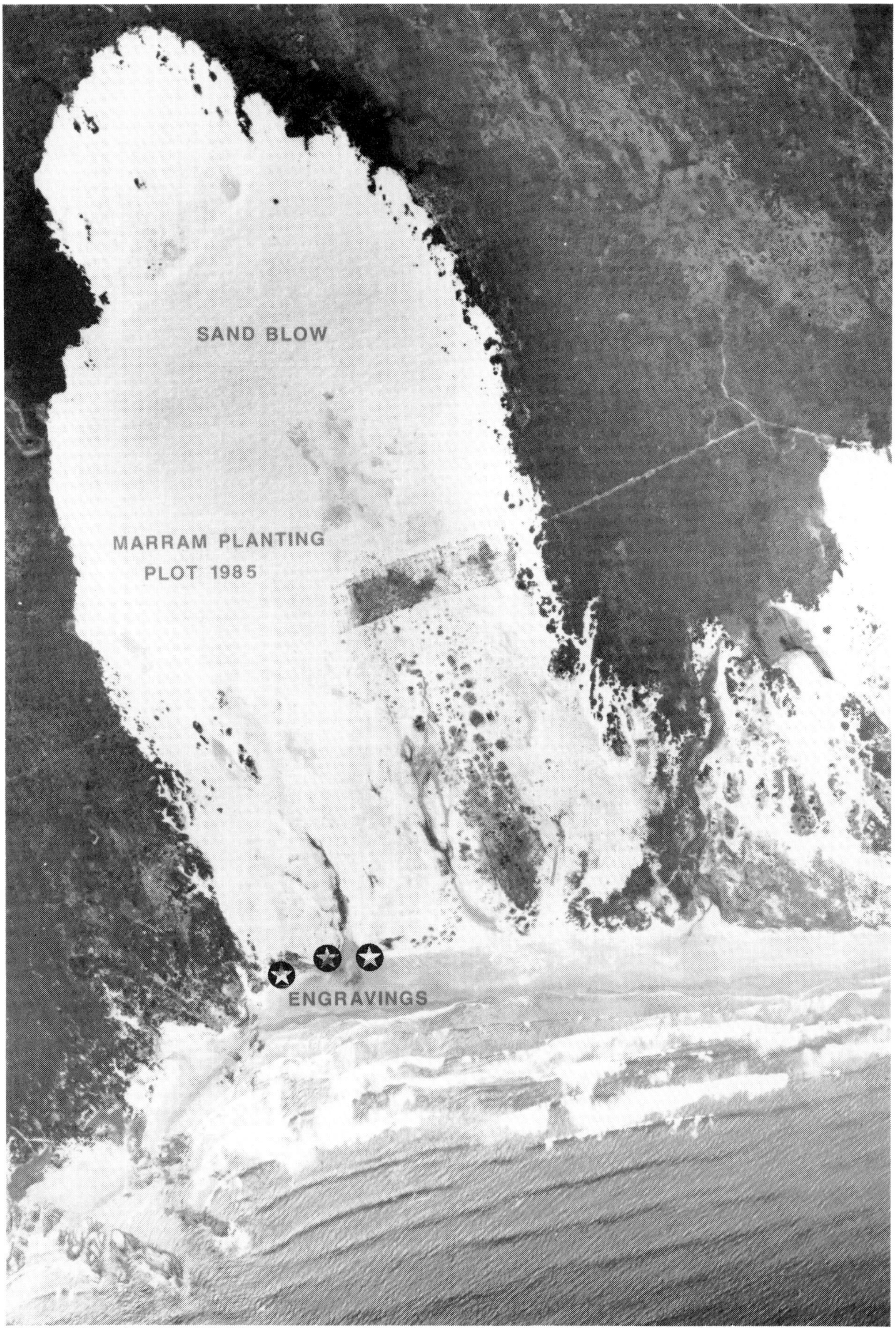

Plate 2. *Aerial photograph showing known locations of rock engravings and extent of sand blow.*

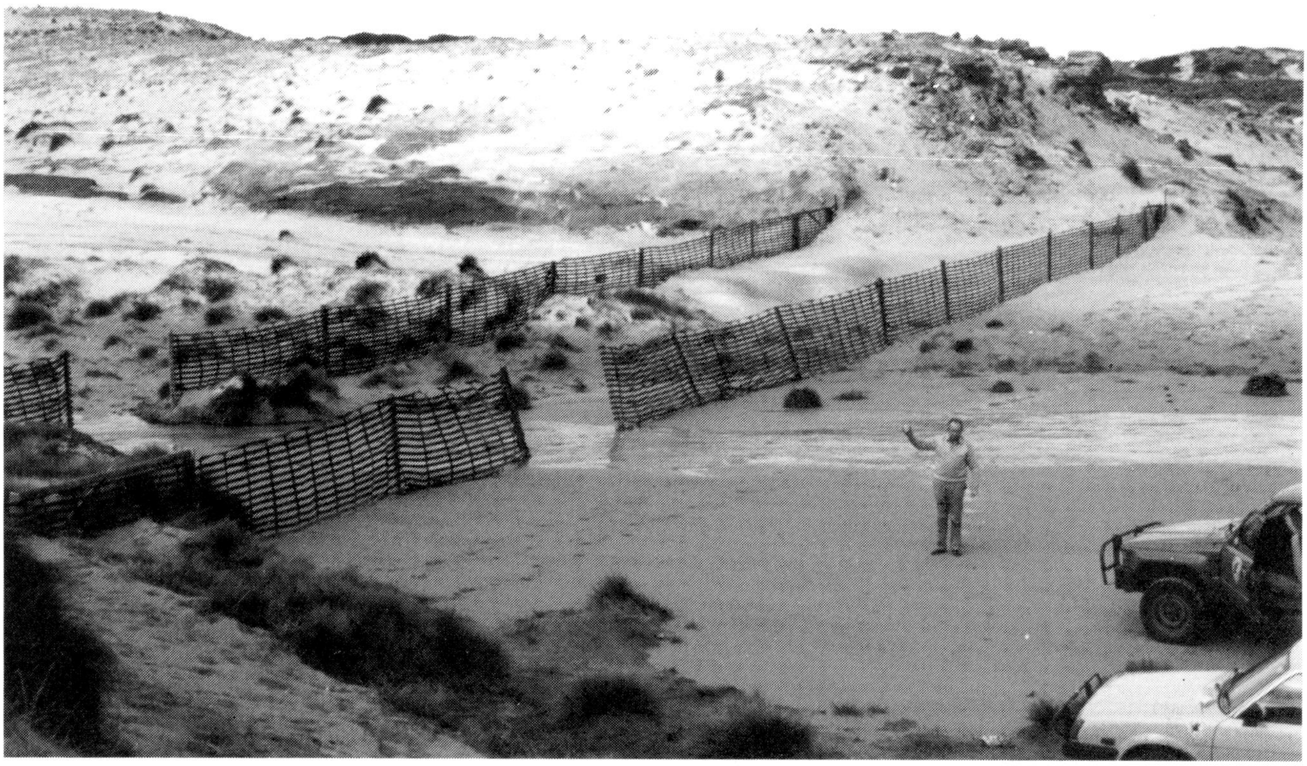

Plate 3. *Drift fences erected in October 1986.*

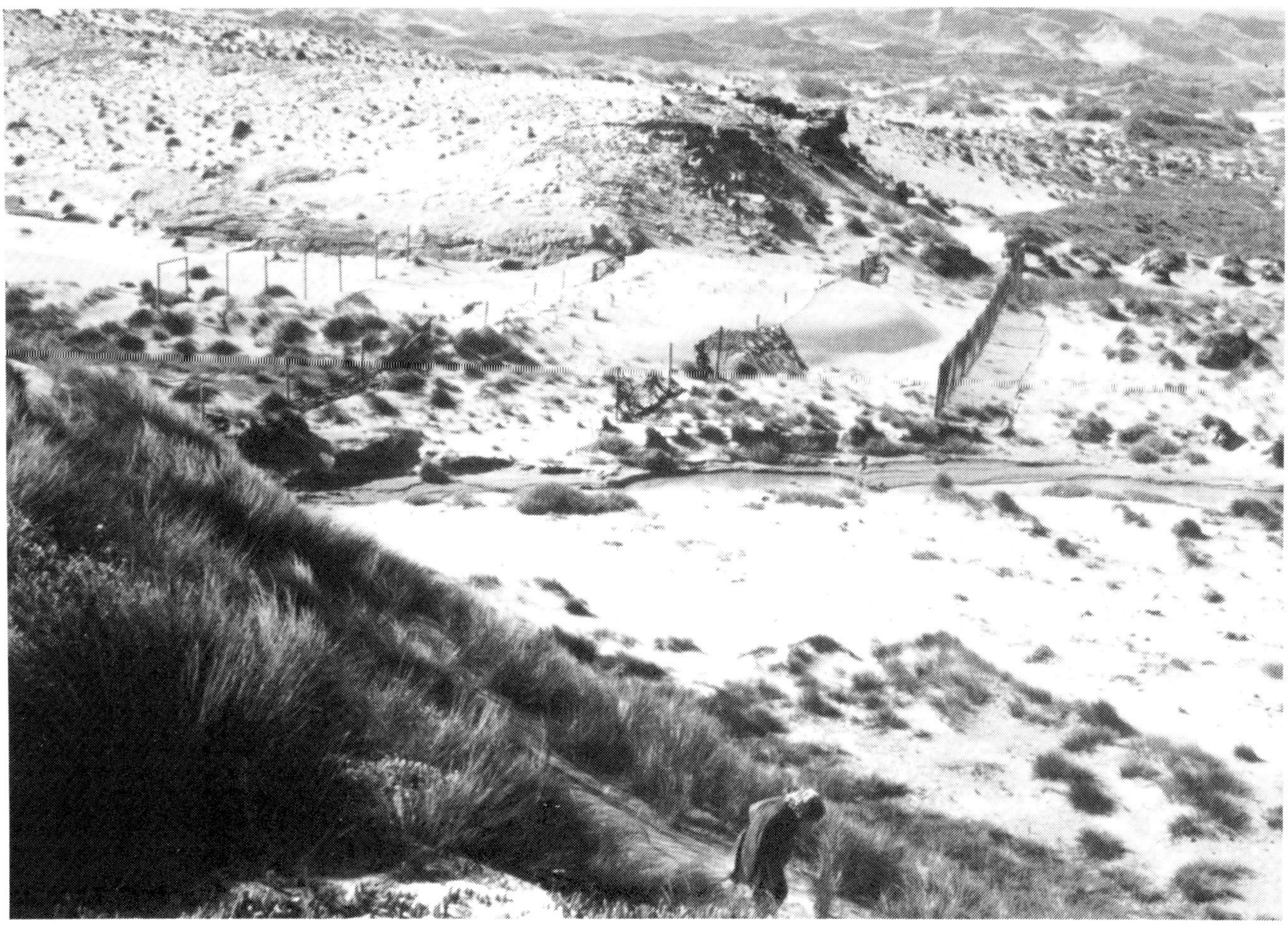

Plate 4. *As plate 3, February 1988. Original drift fences have been largely buried and new fences have been erected.*

9

PADJARI MANU (FORMERLY BUNBURY) CAVE, NORTH-WEST CAPE, WESTERN AUSTRALIA: PLANNING FOR TOURISM

Peter Randolph and Joseph Wallam

Abstract. The North-west Cape region contains few Aboriginal art sites. Tourist pressure to visit Padjari Manu (formerly Bunbury) Cave is growing. The region's other attractions are being promoted and compound the potential impact. No traditional Aboriginal inhabitants are said to be alive, however, neighbouring traditional groups have utilised the region for about 100 years. The information and views of these people and the opinions of Aboriginal people from the wider West Pilbara, Ashburton and Gascoyne regions have been sought and are taken into account. The vulnerability of the art and archaeological deposits are considered as well as the proximity of the cave to other archaeological sites (especially middens). A plan for tourist visitation taking all the above into account is formulated in cooperation with the local Tourist Bureau and the Western Australian Department of Conservation and Land Management personnel.

Introduction

Gradually the wider Australian community is gaining an awareness and understanding of the role Aboriginal heritage has in characterising Australia - its past, present and future. Along with awareness and understanding is an inquisitiveness to see and experience Aboriginal culture. The resulting growth in tourist visits to Aboriginal sites has begun and plans need to be developed to respond to it and anticipate its growth. Andrée Rosenfeld (1985), Fay Gale (1988), Mary Colreavy (1987) and many others have already written about the topic. This specific case study emphasises the need for consultation with Aboriginal people even for presumed pre- historic sites.

Rock art - a scarce resource at North-west Cape, W. A.

Though rich in occupation sites, the North-west Cape region contains only two known art sites; Padjari Manu Cave and a neighbouring cave. The latter contains few motifs and they are of poor quality compared with the art present in Padjari Manu Cave.

The next location of known sites containing rock art is more than 125 km east of Padjari Manu Cave. The most well known is Checkendon Combe, a site on the register of the National Estate - located on Uaroo Station 170 km to the south east. However, art at these sites is on exposed granite boulders and has been produced by a technique closely related to the numerous Pilbara engraving sites.

Site description

Padjari Manu (formerly Bunbury) Cave (W. A. Museum, Department of Aboriginal Suites, site file PO267) is close to the crest of the Cape Range escarpment approximately 6 km south of Vlaming Head (*Padjari Manu*) and just 2.5 km from the Shire (Figure 1).

The cave, situated in ancient limestone deposits, has a level floor with a charcoal rich deposit. There are several grinding bases, artefacts and bones (including macropod, fragments of baler shell and turtle carapace) on the surface. The internal dimensions of the main chamber are 12.5+ m wide and 10 m deep with a ceiling height of 2.5 - 3 m throughout. The mouth is oriented to the north-west, i.e. facing the ocean. At the rear of the cave is a water-worn depression which locals describe as a semi-permanent water source fed by seepage from the ceiling. At the time of inspection the depression was damp but did not hold any water.

The walls and ceiling are coated with a dark brown to black crust (micro-crystals of gypsum and calcium oxalate salts - J. Clarke pers. comm.). Much of this surface has been decorated with engravings which utilise the contrast of the dark coating on the surface and the white limestone beneath. The outer surface has been removed by a series of conjoined punctures, each of which has an irregular perimeter.

[Note: We would like to point out that this paper contains no illustrations of individual rock art motifs - the reason for which will become obvious as the details of the investigations unfold.]

Evidence of increasing tourist pressure

Padjari Manu Cave has been the subject of uncoordinated tourist activity for many years. The local Chamber of Commerce and Tourist Bureau have, through the Western Australian Department of Conservation and Land Management, expressed a wish to formalise visits by tourists. Over the last two years in particular, the region's other attractions (coral reefs, game fishing, natural formations, shipwrecks, flora and fauna) have been promoted. Recently the Ningaloo Marine Park was opened. This is expected to be one of the region's biggest tourist attractions (May et al. 1983 and Department of Conservation and Land Management 1988). An omnibus licence has been issued for a proposed bus tour of the region including places of natural beauty and historic significance. Padjari Manu Cave is listed on sites which may be included in this proposed tour.

The Australian Bureau of Statistics' figures (Gascoyne Regional Profile, p. 24) show a sharp increase in visitor arrivals to the region, e.g. Shire of Carnarvon: 1980 - 59 300 arrivals, 1984 - 92 800 arrivals. The regional trend is reflected in arrivals at the North-west Cape town of Exmouth: 1982 - 15 800 arrivals, 1987 - 25 500 arrivals (pers. comm. N. Nickerson, Tourism Commission).

Aboriginal links to the area and the cave

Tindale (1974) nominates the group which traditionally occupied the North-west Cape and its peninsula as the *Jinigudira* (elsewhere in this report *Yinigudira*) (Figure 2), and suggests they were sometimes considered to be western *Talandji* though informants were sure they had separate identities for a long time. This was confirmed by our investigations (Randolph and Wallam 1986) when Algie Patterson stated that the *Yinigudira* kept very much to themselves and would not share their fires. Hearsay, amongst local pastoralists, is that the whole of the North-west Cape was/is a taboo area and that the traditional inhabitants were wiped out by a natural disaster - some say a tidal wave, others an epidemic of some kind. In fact no *Yinigudira* informants have been identified by recent researchers and little ethnographic evidence of the group is available. Despite this, the suggestion that the area was/is taboo is probably not so straight-forward. Nor is the explanation for the demise of the group. As recently as 1895 Thomas 'Birdman' Carter (Valli 1983: 120) is reported as having sent a letter to his friend Donald Scott complaining that

> the Aborigines of the district were developing a liking for the lights of Carnarvon. He said the few left were content to concentrate in the Bay of Rest area, awaiting handouts from pearling crews in exchange for their women. The murder of a black-birding pearling skipper by avenging Aborigines at the Bay of Rest saw an acceleration of this exodus.

Furthermore, Turner (1985) obtained evidence of recent exploitation. Several of her informants were keen to emphasise that during seasonal breaks from pastoral employment they 'holidayed' along the west coast of the Cape, to catch fish, dugong and exploit other marine resources. However, they described their affiliations as descendants of the neighbouring groups. It should be noted that a strike by Aboriginal pastoral workers in the 1940s and the introduction of award pay for them caused a renewed exodus from the pastoral stations to regional towns and probably made the subsequent exploitation of North-west Cape more sporadic.

In addition to the ethnographic evidence, there is considerable archaeological evidence of occupation of the North-west Cape though the site itself has not been dated. The earliest dating is of a rock shelter in a Cape Range escarpment which has been dated to 25 000 BP (Morse in press), while coastal midden sites have varying dates; for example 8000 BP for the Waroora Midden (Kendrick and Morse) and 5000 BP to about the time of European settlement for middens on the west coast of North-west Cape adjacent to Padjari Manu Cave (Morse pers. comm.).

Consultation with Aboriginal people

Based on the information provided above it would seem that consultation with Aboriginal people would not be necessary. However, Turner (1985) produced evidence which convinced us of the need for consultation.

Specific endeavours were made in the field to ascertain (a) whether any person had knowledge of any engraved or painted cave sites in the region and (b) the general attitude of these people and other Aboriginal people of the region to visits of such sites by tourists. We could find no Aboriginal people with specific knowledge of any engraved or painted cave sites in the region. Indeed apart from those who have come to know the existence of the caves in recent times (e.g. those who accompanied Turner in 1985), no Aboriginal people we spoke to seemed aware that such caves existed. One informant even doubted whether a cave could be found as there were no elders to lead anyone to it (Randolph and Wallam 1986).

In an early reference to the cave, naturalist Harry Butler (1963: 92-95) calls it a sacred cave and states that the figures are very reminiscent of Warburton Desert paintings and petroglyphs. Turner (pers. comm.) indicated that in a conversation with Butler he stated that, though no local Aboriginal people had spoken to him, he believed the cave was associated with male initiation rituals.

More recently and related to the geographically nearby West Pilbara region, Turner (pers. comm.) was told by two elderly informants (regrettably deceased) that similar motifs to those found in the caves in Cape Range were symbols warning Aboriginal women away from an area. Other Pilbara informants have emphasised the particular importance of one motif said to symbolise a headdress. This evidence would certainly go some way to validating Butler's interpretation and highlight the need for caution regarding visits to the caves by Aboriginal people, especially Aboriginal women.

However, the views of the most relevant Aboriginal people - those elders, both men and women, of neighbouring and nearby traditional groups - was clearly that visitors should be allowed. These informants included members of the Baijungu, Ingada, Djiwal/Wariangga, Mardathunera/ Kuruma and Binigura groups (see Figure 2). One informant, while supporting the view that tourist visits would be acceptable, said she had recently heard that the cave was eerie. Similar sentiments were also expressed by an Aboriginal field assistant who visited the cave with Turner in 1985. These expressions of sensitivity reinforce the need for caution regarding visits to the caves by Aboriginal people. Reasons for particular Aboriginal views about tourist visits are therefore not easily summarised, but include:

(a) because the country around the site has no living traditional owners, people can visit it;
(b) everybody (especially non-Aboriginal Australians) should learn more about Aboriginal culture;
(c) people have been going there on an ad hoc basis for many years; they should still be able to visit;
(d) the cave may still have spiritual power so people should be warned of this possibility. While some Aboriginal people would not go into the cave, individual tourists could decide for themselves whether to enter the cave.

Consultation was spread over almost two years including, in May 1988, a field trip to discuss tourism at Padjari Manu Cave. A good part of this trip was spent discussing a management proposal with the people - specifically the wording of a permit pamphlet. Informants consulted on this trip were:

Algie Patterson - Marduthunera/Karuma
George Hughes - Binigura and Joe Butler - Djiwal/Wariangga

Dolly Butler - Baijungu

All informants agreed with the proposed idea of letting tourists visit the cave and with the wording of the proposed permit pamphlet. Two months later we arranged for Mary Cooyou (Baijungu) to accompany us to the cave with John Dale (son of Molly Dale, Ingada, who was consulted in October 1986). As Mary would not enter the cave it was agreed that John Dale, on behalf of his family would actually visit the cave and relay back his views for consideration. The view of Mary Cooyou and John Dale was consistent with the view of other informants, namely: visits could be allowed but visitors ought to be warned of the significance/power of the cave and decide for themselves if they should or should not enter the cave.

This visit completed our consultation. Other people contacted (e.g. Victor Hughes) specifically nominated one or more of those consulted as the most relevant people to speak for the region.

Plan for tourist visitation

Having gathered a range of facts concerning the cave (the Aboriginal viewpoint being of prime consideration), a plan for permitting tourists visits was devised.

A novel approach to the concept of limited access was conceived which will (it is hoped) meet all concerns and yet not be too restrictive. In fact the proposal is not at all overtly restrictive.

The plan is as follows:

Signs:

(i) no sign-posting of the site or overt advertising on brochures or maps;

Information:

(ii) permit pamphlet will only be issued on request and will include a cautionary message highlighting the concerns of Aboriginal people (see below);

(iii) regular monitoring trips will be arranged to clean up litter, report on any desecration and restore any damage;

Access:

(iv) vehicle access will be restricted by fences and windrows at various points. A foot track will be the only access for the final 300 m.

The Tourist Bureau will only issue a permit pamphlet to tourists who specifically ask for information about the cave or enquire about sites of Aboriginal heritage. A person making a general enquiry about attractions in the region will not be told of the cave. The Bureau will keep a register of those to whom permits are issued and ensure, as far as possible, that the permits are returned. In this way it is hoped to minimise the multiplier effect and cause any new visitors to follow the same procedure and hence be made aware of the cautionary message and the vulnerability of the art and the deposit.

These measures would be aimed at providing the greatest possible protection to the site, given the available resources, while catering for Aboriginal opinion and allowing truly interested people to visit. In addition the impact of physical site protection (e.g. signs and fences) is minimised on the site and environment.

Discussion

At a superficial glance this study has simply made provision for the maximum preservation of Padjari Manu Cave. However, it provides both a precedent and justification for the consultation with Aboriginal people even in circumstances where first-hand associations with the site seem non-existent.

It is our view that the management of any Aboriginal site *requires* the involvement of Aboriginal people. It is not valid for archaeologists, state site authorities, the landowner or any other person to claim absolute control and make decisions affecting the future of the site without adequate consultation with relevant or representative Aboriginal people. Bowdler (1987) draws to our attention the event of the Australian Archaeological Association in Hobart in 1987 when Ros Langford

> confronted the archaeologists with views of the Aboriginal people which held that prehistoric archaeological sites in Australia were a part of the Aboriginal heritage, and that heritage was theirs to control.

Bowdler (op. cit.) goes on to say

> Aborigines did not at any time claim to 'own the past'. They claimed to own their heritage, to my mind a very different thing ... It seems to me that when you hear someone say who owns the past it is time to go on full alert: someone is about to be excluded from something, someone is about to proclaim a particular hegemony over something, by proclaiming that something cannot be owned, or is owned by us all.

All sites whether pre-historic, historic or of current use that are categorised as Aboriginal sites are by definition part of Aboriginal heritage. Management decisions of such places should therefore involve Aboriginal people in meaningful input.

REFERENCES

BOWDLER, S. 1987 (unpublished). Who owns what. Paper presented at Conference of W. A. Museum Anthropologists, 1987, Fremantle.

BUTLER, H. 1963 (unpublished). W.A.M. Field Note Book 1963: 92-95. W. A. Museum, Perth.

COLREAVY, M. and CAVANA, M. 1988 (unpublished). Bungle Bungles and National Park and Conservation Reserve visitor survey, April to September 1987. Department of Conservation and Land Management, Perth.

Department of Aboriginal Sites (nd)(unpublished). Padjari Manu Cave site PO267. Department of Aboriginal Sites, W. A. Museum, Perth.

Department of Conservation and Land Management 1988. *Ningaloo Marine Park. Draft Management Plan, May 1988*. Department of Conservation and Land Management, Perth.

GALE, F. and JACOBS, J. M. 1987. *Tourists and the National Estate: Procedures to protect Australian heritage*. Australian Heritage Commission. Special Australian Heritage Publication Series Number 6. Australian Government Publishing Service, Canberra.

HARRISON, D. J. A. 1985. *Gascoyne regional profile*. Joint publication, Gascoyne Regional Development Advisory Committee, Carnarvon, W. A. and Department of Regional Development and the North-west, Perth.

KENDRICK, G. W. and MORSE, K. 1982. An Aboriginal shell midden deposit from the Warrora Coast, North-western Australia. *Australian Archaeology* No. 14. Australian Archaeological Association.

MAY, R. F. et al. 1983. *Ningaloo Marine Park. Report and recommendations by the Marine Park Work Group June 1983, National Parks Authority Report 1*. Wescolour Print Pty Ltd, Perth.

MORSE, K. and KEE, S. 1985 (unpublished). Preliminary report of a survey for Aboriginal archaeological sites in the Cape Range National Park, North-west Cape, Western Australia. W. A. Museum, Perth.

MORSE, K. (in press). Madnu Mandu site PO5646. Department of Anthropology, W. A. Museum, Perth.

RANDOLPH, P. and WALLAM, J. 1986 (unpublished). Investigation at selected Aboriginal sites on the North-west Cape Bunbury Cave - PO267, another cave containing Rock Art - PO751 and Mesa Camp Midden - PO5649. Including recommendations for management and tourist visitation. Department of Aboriginal Sites, W. A. Museum, Perth.

RANDOLPH, P. and WALLAM, J. 1987 (unpublished). Bunbury Cave, N.W. Cape, W. A.: the task of consulting with Aboriginal people. Paper presented to the Australian Archaeological Association 1987, Perth.

ROSENFELD, A. 1985. *Rock art conservation in Australia*. Australian Heritage Commission, Special Australian Heritage Publication No. 2. Government Publishing Service, Canberra.

TINDALE, N. 1974. *Aboriginal tribes of Australia*. University of California, Berkeley.

TURNER, J. 1985 (unpublished). Report on preliminary ethnographic investigations for the area encompassed by the proposed Ningaloo Marine Park. Department of Aboriginal Sites, W. A. Museum, Perth.

VALLI, J. 1983. *Gascoyne days*. St. George Books, Perth.

VALLI, J. 1987 (unpublished). Letter in correspondence file DAS 77131. Department of Aboriginal Sites, W. A. Museum, Perth.

WALLAM, J. 1988 (unpublished). Field trip Bunbury Cave - May 1988. Aboriginal Sites Department, W. A. Museum, Perth.

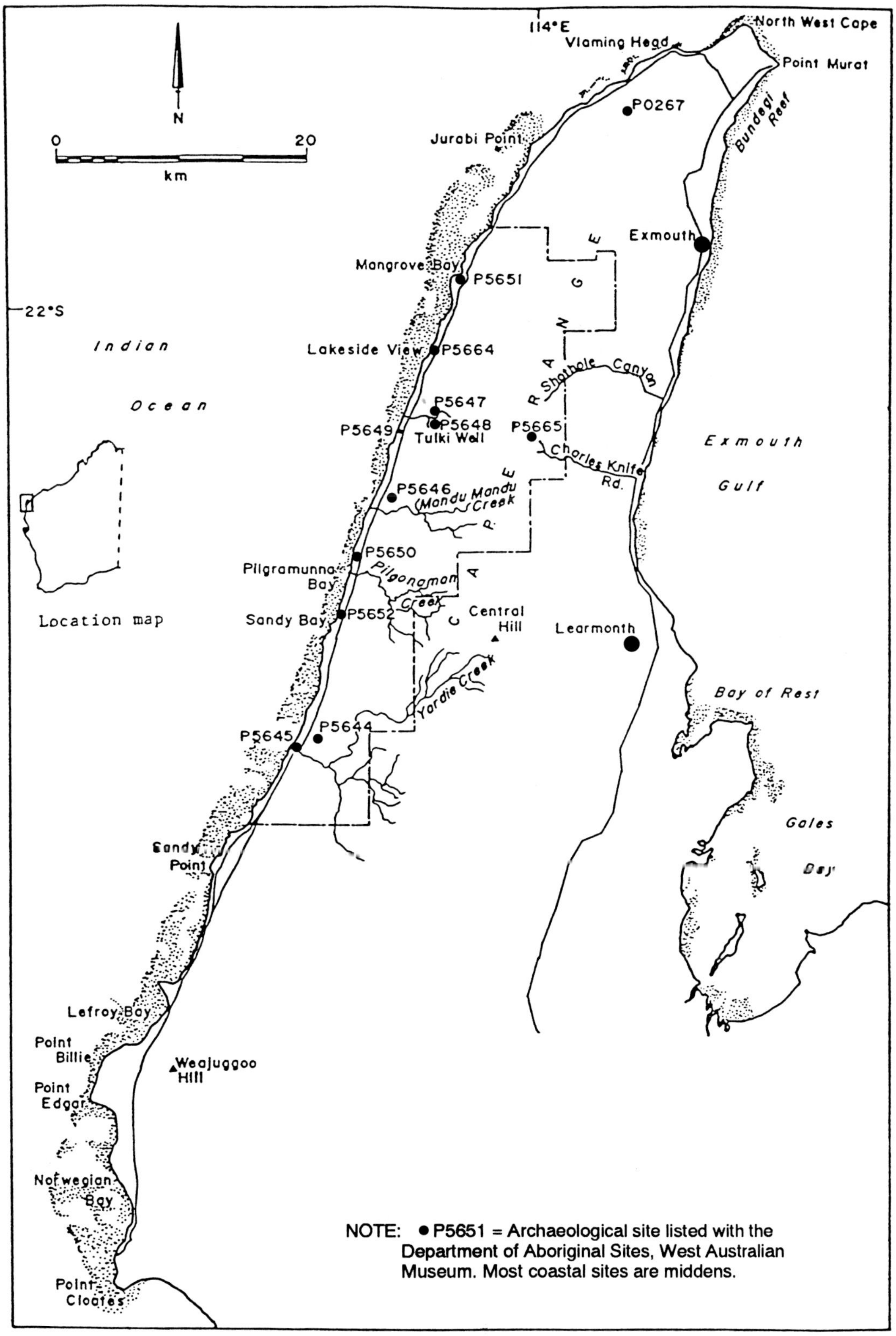

Figure 1. *Archaeological sites in the North West Cape region. Extract from Morse and Kee (1986).*

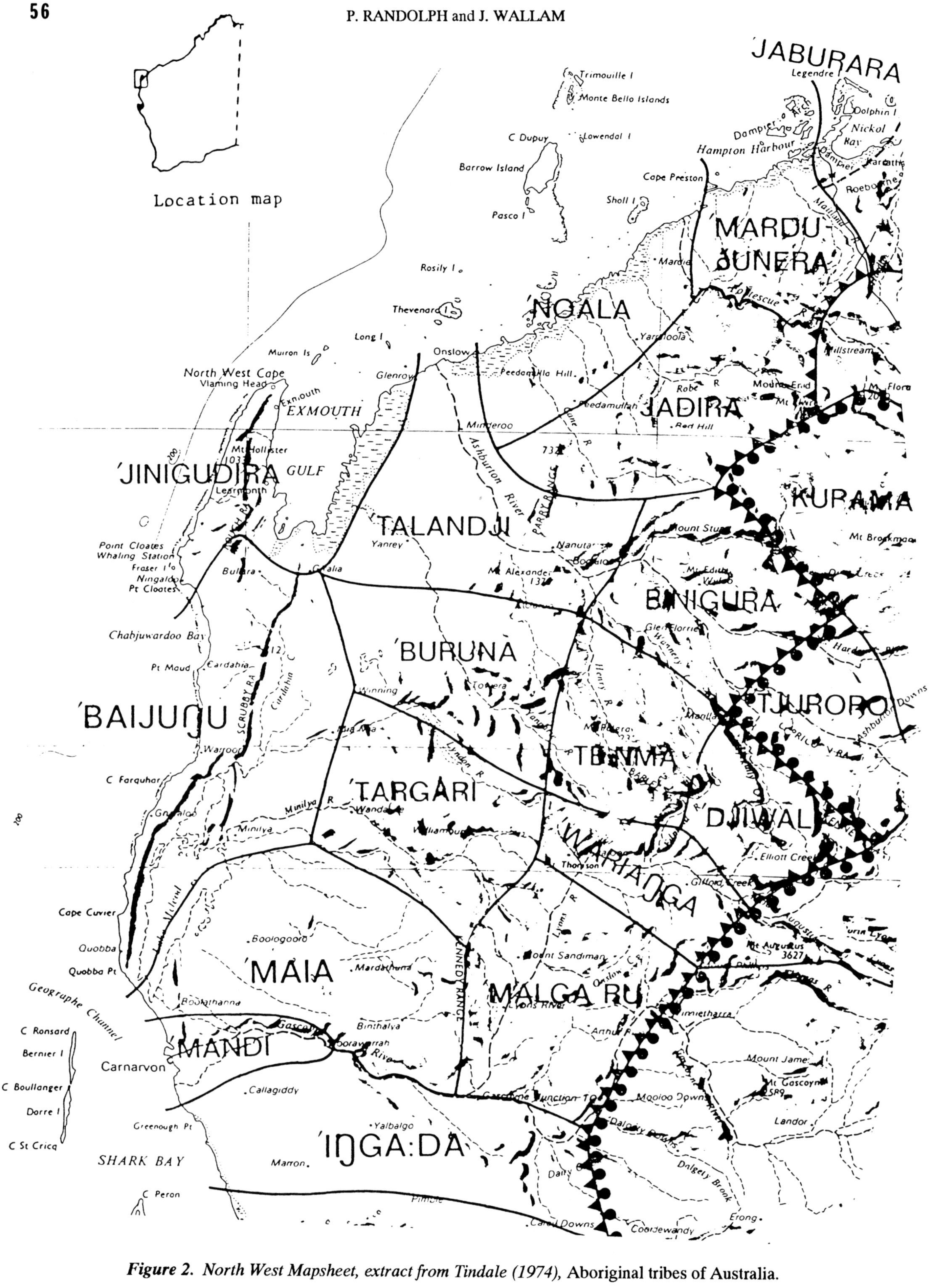

Figure 2. North West Mapsheet, extract from Tindale (1974), Aboriginal tribes of Australia.

10

LES GROTTES DES COMBARELLES (LES EYZIES, FRANCE):

Apports de la connaissance de l'environnement à la conservation des gravures

J. Brunet, J. Vouvé et P. Vidal

Abstract. The immediate surroundings of prehistoric caves are important for the preservation of rock art. The vegetation above the caves regulates moisture infiltration by processes of evaporation and transpiration. That is one of the most important factors determining the subterranean climate. The knowledge about the extent of Combarelles Cave and its geological context allowed us after a field study to define five areas of protection. In order to conserve the engravings of the cave it will be necessary to prevent destruction of forests and fields in the area.

La conservation des oeuvres pariétales préhistoriques du domaine souterrain est un travail à long terme qui demande l'avis, le diagnostic de différents spécialistes et techniciens. Une des phases importantes de ce diagnostic se trouve dans la saisie et la compréhension de l'influence de l'environnement sur les grottes. L'ensemble de Combarelles, comprenant les grottes dites Combarelles I et Combarelles II, servira d'exemple pour illustrer les relations entre l'environnement naturel extérieur et le milieu souterrain.

(1) Localisation - description

Les grottes des Combarelles sont situées sur la commune des Eyzies dans le sud-ouest de la France à environ deux kilomètres à l'est du bourg (Fig. 1). Elles sont creusées dans un massif calcaire qui domine la vallée de la Petite Beune et s'ouvrent au bas d'une falaise au fond d'un vallon. Le massif calcaire est caractérisé par un toit arrondi culminant à la cote + 204 m N.G.F. et sa base est voisine de la cote moyenne + 80 m N.G.F. soit donc une dénivellation de 124 m.

La base de la colline possède une topographie contrastée avec présence de falaises verticales et localement de surplombs comme au-dessus de l'entrée de la cavité. Par contre, cette topographie n'est que peu contrastée à mi-pente et seuls quelques abrupts de 2 à 5 m de hauteur perturbent la surface. Elle devient par la suite très douce au sommet de la colline. Ce phénomène est en liaison avec la nature des sédiments qui s'y sont déposés.

Les cavités des Combarelles appartiennent à la famille des 'grottes topographiquement basses', elles s'opposent ainsi à la famille des 'grottes topographiquement hautes' telles que Font de Gaume et surtout Lascaux (Fig. 2).

La longueur du réseau de Combarelles I est de 240 m et 342 m si on inclut la partie où coule un ruisseau difficilement pénétrable (Fig. 3). C'est une cavité à un seul conduit de faible longueur.

Le réseau karstique auquel elle appartient à peu près horizontal tout au long de son développement souterrain a une altitude peu élevée (cote du toit inférieur à 93.50 m N.G.F., et celle du plancher du réseau actif près de la cote + 89 m N.G.F.).

La hauteur de ce conduit unique est voisine de 2 m après travaux sur la majeure partie du tracé. A la découverte en 1901, cette hauteur était en moyenne réduite de moitié et il y avait des passages 'rampants' et très étroits. Au porche d'entrée, le conduit atteint 7 m de haut et un autre réseau indépendant mineur y aboutit (Combarelles II). Obstrué par un remplissage argilo-sableux, ce conduit a été dégagé par l'homme. Il est long de 104 m environ (Fig. 3) (Aujoulat 1984).

(2) Rappels sur l'art pariétal des Combarelles

La grotte des Combarelles I possède une importante ornementation pariétale sous forme de gravures et de dessins (plus rares) (Photos 1, 2). Le décompte totale n'est pas réalisé et H. Breuil dans son inventaire décrivait plus de 291 gravures auxquelles on pouvait ajouter 'une centaine de débris d'images trop mauvaises ou trop mal conservées pour que l'on puisse les identifier' (Breuil 1952). Parmi les principales figures, on distingue: 116 équidés, 37 bisons, 19 ours, 14 rennes, 13 mammouths, 9 bouquetins, 7 bovinés, 5 cerfs, 5 lions, 3 biches, 1 daim, 4 loups dont 3 douteux, 1 renard, 1 rhinocéros, 1 poisson (?), 1 serpent (?). Il y a, en outre, une main cernée de noir, 4 tectiformes, 39 figurations humaines ou anthropomorphes, 4 signes de caractères sexuels. A cet inventaire déjà ancien, s'ajoute le fruit de nouvelles recherches menées par Claude Barriere et Monique Archambeau (Barrière 1984).

Toujours selon H. Breuil 'les dimensions des figures des Combarelles sont variables; les plus petites n'ont que 10 cm de haut, tandis que les plus grandes ne dépassent guère 1 m. Les techniques sont variées. La gravure prédomine partout; le trait peut être léger ou profondément incisé; parfois les contours sont dégagés par un champ-levé qui donne à la figure l'aspect d'un bas-relief; souvent les poils sont indiqués par des séries de hachures, et le modelé ou les variations de pelage de l'animal sont soulignés par des raclages'.

Pour A. Leroi Gourhan, l'ensemble des Combarelles est du Magdalénien et plus exactement du style III, IV.

Pour Combarelles II, 33 figures ont été recensées réparties en 22 représentations animales, 7 signes et 4 figures indéterminées.

L'Abbé Breuil (1952) mentionnait pour la grotte de Combarelles I un certain nombre de 'débris d'images' trop mauvaises ou trop mal conservées pour être identifiables. Cette appréciation traduit bien l'importance des phénomènes de dégradations des parois qui ont dû se produire naturellement au cours des millénaires. Dans l'ensemble, la surface de la roche a été corrodée puis fréquemment recouverte de coulées de calcite. Le concrétionnement est donc actif, la cavité est soumise aux influences externes.

(3) Etude de l'environnement externe

3.1 Généralités

L'environnement immédiat des cavités préhistoriques joue un rôle capital dans la conservation des oeuvres pariétales. En effet, sur les sols au-dessus des cavités, la végétation influence certains paramètres tels que les infiltrations, vers la profondeur, le ruissellement à la faveur des mécanismes d'évapo-transpiration. Dans ce cadre, la liaison entre le milieu extérieur et le milieu souterrain est en partie assurée par l'eau. Rappelons ici que l'eau capte et transporte une partie du flux thermique (Mangin et Andrieux 1984) apporté à la surface du sol par le soleil en été, la pluie, la neige en hiver; qu'elle sert de vecteur à des substances dissoutes et en suspension; et enfin qu'elle intervient dans les phénomènes de dissolution et de dépôt de sels (sels de calcium en milieu carbonaté).

Notons aussi que dans ces deux derniers cas, la présence en plus ou moins grande quantité de gaz carbonique dans le sol et les terrains, va influencer les concentrations et les précipitations chimiques, elles-mêmes dépendantes entre autre du facteur température.

Mentionnons, en outre, que l'eau d'infiltration peut aider à la migration per descensum d'éléments microbiologiques en liaison avec un milieu géologique favorable (pores de la roche ouverts et interconnectés, microfissures, fissures ...).

Signalons pour mémoire les dégâts et détériorations imputables à l'eau en milieu souterrain: lessivage des tracés à bas de charbon de bois cas du Salon Noir de la grotte de Niaux, transport puis recouvrement par des particules argileuses d'une frise de bisons, cas de la grotte de Font de Gaume, sur - épaisseurs de concrétions de calcite cas des chevaux de la grotte de Villars, dépôts d'amas vermiculés cas des parois de la grotte des Combarelles, coulures de matières noirâtres sur les gravures - cas de la grotte de Chabot.

L'eau jouant un rôle important dans le milieu souterrain, il convient de connaître l'extension des sols et des terrains qu'elle a pu traverser sachant que toute modification externe (naturelle ou artificielle) aura des répercussions plus ou moins importantes et plus ou moins rapides sur la stabilité du milieu souterrain.

C'est donc tout un écosystème qui est responsable directement ou indirectement de l'état de la vie d'une cavité à quelques mètres de profondeur. La végétation au-dessus des cavités, régule les mécanismes d'évaporation, les infiltrations, le ruissellement, l'ensoleillement direct (Fig. 4), et précisons à ce propos que la France de l'ouest est soumise au climat de type océanique tempéré.

A partir de l'étude des terrains composée d'une enquête sur l'état des lieux, de l'identification du contexte géologique et hydrogéologique de la cavité, il a été possible de mener une réflexion pluridisciplinaire sur la protection de l'environnement des cavités de Combarelles. Dans l'esprit de la conservation, rappelons que l'objectif final est de définir sur le terrain un découpage géographique révélateur de la sensibilité. de la vulnérabilité potentielle à l'égard d'actions altéragènes et d'établir des aires de protection (Vouvé et Brunet 1984; Brunet et al. 1987).

3.2 Enquête sur l'état actuel des lieux (Figs 5, 6 et 7)

La grotte se trouve à proximité d'un certain nombre de maisons d'habitation et d'exploitations agricoles. Les deux groupes de maisons des Combarelles (1) et (2) sont construits à quelques distance de la grotte, au pied des abrupts calcaires. Elles n'ont aucune incidence sur l'équilibre climatique, hydrogéologique et bactériologique de la grotte. Au sud-est, nous avons inventorié quatre maisons en ruine, l'une d'entre elles (6) n'était pas cartographiée 'Fontenelles'. Au sud, 'Trémoulède' et 'Le Repaire' sont deux lieux-dits habités importants qui comportent des exploitations agricoles et quelques résidences secondaires restaurées récemment. La maison habitée la plus proche et à l'amont de la grotte, porte le No. (11). Elle est construite en bordure d'un petit chemin. Cet écart vient d'être alimenté en eau potable à partir du réseau communal.

Les informations synthétiques sont rassemblées dans le tableau récapitulatif No. 1.

Les lignes électriques et le réseau communal d'eau potable se développent peu à peu et ceux-ci sillonnent la colline des Combarelles et celles voisines, offrant ainsi des possibilités nouvelles aux propriétaires terriens locaux. Désormais, il sera en effet plus facile de vendre un ou plusieurs lots proches de ces commodités. Le potentiel de vente est d'autant plus grand que la demande extérieure est forte en vue de la construction de résidences secondaires.

A cette enquête, il faut ajouter des renseignements d'ordre administratif concernant la propriété des terrains concernés. En effet il n'appartient à l'Etat français qu'une toute petite surface sous le porche d'accès à la grotte (en plus du réseau orné lui-même), et sont du domaine privé les surfaces boisées et herbeuses qui surmontent le porche, le réseau souterrain et au-delà en amont.

La gestion de ces surfaces boisées et herbeuses est quasi nulle depuis plusieurs décennies et leur entretien arrêté depuis la disparition de l'exploitation du charbon de bois (charbonnières). La qualité des peuplements végétaux correspond à la réimplantation naturelle après la déforestation du début du siècle rendue nécessaire pour l'alimentation des fours des forges des Eyzies. Ce peuplement végétal est associé étroitement à la nature des affleurements géologiques.

C'est ainsi que sur les calcaires à sol peu épais ou quasi inexistant, on note la présence de chênes; sur les argiles plus ou moins sableux ce sont les châtaigniers qui se développent, alors que sur les placages argilo-calcaires nous avons identifié chaque fois la présence de pins sylvestres réunis en petits bosquets.

Le tableau ci-dessous donne la répartition de la mosaïque forestière et agricole pour la région correspondant au bassin hydrogéologique associé à la grotte ornée (superficie = 983 750 m^2). Il montre la prédominance de la chênaie sur tous les autres types de végétation.

3.3 Synthèse géologique (Fig. 6)

Sur la base des prospections et analyses il apparaît que les grottes des Combarelles I et II sont creusées au sein d'un puissant horizon calcaréo-gréseux jaunâtre à beige du Crétacé supérieur.

Celui-ci est sub-horizontal et affleure largement de part et d'autre des deux flancs de la vallée de la Beune.

Cet horizon est composé:

- d'une fraction sableuse dont le pourcentage varie légèrement d'un point à un autre, la taille des grains de quartz étant également variable;
- d'un ciment calcaire contenant très peu de débris de

fossiles à l'exception de quelques lieux bien précis.

Des variations rapides de faciès se produisent çà et là au sein de la masse rocheuse provoquant l'apparition des corps sédimentaires de formes lenticulaires et de tailles décimétriques et métriques.

Pétrographiquement, le réseau des Combarelles se trouve dans des niveaux calcaréo-gréseux, et la dissolution chimique qui s'y est exercée est responsable de la structure ajourée de certaines portions de la voûte. La mise en place de cette cavité par le jeu de la karstogenèse a été favorisée par la présence de joints horizontaux et de diaclases. Deux directions principales de diaclases apparaissent dans le plateau calcaire et le tracé très anguleux de la galerie en traduit l'influence.

Une prospection plus fine a été faite sur place et complétée par un certain nombre de sondages à la tarière à main.

De cette enquête, il ressort que les placages argileux, argilo-sableux et sablo-argileux sont fréquents à l'amont immédiat et rapproché de la cavité et qu'ils sont d'épaisseur variable (entre 0.30 et 1.05 m). Et dès l'instant où l'on atteint les dépôts les plus importants cette épaisseur devient très vite supérieure à 1.20 m ce qui n'est pas à négliger dans l'esprit de la conservation naturelle de la grotte.

3.4 Hydrogéologie générale

Le bassin versant qui se dessine à l'ouest de Trémoulède constitue l'impluvium normal à travers lequel une fraction de l'eau de pluie s'infiltre, l'autre étant évacuée rapidement par ruissellement superficiel.

Dans la mesure où il n'existe pas d'anomalie karstique aérienne connue à ce jour, la couverture argilo-sableuse est présente pour partie, la densité de cassures qui affecte les affleurements calcaires est faible, il apparaît vraisemblable de penser que l'infiltration est de type mixte.

Elle est directe au droit des affleurements calcaires et retardée là où la couverture sableuse existe. L'eau infiltrée circule le long des petites fissures et joints, elle gagne plus ou moins rapidement un collecteur unique (chenal), de taille modeste. Ce chenal se subdivise à l'aval comme semblent l'indiquer les résultats du traçage chimique.

Le collecteur a une géométrie différente de celle du réseau orné. Il émerge dans les alluvions calcaires de la vallée.

L'étude hydrogéologique a montré que:

- la grotte ornée des Combarelles n'est pas affectée par des venues d'eau importantes en débit. Cependant, il n'en a pas toujours été de même à des époques plus ou moins anciennes comme le révèle le concrétionnement calcique;
- la fraction du réseau la plus active se situe tout au fond et à l'intérieur du conduit transverse menant au réseau inférieur;
- la liaison hydrogéologique verticale ou sub-verticale entre la surface et la cavité se fait par l'intermédiaire d'un réseau de microfissures;
- la mise en réserve de l'eau en période d'alimentation des nappes, dans les fissures ouvertes au sein des calcaires, n'est pas négligeable compte-tenu du retard à l'écoulement qui atteint 71 jours. C'est l'écart de temps entre la reprise de la pluie efficace et la croissance du débit des suintements à l'intérieur de la grotte;
- ce retard à l'écoulement décroit sensiblement dès l'instant où les fissures du réseau ont été envahies par l'eau.

En résumé, à la suite de la synthèse géologique détaillée et du bilan hydrogéologique effectué sur un an de contrôle en continu nous savons que toute la partie de la colline qui n'est pas recouverte par les dépôts argilo-sableux participe directement et rapidement à l'alimentation de la cavité ornée.

La fraction la plus proche du vallon, c'est-à-dire celle située au nord-est, est composée de calcaires affleurant ou sub-affleurant masqués par une mince couche d'éboulis calcaires.

Plus au sud-ouest, les dépôts sont peu épais (jusqu'à 1.20 m au maximum) et sont de nature argilo-sableuse, sableuse ou argileuse.

(4) Protection de l'environnement externe: recommendations

La France possède une législation qui confère à l'Etat, par des limitations du droit à la propriété des biens immobiliers, les pouvoirs nécessaires à la protection du patrimoine archéologique. Citons parmi ces textes la loi du 31 décembre 1913 sur les Monuments Historiques permettant la protection des immeubles dont la conservation présente du point de vue de l'histoire ou de l'art un intérêt public; la loi du 27 septembre 1941 sur les fouilles archéologiques, validée par ordonnance du 13 septembre 1945; la loi du 2 mai 1930 sur la protection des monuments naturels et des sites, grâce à laquelle les cavités décorées peuvent bénéficier d'une protection lorsqu'elles sont incluses dans le périmètre d'un site naturel ayant fait l'objet d'un classement ou d'une inscription; la loi du 10 juillet 1976 relative à la protection de la nature prenant en compte 'la préservation des sites présentant un intérêt particulier pour l'étude et l'évolution de la vie et des premières activités humaines' (Geneste 1984; Pattyn 1981).

Ce cadre législatif et administratif doit être renforcé par des données de terrain.

Comme nous l'avons déjà évoqué (cf. 3.1), les découpages géographiques symptomatiques de la vulnérabilité à l'égard des altéragènes seront faits en fonction de la nature des terrains et de la dépendance hydrogéologique (infiltration directe ou retardée au profit du micro-aquifère drainé par la grotte ornée). Ce partage en zones délimite en fait les surfaces sous lesquelles l'alimentation au profit de la grotte sera plus ou moins directe.

En général, nous les définissons à trois modules et pour Combarelles deux ont été dissociés en sous-unités.

- L'aire de protection immédiate occupe au sol la surface la plus restreinte; elle englobe en général le territoire immédiatement circonscrit au monument. Les règles qui la régissent sont des plus contraignantes.
- L'aire de protection rapprochée intercepte au sol la surface altimétrique plus élevée que le monument, et en-dessous de laquelle une pollution est susceptible de se transmettre rapidement au niveau des peintures et/ou des gravures et sculptures. Le délai peut varier entre quelques heures et quelques jours (8 à 10 environ). Les contraintes associées à cette protection comporte des interdictions et des autorisations sous conditions ou assorties de conseils selon la nature et la géométrie de

l'environnement aérien et souterrain.

- L'aire de protection éloignée englobe toute la surface au sol intégrée dans le bassin versant topographique ou impluvium. Il s'agit en fait de la surface au-dessous de laquelle la pluie après ruissellement et infiltration est susceptible d'émerger un jour au contact des peintures et des gravures après un temps de transfert souterrain plus ou moins long et inférieur en général à 12 mois. Les contraintes associées à cette protection ne comportent que des règlementations et des recommandations.

Appliquée au cas particulier des Combarelles, cette protection permet de différencier cinq zones (Figs 5 et 7).

Zone No. 1

Elle surplombe directement la cavité et s'étend immédiatement à l'amont de celle-ci. Elle est boisée; la nature du sous-sol est essentiellement calcaire.

Cette surface est sensible à la pollution et il serait souhaitable que:

- la construction soit interdite
- le déboisement soit réglementé

Ces coupes pourraient être faites sous réserve que les plantations ultérieures soient programmées et réalisées dans des délais raisonnables, ne dépassant pas deux ans.

Zone No. 2

Elle constitue la fraction la plus élevée de l'impluvium qui profite à la grotte. Le sous-sol est argileux sur une certaine épaisseur, non connue à ce jour; en-dessous se trouvent les calcaires. Elle est boisée, cette surface sera peu sensible à la pollution si la puissance du revêtement argileux est supérieure à 2.50 m.

La construction pourrait donc être tolérée sous réserve d'un contrôle préalable de la puissance du revêtement. Le déboisement pourrait également être toléré sans contrainte particulière.

Zone No. 3

Il s'agit d'un vallon à fond plat constitué de parcelles boisées à l'aval et cultivées à l'amont; c'est le collecteur naturel du ruissellement aérien. Cette surface nous paraît sensible à la pollution, dans ce sens, il serait souhaitable que la construction soit interdite.

Zone No. 4

Elle constitue le flanc oriental du vallon qui débouche aux Combarelles; elles est couverte de bois dans sa grande majorité. Quelques prairies et chaumes ont été repérés autour de la maison en ruine des Combarelles-Haut. Cette surface ne nous paraît participer que pour une part très faible à nulle à l'alimentation du réseau. Pour cette raison, il n'y aurait pas de contrainte particulière à envisager ici.

Zone No. 5

Cette surface représente une fraction topographiquement haute de l'impluvium des Combarelles. Le sous-sol de la partie inférieure est constitué de calcaire; celui correspondant à la partie haute est couvert par un placage argileux dont la puissance nous est inconnue. La protection naturelle nous paraît acquise en haut. Elle ne l'est sans doute pas complètement en bas. La constructibilité serait donc envisageable sous réserve, pour la fraction topographiquement basse, d'un contrôle préalable de la présence du revêtement (nature et épaisseur jusqu'à concurrence de 2.50 m).

A ce stade de l'étude, il est possible de dresser un tableau synthétisant les données en notre possession sur la notion de vulnérabilité (Tableau 3).

Conclusion

Ces zones de protection aux abords de la cavité des Combarelles sont basées sur une sensibilité différente à des activités pouvant avoir des répercussions sur les conditions internes du milieu souterrain (pollution domestique, mutation des cultures et des forêts) et sur la nature et les caractéristiques hydrogéologiques des terrains. Pour ce secteur particulier du Périgord - qui est très touristique - elles ont pu être présentées à l'Administration du Ministère de la Culture pour être incluses à une protection plus globale au titre des sites et de la vallée de la Vézère qui est un vaste ensemble préhistorique. Les prescriptions énoncées zone par zone nous paraissent capitales dans l'état actuel de nos connaissances sur le sujet.

Dans un souci de protection de la cavité, de ses abords, il conviendra à l'avenir de tenir compte des ces conclusions pour ne pas introduire de bouleversements dont les conséquences seraient préjudiciables à court et long terme au bon état de ces oeuvres pariétales.

REFERENCES

AUJOULAT, N. 1984. Grotte des Combarelles II. In *L'art des cavernes. Atlas des grottes ornées paléolithiques françaises*, pp. 114-8. Imprimerie Nationale, Paris.

BARRIERE, C. 1984. Grotte des Combarelles I. In *L'art des cavernes. Atlas des grottes ornées paléolithiques françaises*, pp. 109-14. Imprimerie Nationale, Paris.

BREUIL, H. 1952. Les Combarelles. In *400 siècles d'art pariétal*, pp. 91-106. Editions Max Fourny, Art et Industrie.

BRUNET, J., P. VIDAL et J. VOUVE 1987. *The conservation of rock art*, Studies and Documents on the Cultural Heritage No. 7. UNESCO, Paris.

GENESTE, J. M. 1984. Le poisson, législation. In *Découvertes de l'art des grottes et des abris*. Périgueux.

MANGIN, A. et C. ANDRIEUX 1984. Les conditions hydrogéologiques et climatiques d'environnement des oeuvres pariétales préhistoriques. In *L'art des cavernes. Atlas des grottes ornées paléolithiques françaises*, pp. 53-7. Imprimerie Nationale, Paris.

PATTYN, C. 1981. Protection et conservation. *Revue des Monuments Historiques 'Les grottes ornées'* 118: 2-11.

VOUVE, J. et J. BRUNET 1984. A propos de la protection de l'environnement des monuments rupestres ornés du patrimoine mondial mythe ou réalité? *Preprints of the 7th Triennial Meeting of the ICOM Committee for Conservation, Copenhagen*, pp. 24.17-24.20.

N°	Sites	Codes	Position géographie par rapport à la grotte	Appartenance Hydrogéologique	Distance en m	Pollution potentielle estimée
	LIEUX CONSTRUITS	1			2	3
1	Les Combarelles	N.H.	Aval	Indépendante	5	Nulle
2	Les Combarelles	H.	Aval	Indépendante	110	Nulle
3	Les Combarelles	H.	Aval	Indépendante	275	Nulle
4	Les Combarelles-Haut	E.R.	Amont	Indépendante	225	Nulle
6	Fontenelle	E.R.	Amont	Indépendante	475	Nulle
7	Trémoulède	H.	Amont	Dépendance indirecte	475	Très faible
8	Le Repaire N.	H.	Amont	Dépendance indirecte	1000	Très faible
9	Le Repaire S.	H.	Amont	Dépendance indirecte	1150	Très faible
10	Fontenelle-Bas	E.R.	Amont	Directe	400	Notable
11		H.	Amont	Directe	400	Notable

Tableau 1.

(1) H = habitée, N.H. = non habitée en 1979, E.R. = en ruine
(2) Plus courte distance calculée par rapport à l'entrée de la grotte
(3) Pollution potentielle estimée dans l'hypothèse de projets de construction de maisons secondaires ou d'exploitations agricoles

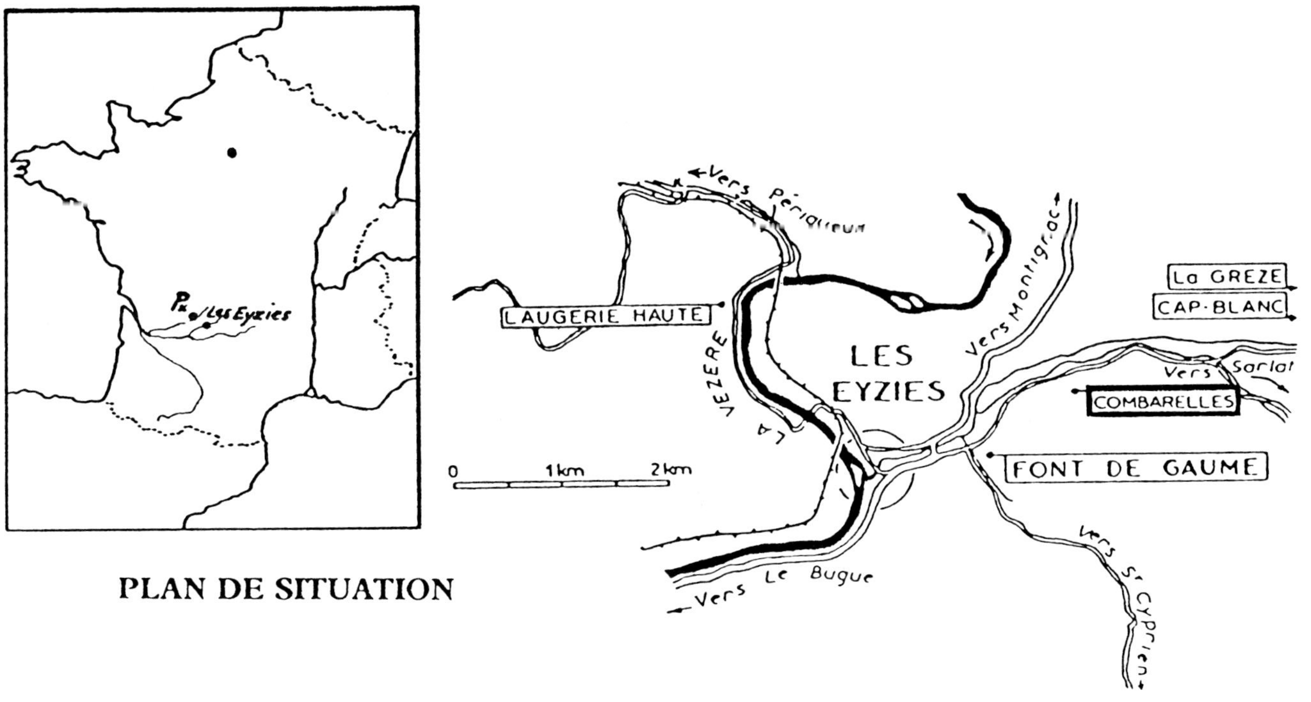

Figure 1. *Localisation des cavités des Combarelles (Les Eyzies, Dordogne, France).*

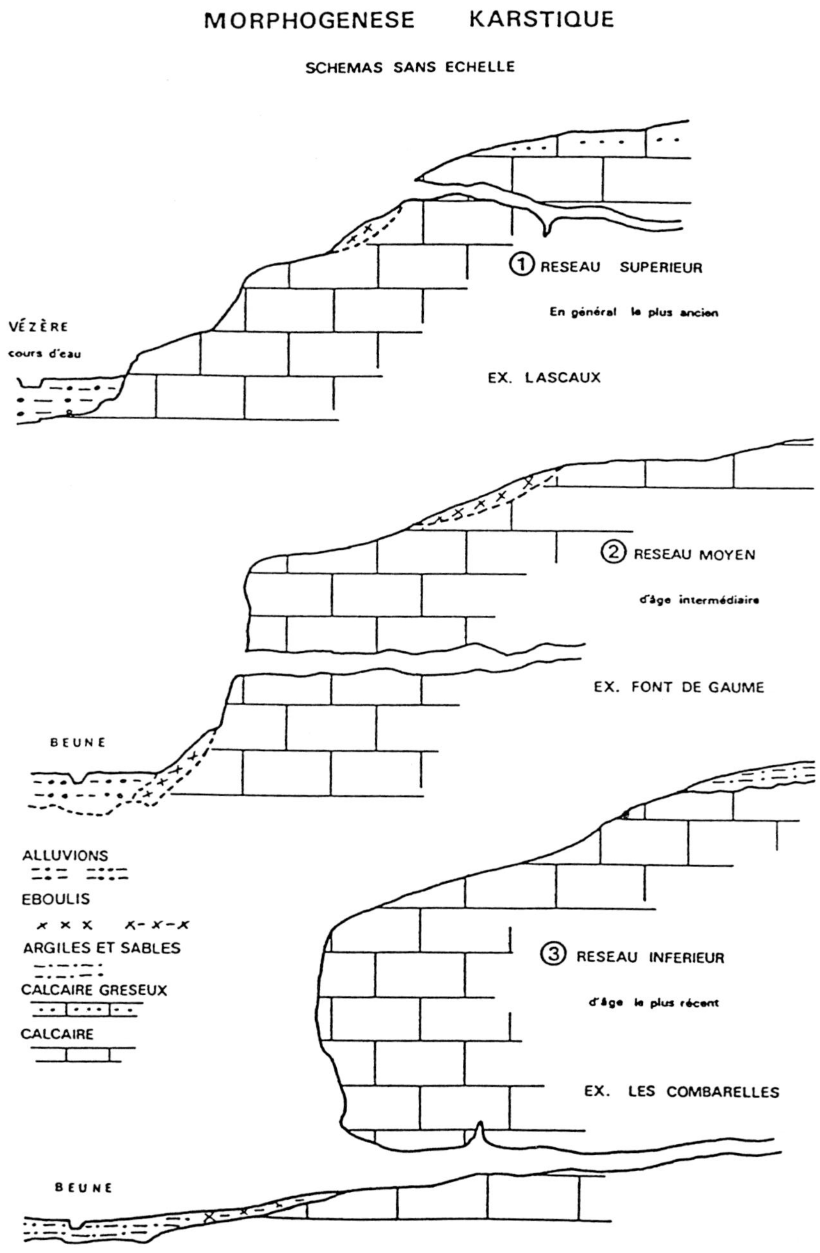

Figure 2. Exemples de morphologie karstique.

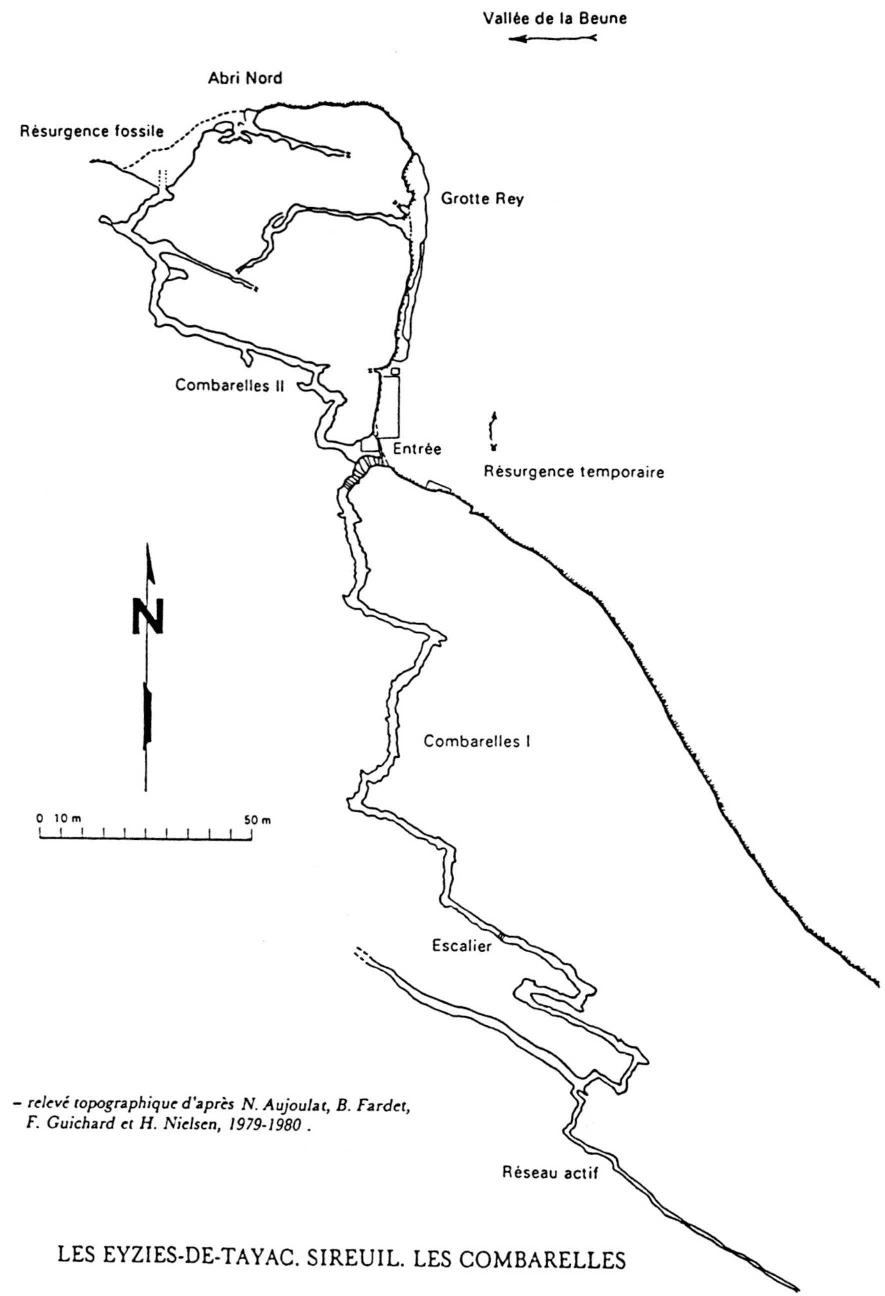

Figure 3. Carte du réseau des Combarelles.

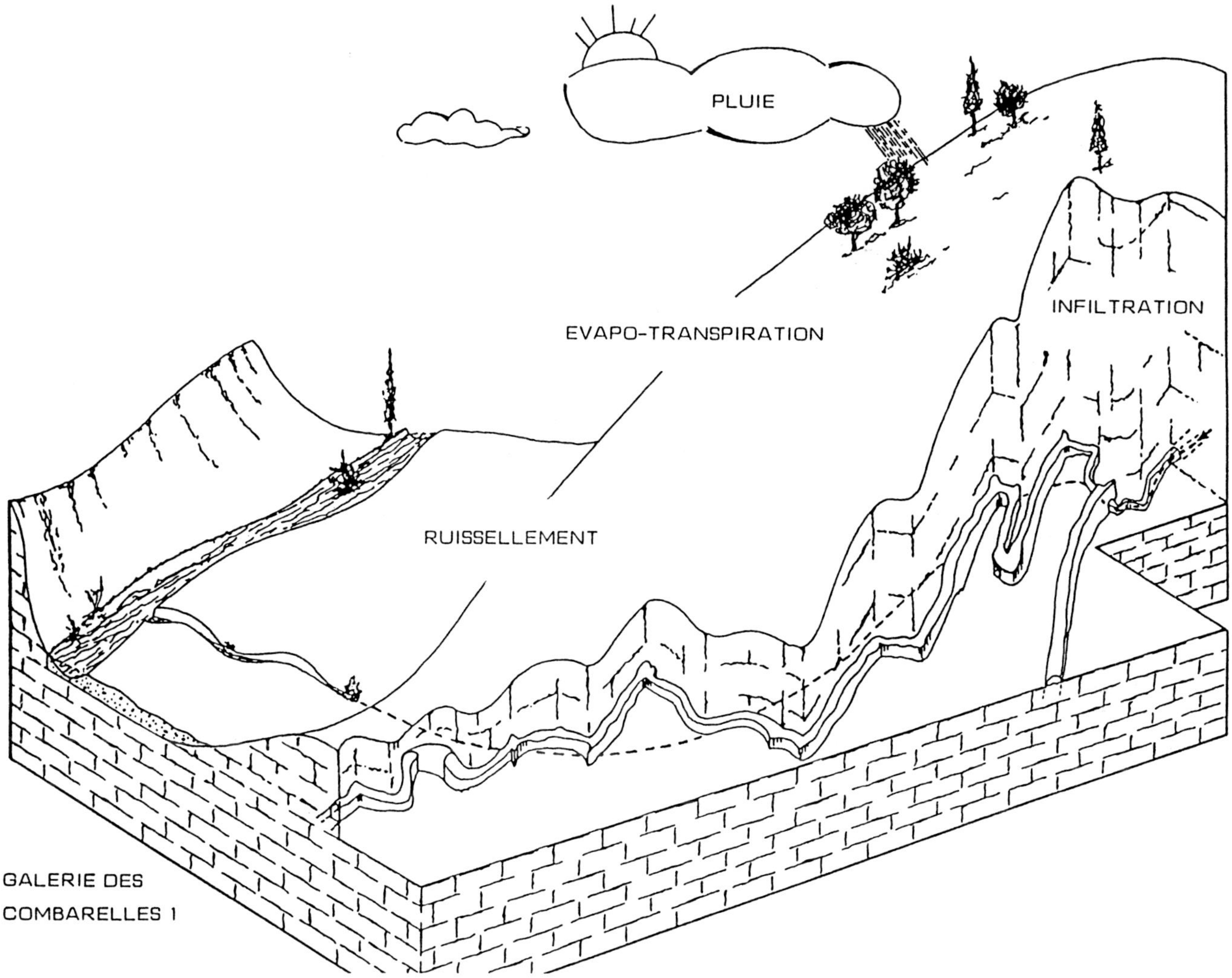

Figure 4. *Représentation schématique de l'action de l'ensoleillement, de la pluie, de l'évapo–transpiration, du ruissellement, de l'infiltration en fonction du couvert végétal sur le site des Combarelles.*

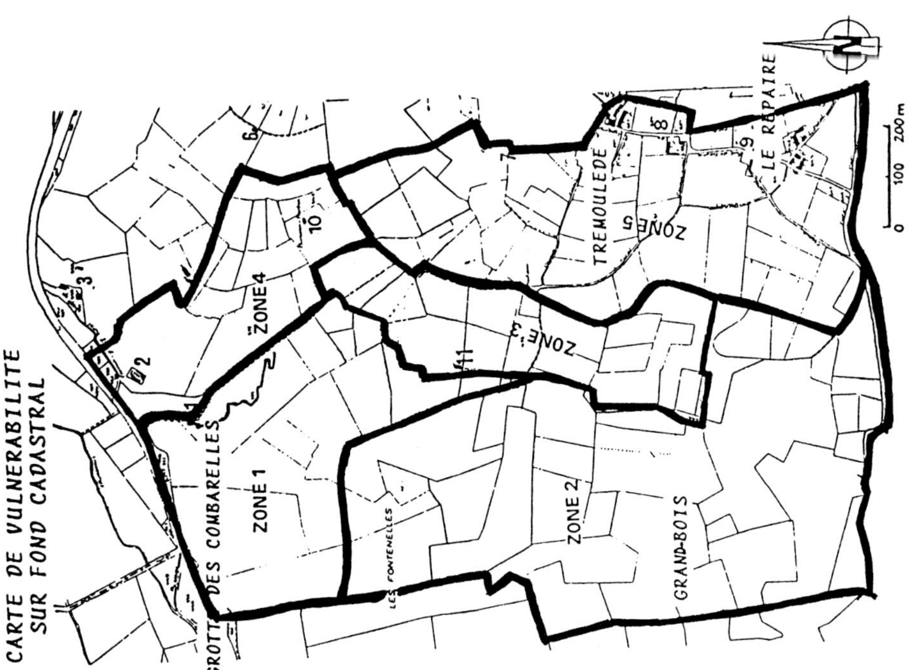

Figure 5. *Répartition des parcelles de terrain sur le site des Combarelles. Le morcellement en nombreuses parcelles est une des caractéristiques du Périgord. Report des zones vulnérables.*

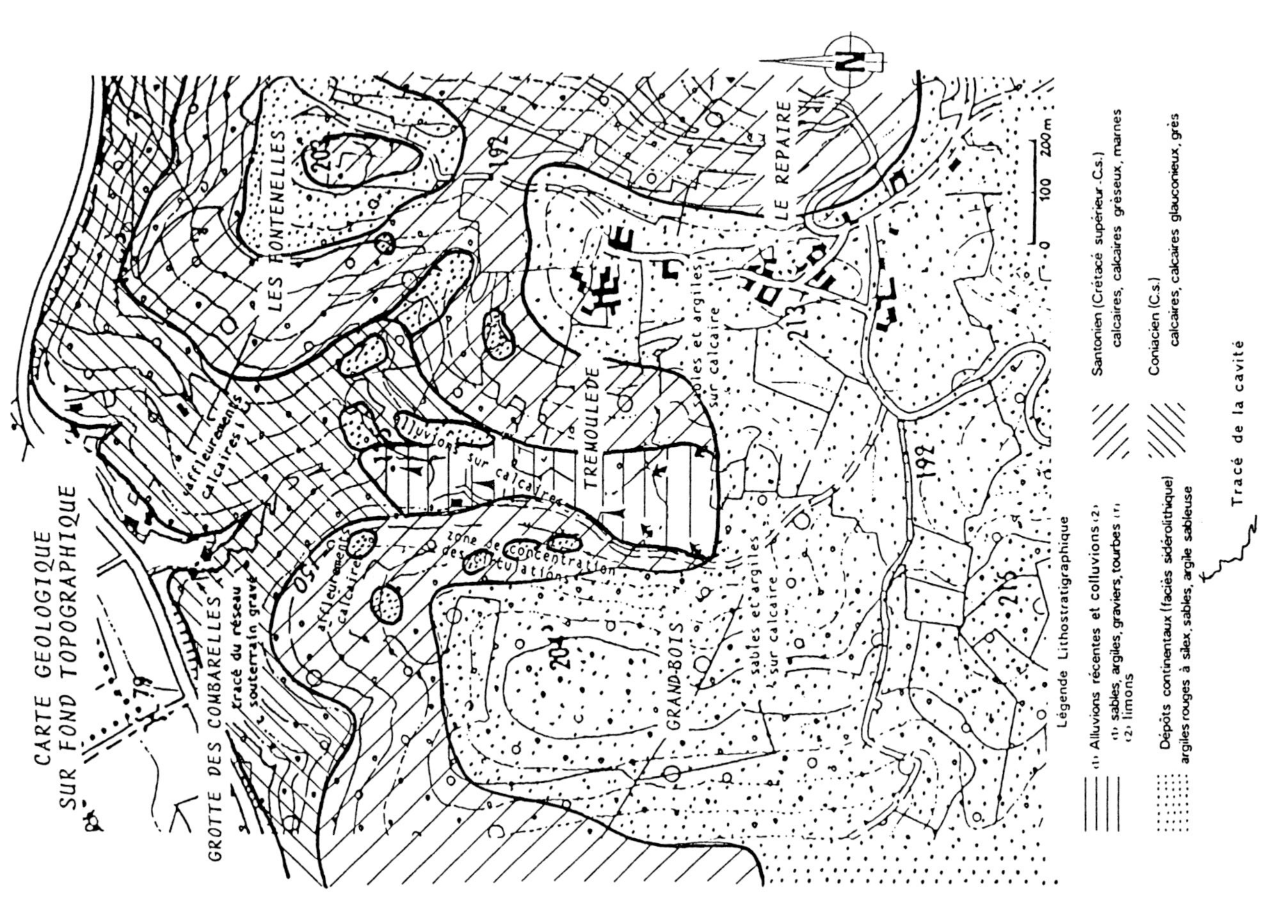

Figure 6. Site des Combarelles: carte géologique.

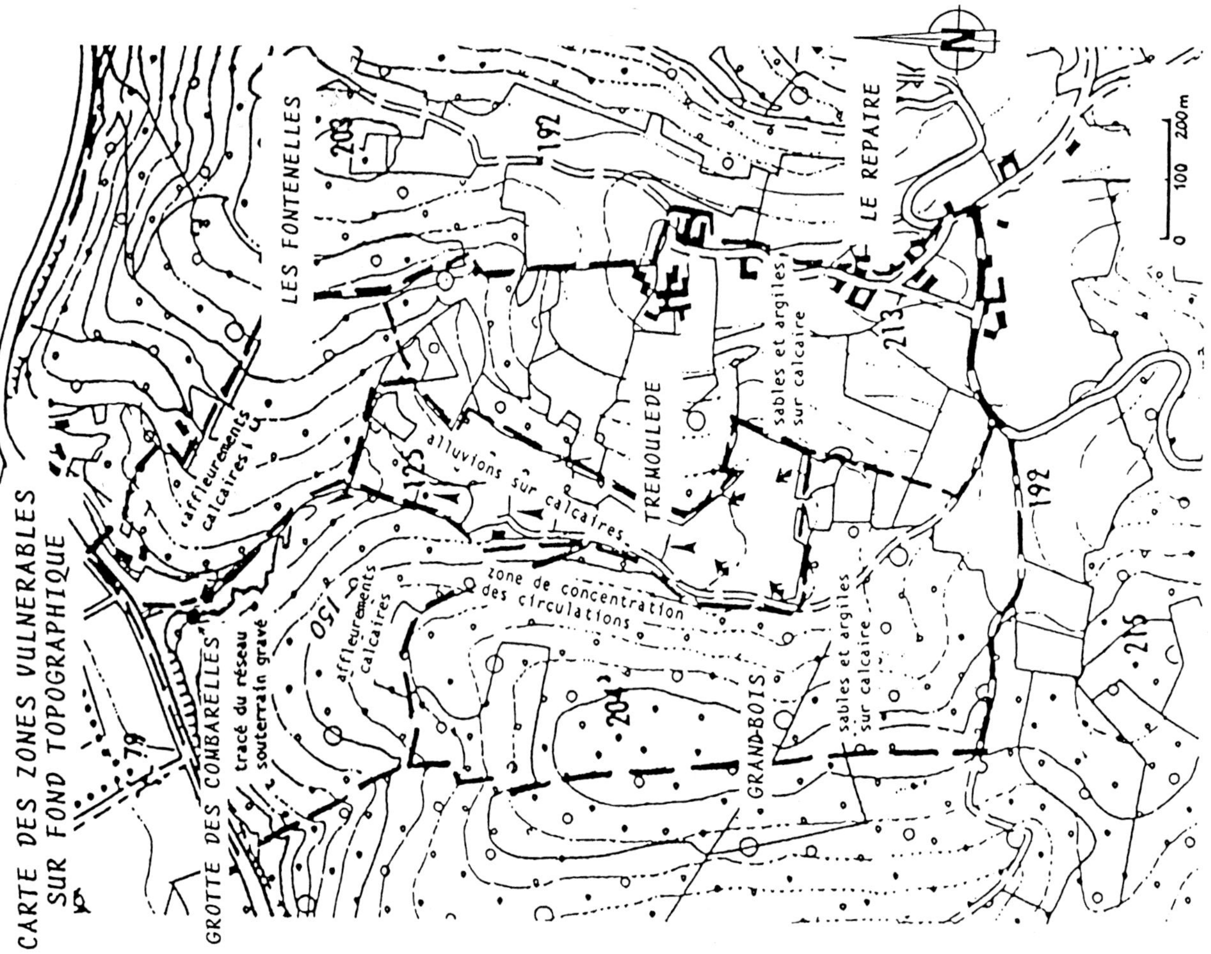

Figure 7. Carte de vulnérabilité sur fond topographique.

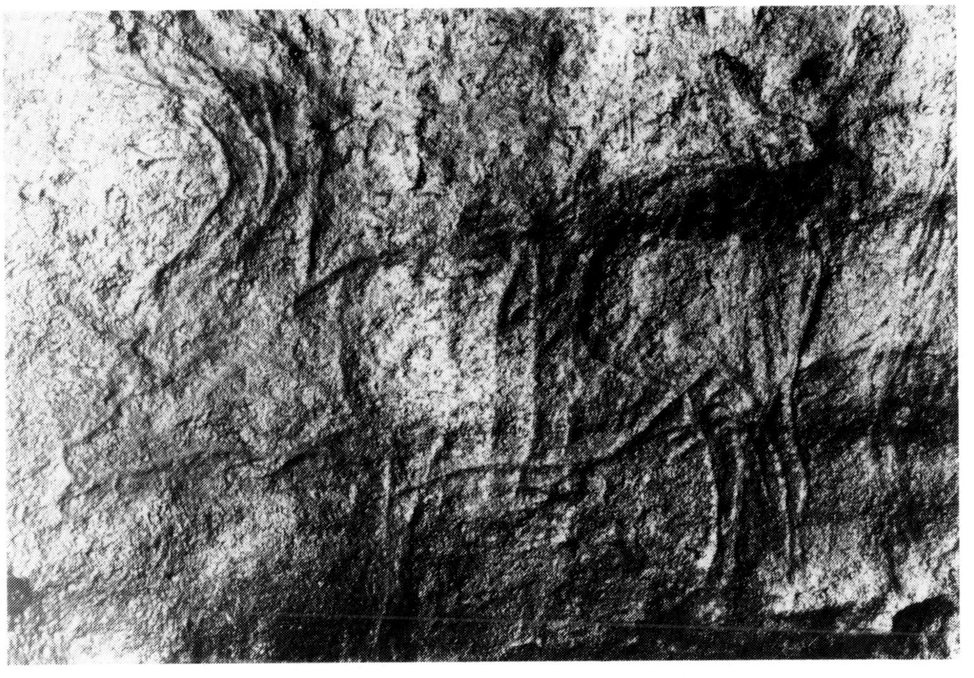

Photo 1. *Gravure de renne. Combarelles.*

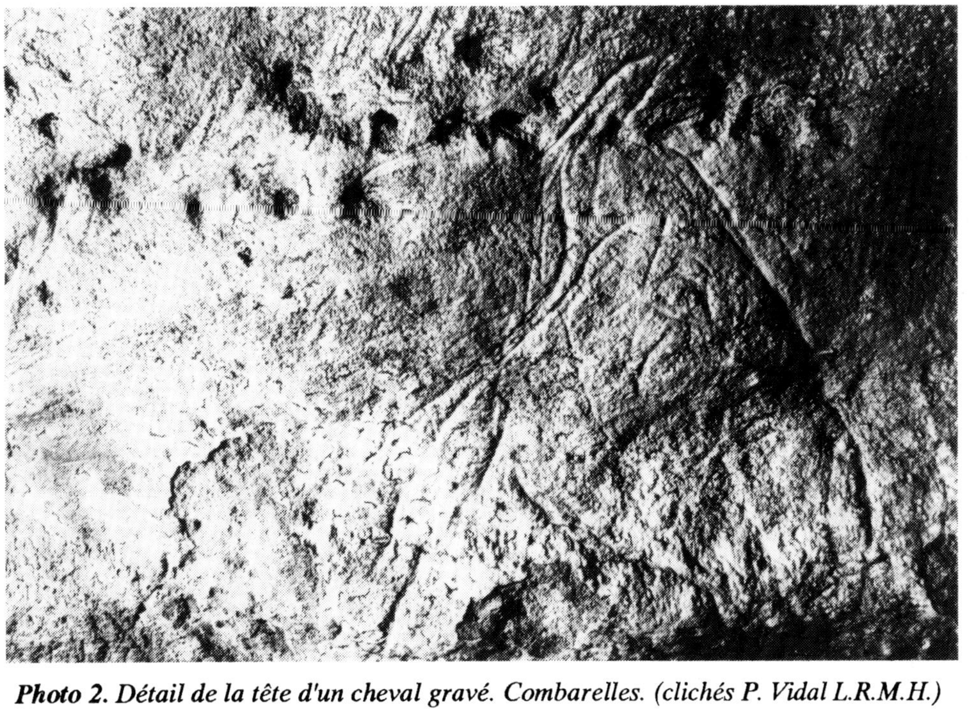

Photo 2. *Détail de la tête d'un cheval gravé. Combarelles. (clichés P. Vidal L.R.M.H.)*

11

THE PAROONG CAVE PRESERVATION PROJECT

Robert G. Bednarik

Abstract. The main objects of the Paroong Cave Preservation Project were to alleviate conservation hazards that had been introduced by the European occupation of the district, to halt the deterioration they were causing to the cave art, and to effect measures of active and passive visitor control. The many conditions that had to be met in the execution of this project, and the way in which they were met, are described. Through a detailed study of the limestone cave, its hydrology, meteorology, pedology and other relevant factors, the processes responsible for deterioration were identified. The comprehensive preservation measures of early 1988 as well as the long-term work are described, and a preliminary assessment of their effectiveness is presented.

Introduction

The Paroong Cave Preservation Project is the first systematically planned rock art conservation project undertaken in Australia that relates specifically to a limestone cave. It has been noted by several writers (e.g. Rosenfeld 1985: 4) that the conditions applicable to conservation work differ significantly between cave sites and open air or rockshelter sites. The meteorological regime even of a rockshelter is substantially regulated by a region's general climate, and by its seasonal and diurnal oscillations. A speleo-climate, by contrast, remains almost unaffected by external conditions and has very site-specific characteristics. On the other hand, factors that have practically no effect on an open air site, such as the mere presence of humans, can dramatically influence the micrometeorological environment of cave art. Moreover, the limestone most caves occur in presents conservation issues that are not relevant to other art sites, even to those on open limestone pavements. The only major cave art conservation studies so far conducted are those in France and Spain, and French specialists have in fact presented some of their results at this conference. Here I shall relate the first intensive cave art management project outside of western Europe.

Description of the Site

Cave art is a rare phenomenon in Australia, where less than 40 such sites are known today. They have all been located only in the last ten years, except Koonalda Cave, which was first published in 1968, and Orchestra Shell Cave, reported three years later. The largest regional concentration, not only in Australia but world-wide, exists at Mount Gambier, South Australia, where presently 29 sites are known (Bednarik 1990). One of the visually most striking of these is Paroong Cave. Its petroglyphs are particularly deeply carved, 30 to 40 mm, densely covering the vertical walls of the 15 m long Art Passage up to a height of 4 m. This art is stylistically so similar to that of Mount Cameron West (see Mt Cameron West Rock Art Protection Project in these proceedings) that the two seem to be twin sites, though they are separated by Bass Strait. The age of the Paroong petroglyphs is unknown, but their stylistic similarity to the early part of the rock art at Olary, around 30 000 years old (Nobbs and Dorn 1988), is obvious, and the land bridge to Tasmania became submerged towards the end of the Pleistocene.

The cave (Figure 1) consists of a horizontal passage formed mostly along a vertical fault, oriented roughly east-west. It reaches the phreatic water level at several

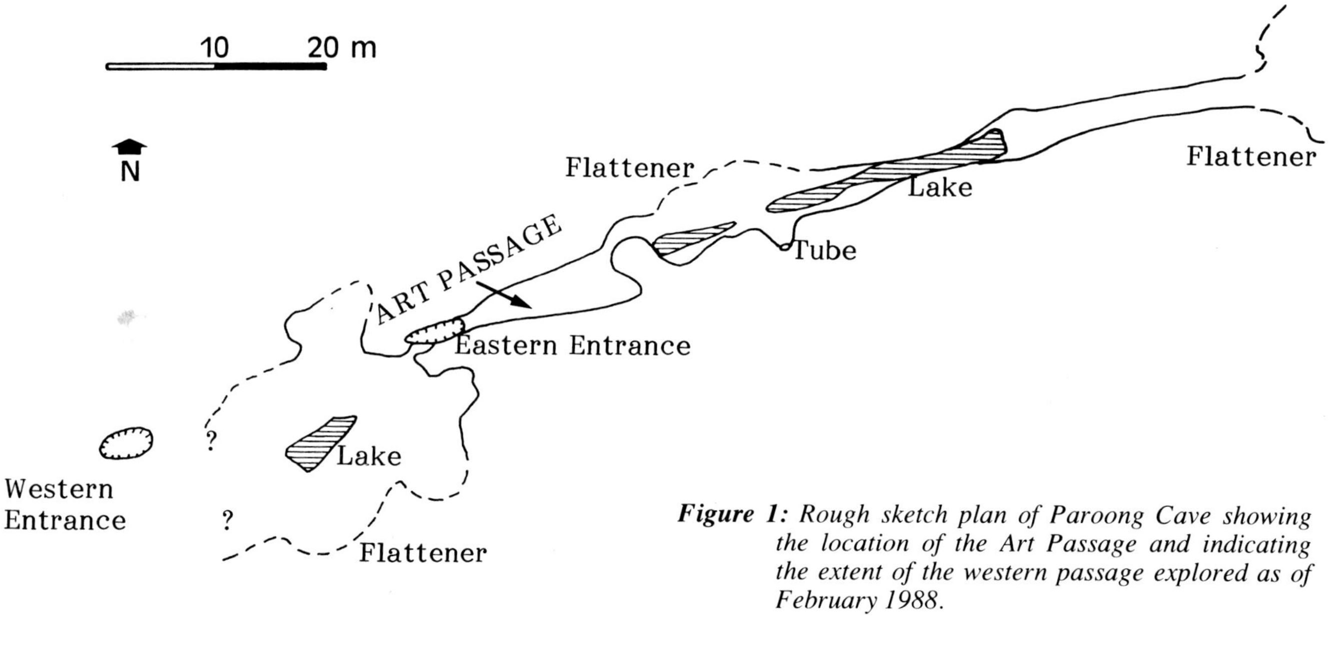

Figure 1: *Rough sketch plan of Paroong Cave showing the location of the Art Passage and indicating the extent of the western passage explored as of February 1988.*

points. The upper aquifer, which characterises the area's macrohydrology, is about 8 m above sea level here and drains in a due southerly direction. There are presently two entrances from the surface and landowners have attempted to fill in both with boulders and agricultural refuse. The western entrance was successfully closed and has remained so. The eastern entrance was only partly obscured by rocks, which had several effects: it closed off access to the western part of the passage and thereby influenced micrometeorology profoundly; it allowed soil sedimentation in the entrance, as well as the establishment of vegetation; it indirectly affected the hydrological regime; and it concealed some rock art.

The petroglyphs were discovered in 1984 by Geoffrey Aslin, a member of the Parietal Markings Project, during routine reconnaissance. Planning for the site's protection began in the same year. It is the Project's established practice to keep site locations confidential, but in this case several people, including the landowners and some cavers had become aware of the find and visitation of the site soon increased. Within three years it became evident that rock art was being destroyed rapidly along the steep climb from the entrance to the passage floor. Here the passage is narrow and visitors found it almost impossible to avoid contact with the densely decorated wall. Especially just above floor level, where the boots of visitors rubbed against the moist and fragile wall, the rock surface receded by about 20 mm in just three years and some petroglyphs have been worn away almost completely. By 1987 I realised that much of this outstanding corpus of cave art, which I have described as one of the 20 most important sites in Australia, would not survive the next 20 or 30 years, unless an effective preservation effort commenced immediately. I secured a seeding grant of $4200 from the Rock Art Protection Program of the Australian Institute of Aboriginal Studies, intended to meet the cost of material only and began organising the project. The main works took just under one month and were executed in January 1988. The project has been such an outstanding success in that I consider the perpetual survival of the cave art as assured, that its methodology is worth relating in some detail.

Preservation Project

The project was subject to a number of basic guidelines or ground-rules: No structural changes to the cave, nor modifications to the rock mass were to be undertaken; all adopted measures had to be fully reversible; costs had to be kept within the limited resources available; the results of the project had to be monitored in the long term; preference was to be given to conservation measures that would control the environment by indirect or natural means rather than by calculated direct intervention; preservation measures had to be compatible with the landowners' perceived management plan and in addition had to have the approval of the Aboriginal Heritage Conservation Branch of South Australia; they had to be conducted without the risk of an accident and had to be within the capability of a very small work crew. But perhaps most importantly, the design of the program had to be based on scientific, quantitative assessment of all crucial factors, not on mere guess-work or opinion, and each measure to be adopted had to be carefully considered before implementation, and its effects, or potential effects, on other measures forming part of the program had to be assessed.

All these conditions were met fully. The final management plan was based on extensive observation and analytical work that began in 1984, and was designed after carefully reviewing several options. The factors considered included surface rainwater run-off, capillary moisture, condensation, mosses, algae, fungi, bacteria, speleofauna, domestic animals, human visitation, physical weathering, structural aspects, salt deposition, tectonic stresses and the speleoclimate. Here I shall only consider the most important aspects, which are those presenting the major threats.

The cave art had obviously survived well for many millennia and it became apparent that practically every type of threat to it is in some way connected to the European occupation of the region. Clearing the former dense stringybark forest had significantly altered the hydrological regime, increasing rainwater run-off into the cave entrance. The deposition of organic matter in the cave had attracted fungi and the deposition of boulders and refuse in the entrance shaft had facilitated the establishment of washed-in sediment. The high level of moisture well within the reach of daylight had generated a dense cover of mosses and other plant species, including grasses. This had created a vicious circle: the plant growth, through the micro-organisms it introduced, produced carbon dioxide which, combined with water to form carbonic acid, attacked the limestone rock. The moss cover assisted in retaining moisture longer, and the by-products of the microchemical attack on the rock probably further assisted in keeping moisture from evaporating. To make matters worse, the partial infill of the cave entrance had reduced the opening size, thereby impeding convective temperature exchange. Perhaps more importantly, this assisted in maintaining a higher air humidity near the cave floor, facilitating the survival of the moss community during the dry summer months.

In attempting to determine the cause of moisture retention in the entrance shaft I had to establish the nature of the relationships between interstitial rock moisture, moisture in the sediment and air humidity. The rock can absorb an average of 27.6 per cent moisture of its own bulk weight, and by determining the bulk volume with the wax coating method (simply coating a sample of rock with wax and determining its volume by water displacement), I ascertained a porosity of 50.8 per cent of volume, of which 73.1 per cent is occupied by water within 24 hours. The rock's porosity is therefore extremely high but the bulk of the pores is readily accessible and the problem did not appear to be the rock's own moisture retention characteristics. It seemed that the water-logging, which was so detrimental to the survival of the petroglyphs, was attributable to buffering by the soil rather than vice versa.

My analyses of the sediment identified it as a local rendzina soil of dark colour, changing from 10YR 3/1.5 to 7.5YR 4/5 (Munsell) upon removal of the 4.1 per cent organic matter component, i.e. its hue shifts towards red, while both value and chroma increase. The coarse fraction (Figure 2) consists entirely of carbonates, a ferruginous component only becomes prominent at 0.4 mm and occurs mostly as the ferric mode. The lowest fractions are dominated by water-insoluble calcium humates, as is typical for rendzinas. The complete absence of drying cracks in the deposit is noteworthy in view of the extremely high linear shrinkage of 14.8 per cent, because it suggests that the soil deposit never experienced significant desiccation. The natural moisture content of the sediment is 33.4 per

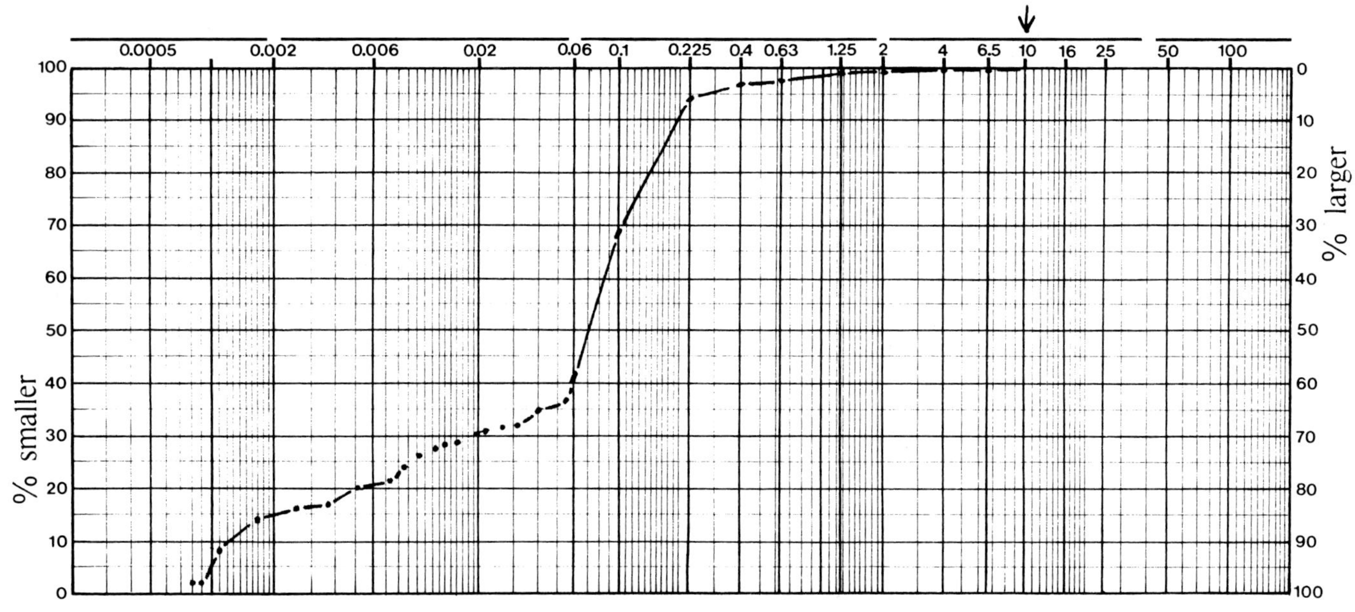

Figure 2: *Cumulative grain size distribution curve of the recently deposited soil in the Eastern Entrance of Paroong Cave.*

cent of mass, which at an assumed s.g. of 2.6 means that an incredible 56.6 per cent of the sediment's bulk volume is occupied by water. This is particularly remarkable in view of the rather modest clay fraction of about 15 per cent, the already mentioned low organic matter content and the marked steepening of the cumulative grain size distribution curve in the fine sand fraction (Figure 2).

This high susceptibility to prolonged moisture retention explains how the moisture is buffered in the rock, and this relationship is assisted by the micrometeorological regime: closeness of the water table prevents the intake of warmer air, cool air is trapped, and buffered by both the water body and the rock mass, allowing its humidity to remain equally stable. As soon as this equilibrium is threatened by convective heat in the entrance, the evaporation rate there would increase, evaporation would effect a lowering in temperature and thus protect the cool air below. In this system equilibrium is maintained permanently, the dew point is not exceeded and consequently condensation does not occur.

During three weeks in January 1988, the air temperature near the floor of the cave immediately below the eastern entrance never strayed outside the range from 11 to 14 degrees centigrade. The air relative humidity just 10 m from external conditions remained continuously above 97 per cent. Both temperature and relative humidity showed only negligible diurnal changes, if any. For instance, external relative humidity averaged 21 and 18 per cent at 9 a.m. and 3 p.m. respectively, it ranged from 30 to 43 per cent just outside the entrance, reached 52 per cent immediately below the ground line, while remaining static between 97 and 99 per cent near the floor of the entrance shaft. Whenever wind speed values were taken, no discernible movement could be recorded a few metres into the shaft. Though no wind velocity greater than 8 m/sec actually occurred during the period of observation, much more extensive studies at other sites in the region suggest that even much stronger winds would be ineffective in producing an air turbulence in the cave.

Conservation Plan

The need for a heterogeneous conservation plan was clearly established by these studies: the amount of rain water discharging into the cave entrance had to be controlled, the deposited rubble and soil had to be removed to return to pre-European hydrological and micrometeorological conditions. The latter measure would create the additional benefit of major spatial changes in the entrance shaft, allowing the relocation of visitor traffic away from decorated panels. The erection of a structure over the entrance shaft would also bring two benefits: the possibility of suspending from it a heavy access ladder designed to keep visitors away from walls, and that of being able to install a grid controlling the access of humans and animals.

I considered several options to achieve each of these main objectives and eventually designed a complete program of preservation measures. Accordingly the boulders deposited in the entrance were removed by mobile crane, taking great care not to damage the walls. This was an extremely difficult and delicate task, requiring considerable experience in the operation of such equipment. The shaft was cleared of the recent deposit, which was incorporated in the construction of a levee outside the cave. The potential carbon dioxide catchment area was determined on the surface, and a 73 m-long drainage channel constructed outside the levee. This effectively reduced the natural water catchment area draining into the cave by 98 per cent, from about 18 150 m^2 to 345 m^2. Two rows of trees of stringybark and casuarina were planted outside this channel. These were the main species in the area prior to European arrival and both are noted for their effectiveness in absorbing moisture with their root systems. They will be located far enough from the Art Passage to prevent any of the mycorrhizal carbon dioxide that will be produced by the symbiotic micro-organisms from reaching the art. The entire complex has been surrounded by an electrified stock fence which will prevent the raising of soil nitrate levels through domestic animals (Figure 3).

To meet the requirement of active and passive visitor control without attaching a grid to the rock I constructed a steel structure of such weight that it could not be lifted without a crane. To prevent its being dragged by a truck I

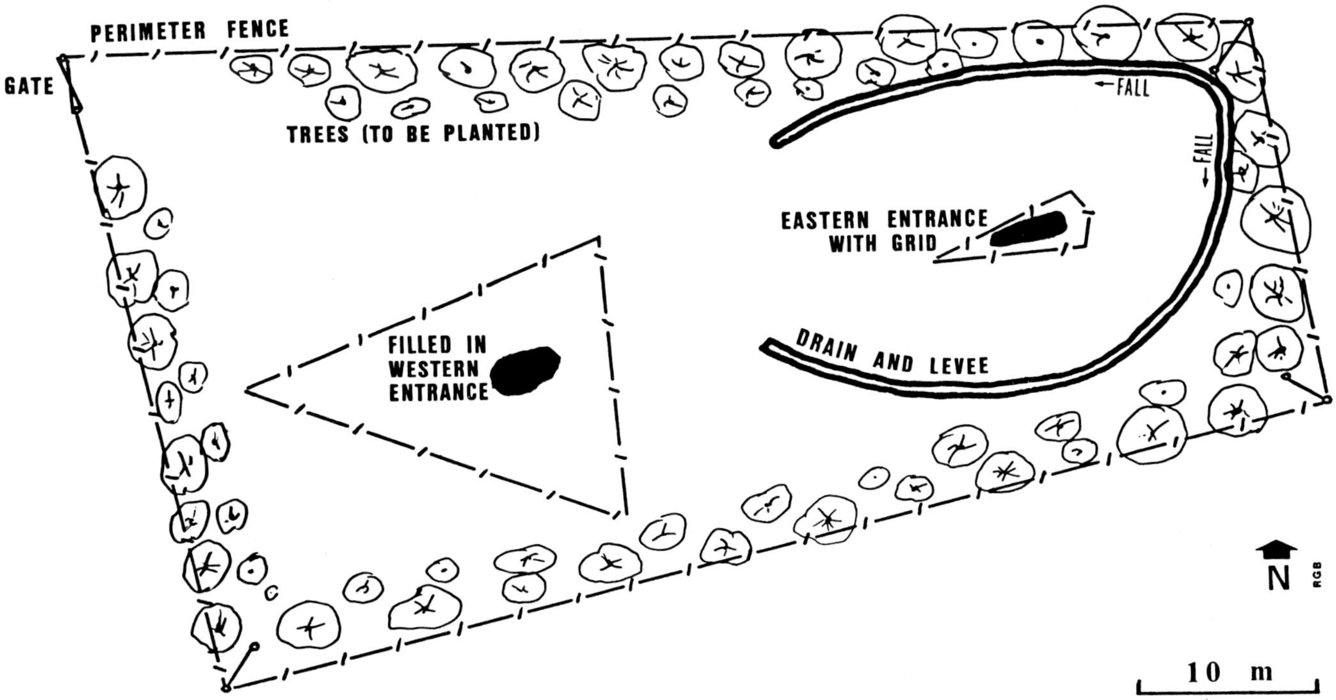

Figure 3: *Sketch plan of the Reserve surrounding the entrance of Paroong Cave.*

exposed the solid rock ledge and shaped the infrastructure component to fit snugly to the rock contour. This grid consists of four basic elements: the infrastructure distributing the load evenly, a set of heavy crossbeams, tubular decking providing the grille, and several superstructures. The finished grid structure is about 8.5 m high and weighs well over a ton. It represents almost 100 hours of welding work. Many structural and practical considerations had to be included in its design, for instance it is intended to discourage cave dwelling swallows from entering, while permitting bats to pass.

Passive visitor control measures are generally less obtrusive than those of active control, they function primarily at the psychological level. They may be 'camouflaged' as measures providing the visitor with comfort, as in this case the 7 m long galvanised steel ladder I suspended from the steel structure, immediately below the lockable trap door (Figure 4). It is a most effective means of confining the visitor to the intended path, and is well clear of petroglyphs. Several other measures of passive visitor control were adopted and are mostly based on the rule that most cave visitors tend to keep to level ground when moving around and shun slopes, piles of rubble and any area that looks unstable or dark. Various specific methods were employed to discourage the visitor from stepping close to the walls. These expedients should provide adequate protection, considering that casual or unappreciative visitors are automatically excluded by an extremely stringent access permit system, and that all visitors are closely supervised throughout their presence in the cave. Their attitude is further conditioned by the requirement that each visitor must study, complete and sign a visitor form before entering the site.

Monitoring of the visitors is achieved by the following means. The access door is locked by a system of three separate locks, requiring the presence of two wardens to open it. This ensures that access is only available to determined and genuinely interested people, and that two wardens are at hand, for instance in case of accident or disability. In principle, no category of visitors is excluded, for example cave divers are welcome to apply for access, but they will need to convince the warden that their diving equipment will not be handled close to petroglyphs. Other rules are the complete banning of smoking, fires and combustion engines in the cave and within its carbon dioxide catchment area. Vehicles are restricted to a designated area, no organic matter other than clothing is to be taken into the cave, and no rock wall is to be touched at any time. Groups permitted to enter the cave are restricted to no more than five people and children must be strictly supervised. The duration of visits is kept to a minimum and after each visit the cave atmosphere is given time to recover to its correct temperature, humidity and carbon dioxide level.

As a result of these stringent requirements only 60 people have visited the cave in the six months from February to July 1988. The accumulation of these records is only one aspect of the ongoing monitoring program. Every completed project aimed at the preservation of rock art needs to be followed up by systematic monitoring of the effects. Quite apart from the obvious benefits to the site in question, through the opportunity of modifying the measures where this is found to be necessary, there is the need to learn from the long-term results of each project. Without such monitoring our understanding of the effects of specific measures would remain very limited indeed, we would tend to repeat mistakes and we would encourage others to duplicate our mistakes. I therefore suggest that the relative value of any rock art preservation work should be largely measured in terms of the extent of monitoring taking place after its completion.

In the case of the Paroong project there is an additional reason for comprehensive a review of results: it is a pilot project for similar work that will be tackled in future years, which will benefit from a reliable base of data and experience. I have listed a program of eleven factors requiring monitoring over future years and expect that the monitoring period will be up to 20 years.

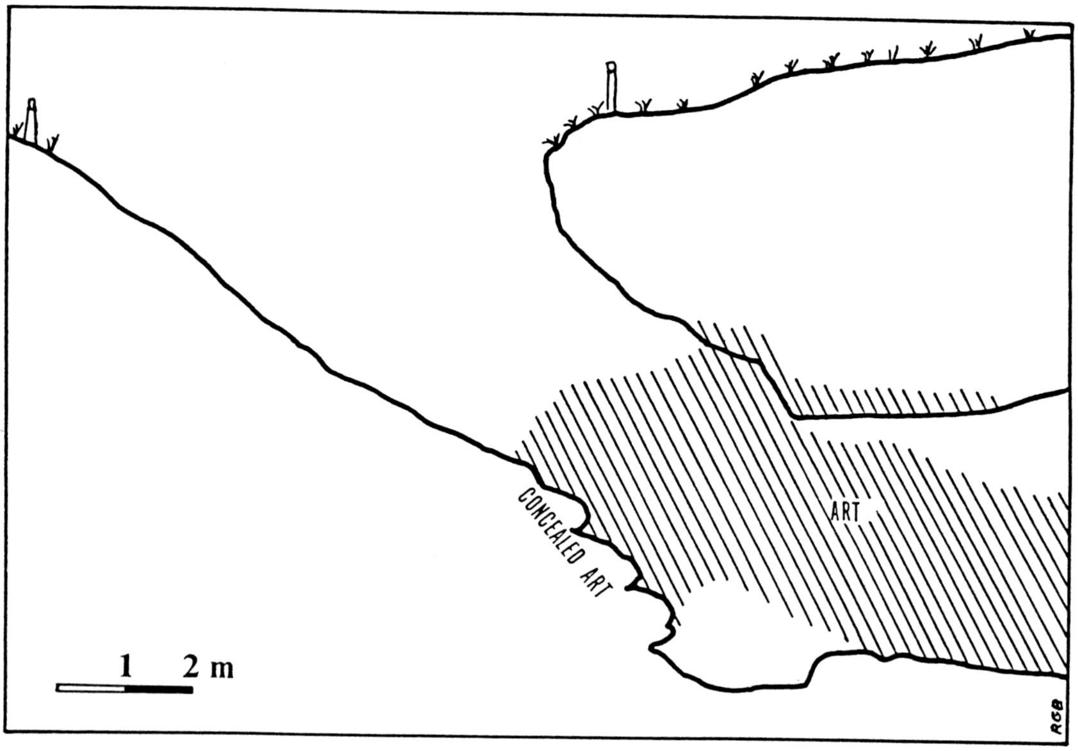

Figure 4a: *West-east section of eastern entrance, Paroong Cave, looking north, as it appeared before commencement of the preservation project.*

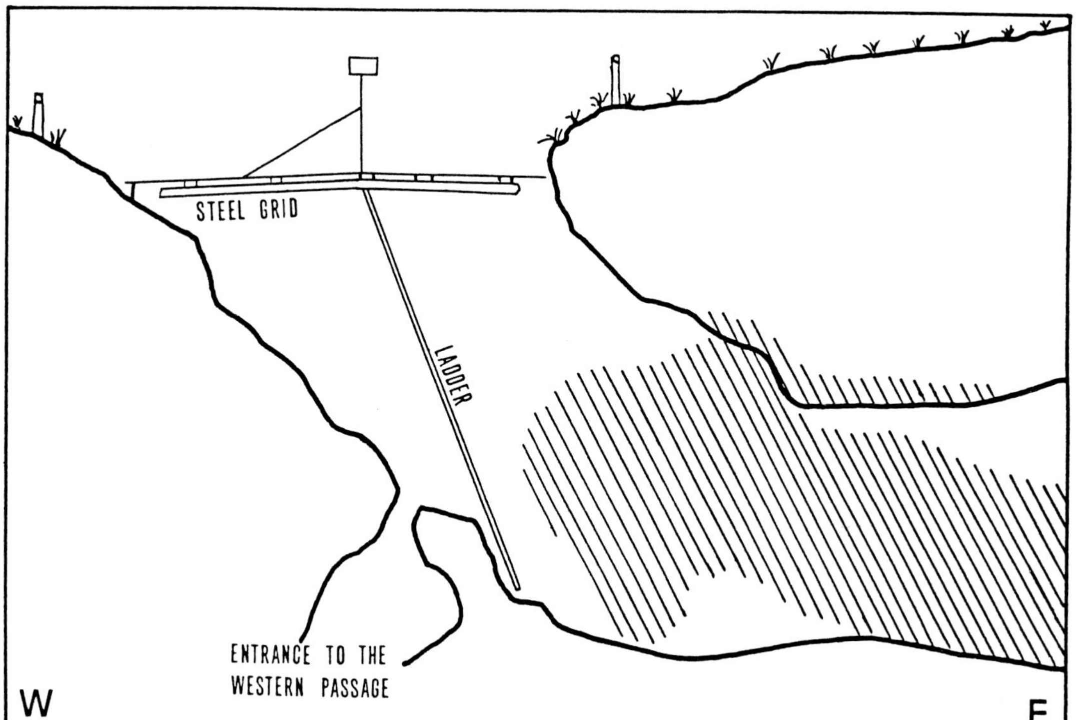

Figure 4b: *Entrance after completion of preservation project: decorated areas are well clear of the suspended access ladder and the western passage has become accessible.*

Acknowledgments

I thank the Australian Institute of Aboriginal Studies for supporting this project financially, and the Aboriginal Heritage Conservation Branch of South Australia for co-operation and for contributing the cost of the fence around Paroong Cave reserve.

REFERENCES

BBEDNARIK, R. G. 1990. The cave petroglyphs of Australia. *Australian Aboriginal Studies* No. 2: 64-68.

NOBBS, M. F. and DORN, R. I. 1988. Age determinations for rock varnish formation within petroglyphs: cation-ratio dating of 24 motifs from the Olary region, South Australia. *Rock Art Research* 5(2): 108-46.

ROSENFELD, A. 1985. *Rock Art Conservation in Australia.* Special Australian Heritage Publication Series No. 2. Australian Government Publishing Service, Canberra.

12

THE REMOVAL OF RECENT OVERPAINT FROM THE IMAGE OF BUNJIL

Andrew Thorn

Abstract. Early in 1988 the rock art at Bunjils Cave was overpainted with unbound pigments. The local community were distressed to find this alteration to the painting and sought advice from the Victorian Archaeological Survey to have the recent additions removed. The overpainting consisted of highlighting the existing figure of Bunjil and his two adjacent dingoes in earth colours and charcoal, quite out of keeping with the original colour range. Tests revealed that the overpaint could be removed, albeit cautiously, without causing damage to the underlying original. This paper outlines the methodology of paint removal with particular reference to the techniques employed at Bunjils Cave, emphasising that the removal of paint, whether it be graffiti or any other form of paint, must be assessed in a scientific, and case-specific manner.

Initial inspection

Bunjils Cave (Plates 1 and 2) is situated in a low range of hills to the east of the Grampians. It lies approximately 8 km south of the township of Stawell, Victoria. The site was first visited by conservators in February 1988, within a month of the paintings being over-painted. A brief inspection of the site, alcove and painting was made and the findings of that inspection are covered in an earlier report by Kosinova and Thorn 1988.

A brief extract of that report is included here to summarise the findings of the paint removal trials.

> The painting at Bunjils Cave, unlike most other sites visited, is a thick layer of paint in most places. The white outline of the body is quite thick. When dry the white paint appears of uniform whiteness but when wetted with water changes into two slightly different hues. Some parts remain unchanged while others become a pink/brown.
>
> This is a reaction that would be expected if the white were made of two applications using pigments with different refractive indices.
>
> It is believed that the paintings were reworked early this century using calcium carbonate pigment whereas the original is believed to be kaolin. If it is the case that the reworking was done with whitewash then the colour changes on saturation support this. One or other of the pigments is becoming more transparent when wet. Calcite will do this, in our experience, particularly when bound in animal glue.
>
> The original red paint is much thinner than the white and also more stable.
>
> All of the pre-1988 paint is broken up and unstable to some extent. The granite surface is quite irregular due to the protruding particles and most of the paint has been removed from the high spots.
>
> In the depths the paint remains quite thick but does not have complete adhesion at the edges.
>
> This loss of adhesion does not call for immediate action however, but should be noted and monitored.
>
> Solvents were tried to determine whether the over layers could be dissolved without disturbing the original paint. The recent overpaint was completely soluble in water and to a lesser extent in other solvents. The majority of the pigment could be removed with a wet cotton wool swab. A residue of pigment remained after this initial cleaning and could be removed with further action.

MICROCLIMATE

The site was assessed to determine as much as possible within the space of one week, the potential threat to the paintings from the environment.

This analysis can only be considered a spot check of a process that is operating not only diurnally and seasonally but with great variation in annual and longer term climatic patterns. The following findings covered the conditions observed from 16-20 May 1988.

These findings should serve at least one useful purpose and that is the need to monitor the alcove over a longer period.

Daily weather pattern

The daily development of atmospheric moisture was observed to be of two types:

Clear nights

When the sky was clear at night, the following day displayed a regular pattern of atmospheric moisture movement.

At 7.00 a.m. (the first observation time) the low lying land around Bunjils Cave was encased in fog.

At 8.00 a.m. (arrival at site) the fog was confined to the low lying areas below the alcove.

At 9.00 a.m. the rising temperature from direct sunlight caused the fog to lift and roll up the hill and completely engulf the alcove. This would inevitably cause air-borne water to deposit on the painting surface and in particular be absorbed by the paint. The relative humidity at this time was measured to be extremely high (>90%).

At 10.00 a.m. the fog had lifted and the site was clear with bright sunshine. It could be seen in the low lying areas to the west that the fog lifted to form a band covering the Grampians except for the very peaks. This cloud usually remained for several hours but had no affect on the alcove.

Overcast nights

When the sky was overcast at night the fog did not appear in the morning. The relative humidity was still very high but there was no direct moisture contact with the painting.

Direct sunlight

The site opens almost directly to the south and during the conservation program received no direct sunlight at any time of the day. This would, however, not be the case in mid-summer when the sun sets 30° S of west. Figure 1 shows that the sunlight could penetrate into the east side of the alcove to a position 200 mm beyond the emu pictograph (see also Plate 2).

Taking the emu pictograph as a reference point in the alcove the following amounts of sunlight would reach the walls.

1. Sunlight could fall on the emu for a total of 45 days in the summer from approx. 1 Dec. - 14 Jan.
2. The maximum sun penetration would occur at the summer solstice (22 Dec.) and would strike the emu pictograph for a maximum of one hour prior to sunset.

The penetration of sunlight does not appear to be significant but may have an impact in two ways on the condition of the painting within the zone identified. Firstly it must be noted that there are several pictographs recorded within the zone of insolation and these pictographs have almost completely disappeared. This may be a result of several factors.

Fading by the sunlight

This would not be significant due to the inherent stability of the pigments used and the low UV content of sunlight prior to sunset. Nevertheless, 45 hours per year over a thousand year period amounts to a high level of radiation. The loss of the pictographs appears to be a complete loss of paint rather than an alteration to existing material.

Heat from sun

The rise in temperature due to the sun would be more significant than the effect of light.

The rise in temperature may cause expansion between minerals within the granite leading to spalling, but the most significant effects would be:

1. promoting greater capillary rise in groundwater (see later); and
2. accelerating chemical reactions.

The moisture survey revealed that the eastern wall has ground moisture rising to a greater height than anywhere else in the alcove (Fig. 2). There is no other obvious cause of this. The slope of the ground in the opening of the alcove would lead one to expect a moisture build up along the west rather than the east wall but this is not the case.

The sun in winter

The sun does however act to protect the paintings by heating the tor to a point where the granite surface within the alcove is kept slightly warmer than the average winter air temperature (Hough 1986). The data in that paper show only monthly averages which cannot be considered reliable for predicting condensation. However, the figures do show that the period May - September should not be prone to condensation. In fact January is the most critical period if there is no significant rise in the surface temperature of the granite within the alcove.

The only way of assessing this situation is to install an appropriate long-term monitoring system within the alcove. During the conservation program (16-20 May 1988) the tor was measured with a probe thermometer placed in a natural fissure and sealed with paper. The probe showed that the tor maintained a constant temperature of 14°C +/- 0.5°C. The rock surface was always warmer than the air temperature for the whole observation period.

The relative humidity observed in the alcove was always in the range 80-95%. At these high humidities the potential for condensation is very great. If the air at 95% contacted a surface only 1°C lower, condensation would occur.

At 80% the surface would need to be 4°C lower. This assumes the rule of thumb that for every 1°C drop in temperature the RH will rise 5%. (The author has confirmed the RH rise to be 4-5%/°C at normal ambient temperatures.)

At Bunjils Cave the temperature of the tor was such that even with the RH at 95%, condensation would only occur when the air temperature is at or above 15°C. This is not a common occurrence at Bunjils in winter but a thermohygrometric program would confirm the likelihood of this occurrence.

Surface water movement

Rainfall during the conservation program was insufficient to saturate the south face of the tor and thus best illustrate the surface water movement. However, the two or three light showers did give sufficient information to support the findings of the electronic moisture survey, combined with a survey of the rock morphology. Figure 4 illustrates the significant features of the tor which have a direct influence on moisture conditions within the alcove.

The southern tor surface

The southern tor surface has been divided into four zones with two significant fissures marked A and B (Fig. 4, also Plates 1 and 2).

Zone 1. Is delimited by the upper end of fissure A. Water falling to the west (left) of point A will tend to run down the surface and fall clear of the alcove. Strong westerly winds are unlikely to drive the surface water into the alcove due to the curvature of the tor away from the entrance.

Zone 2. Lies between point A and the outer most curve of fissure B. Water entering this zone will either run directly into the entrance of the alcove or, into fissure A, where it can enter the roof of the alcove and reach the terminal point of the fissure in the centre of the alcove roof. Water reaching the ground will then flow downhill and collect around the western side wall.

Zone 3. This area is bisected by fissure B and lies between the outer curve and terminal point B. Water falling above the fissure will run into it and down, partially spilling onto the surface above the east wall of the alcove before reaching terminal point B. Whereas water falling below the fissure will run directly into the alcove to the natural drip line.

Zone 4. This area lies to the east of fissure terminal B.

Water falling in this zone will reach the ground and proceed downhill, contributing to the build up of ground water around the alcove.

The fissures

The two fissures A and B are the most significant path for water movement into the alcove from the tor surface. Fissure A deposits water into the depths of the alcove allowing the natural drip line at the entrance to be breached.

Fissure B, although draining water from above the alcove, fails at a critical point above the east wall allowing excessive amounts of water to spill onto the entrance and surrounding surfaces. Water collected into this fissure is deposited uphill from the entrance thus increasing the volume of groundwater entering the soil within the alcove. The rainfall was not heavy enough to understand the full contribution from these two fissures, but measurements using a Protimeter resistance moisture meter showed that the water migrated, during the study, to a distance of 100 mm beyond the fissure. This pattern was also noted on the drip line itself. This distance could be quite misleading due to the mild weather and a more severe migration may be prevalent during higher rainfall. Fissure A may even permit water to flow from the terminal point during heavier rain periods.

The natural drip line

The entrance of the alcove forms an arch which has its lower edge near the entrance, well away from any existing pictographs. This lower edge forms a natural drip line which prevents surface water entering the body of the alcove except through fissure A.

The drip line displays a zone of water migration which can be measured electronically and is manifest as a leached band of clean granite surrounded by a consistent border of a black deposit with a white inner edge, consistent with gypsum formation. This leached zone suggests that water does not migrate beyond the 100 mm recorded by the Protimeter, however infrequent heavy rains may penetrate further into the alcove without altering the zone of deposition.

Lichen growth

The presence of lichens, particularly in zones 2 and 3, must be considered as potential agents of deterioration due to their production of acids. Although the lichens proliferate in areas well away from the art, water running through them will be more acidic and more reactive. This should not be considered a situation that demands removal of the lichens, however, as the natural drip line would prevent the acidified water reaching the art. The drainage of fissure B could be corrected and thus prevent any contact of lichen-enriched water with the alcove.

Groundwater movement

The movement of groundwater is closely related to the water running off the tor surrounding Bunjils Cave. This shed water is indeed a real contribution but may be quite small compared to the drainage of the hill running up to the east of the tor.

Water movement in the soil

Run-off from the tor combined with any run-off from the hill during the conservation program presented a stable line of wet surface soil as seen in Figure 1.

A survey using the Protimeter revealed a second contour of high moisture at a depth of 50 mm below the soil surface as can be seen in Figure 1. This revealed that the moisture, although seemingly kept out of the alcove, did in fact flow into it at lower levels. Further readings were attempted at 100 mm depth. There were too many rocks in the soil at that depth to build a consistent contour. It was noted however, that there were areas of high moisture content at 100 and 150 mm at various points within the alcove.

The pattern shown in Figure 1 illustrates three very important points.

1. The wet soil is no indication of water movement.
2. The water moves into the alcove in a descending pattern and was detected at the back wall at a depth of 150 mm.
3. Subsoil water accumulates more around the east (uphill) side of the alcove. It is this area that shows the greatest capillary moisture movement within the granite (Fig. 2).

Capillary rise

Figures 2 and 3 illustrate the moisture distribution along the two pre-defined sections (Fig. 1) within the alcove. These illustrate the general conditions within the alcove during the survey period only. The numbers are for comparative purposes only; the higher the figure the wetter the surface.

Section B-B' running approximately east west passes through the emu pictograph. The gradient of the floor is not indicated.

Section A-A' runs from the centre of the Bunjil figure to approximately the centre of the opening.

The following observations can be seen from this survey.

1. The highest readings occur on the high side around the emu with a gradient of decreasing moisture with height.
2. A lesser gradient can be seen on the west wall.
3. A slight peak can be seen in the fissure in the roof of the alcove.

The consistent diminishing moisture gradient with height is typical for rising moisture through capillary attraction. The highest readings around the emu, although surprising when the site is viewed superficially (high side of alcove, furthest from wet soil), are explained quite clearly from the sub-soil survey which shows an accumulation of moisture around the base of this pictograph (Plate 2).

Furthermore, the sun falling in this area during the summer would further exacerbate the situation. The tor surface as it is heated will allow rising water to move higher up due to a higher evaporation rate. This can best be observed in buildings where the wall facing the sun will often have rising damp to a greater height than its counterpart on the shaded side.

From the condition of the alcove, particularly the disposition of old spalling (Gunn 1983: 24) and existing flaking areas, it is clear that most losses have occurred in the lower parts. This may be the result of physical contact but the survey demonstrates the potential problem of rising moisture and this must be considered a significant contribution to the deterioration within the alcove.

CONDITION OF BUNJILS CAVE

The physical condition of the alcove was thoroughly inspected with particular reference to those factors that influenced the condition of the pictographs themselves.

The Support

The granite supporting the painting has been studied more intensively and as a result its condition can be considered to be moderately unstable. Of particular note is the extent of the spalling below the figure of Bunjil (Plate 2) which is a continuation of a problem that has been active for some time. Gunn (1983) has documented the loss of granite as a function of the loss of the pictograph of Bunjil. These areas correspond with the present author's survey indicating the current loss of adhesion within the granite. This loss of adhesion can safely be attributed to the weakening of the stone due to rising groundwater.

There are, however, isolated areas of spalling higher up in the alcove which cannot be a result of rising moisture. These isolated spalls are well away from any pictographs and should not therefore be considered a threat to the stability of the pictographs. There is no obvious cause of these losses of adhesion and without a complete study of the microclimate and the granite no explanation is attempted.

The position of each area has been recorded using two measurements taken from two reference points (Plate 2). Co-ordinate A is the centre of Bunjil's top left eye (right eye to the observer). Co-ordinate B is the west ear of the closest dog to Bunjil. The measurements represent the distance in centimetres from each point and define a unique position on the back wall of the alcove. These co-ordinates are measured to the nearest cm and should assist in relocating the spall sites.

Site	from A cm	from B cm
01	120	135
02	119	133
03	120	115
04	123	116
05	113	103
06	112	95
07	123	105
08	117	90
09	108	87
10	110	80
11	106	72
12	113	74
13	109	67
14	121	81
15	125	78
16	117	62
17	121	55
18	135	66
19	130	57
20	135	61
21	112	19

The painting

The condition of the paint was examined thoroughly during the initial inspection in February 1988 by Kosinova and Thorn (1988). The following observations were made at that time.

1. The pre-1988 paint survives in a relatively stable condition with the exception of the white which is poorly adhering in most places. The white paint is much thicker than any other colour which perhaps suggests that it is a more recent application.
2. The original paint tends to lie in the depths of the granular surface of the coarsely textured granite. The recent paint covers the whole surface of both the original paint in the depths and the bare peaks.
3. The white paint is soluble in water but not in ethanol or acetone.
4. The red paint is much more stable to the solvents tested and is much thinner than the white. It is also appreciably thinner and more broken up than the similar red applied as graffiti outside the alcove. This is confirmed by other analyses (McConnell n.d.).

PAINT REMOVAL

The overpaint

The recent paint was executed in a deliberate attempt to copy the main pictographs of Bunjil and the dingoes. The painter used an aqueous slurry of unbound pigments to enhance the existing painting without making any new pictographs. The colours, ground on site, are charcoal and orange and yellow earths.

The decision was made to have the recent paint removed and in February 1988 a test program was undertaken by the authors to determine the feasibility and best approach for the paint removal. In the time between the test program and the conservation program the author was able to discuss the proposals in an international forum at the Getty Conservation Institute, Los Angeles during the 'Conservation of Rock Art' course held in April 1988.

A separate report for the test program was prepared at that time (Kosinova and Thorn 1988) and the following extract outlines the findings:

> *Paint removal*
>
> Due to the uneven surface of the granite the range of paint removal options is limited. Any mechanical removal other than dry removal with a scalpel under a microscope would cause a great deal of damage to the painting. Mechanical abrasion acts more severely on the high points leaving the lower depths untouched.
>
> Air abrasive techniques can overcome this problem to some extent by the correct selection of abrasive hardness and to a lesser extent the air pressure. However, the paint surface at Bunjils is too vulnerable for this approach. Solvents were tried to determine whether the over layers could be dissolved without disturbing the original paint.
>
> It was noted that the original white paint was slightly soluble in water but with care the overpaint could be successfully removed. This had to be carried out under a microscope to ensure that the original was not being activated by the water.
>
> Acetone and ethanol were also tried but had little effect on the overpaint. They were, however, quite safe to use on the white paint. The red paint was very stable to all solvents and a slightly more aggressive cleaning could be tolerated. The recent overpaint was completely soluble in water and to a lesser extent in the other solvents. The majority of the pigment could be removed with a wet cotton wool swab. A residue of pigment remained after this initial cleaning and could be removed with further action.
>
> This second phase is the critical one where most damage could potentially be done. Due to the prolonged wetting and extra abrasion the original surface easily

becomes disturbed as well. The contact time is critical and the use of a microscope essential. By deliberately prolonging the cleaning action it is possible to establish a safety time zone for the cleaning but it is essential that the process be closely watched at all times.

The cleaning of the painting can be divided into three parts:
white paint - requiring meticulous attention
red paint - permitting slightly more aggressive cleaning
granite support - may be cleaned readily

Paint removal

Solvents

It was noted from the trial program that the recent overpaint was readily soluble in water but sparingly soluble in other solvents. The pigment removed with other solvents was attributed to the abrasion of the cotton wool swab. It was further noted that the white original and to a much lesser extent the red were also soluble in water. Having established that solvent removal was the safest approach the first requirement was to modify the solvent to remove the overpaint yet have a minimal effect on the original.

The addition of acetone was chosen for several reasons:

1. Non-solvent for the original paint.
2. Its volatility allowed the solvent solution to evaporate much more quickly thus reducing the contact time. This could be measured with the Protimeter to show that whereas water alone kept the surface damp for the several hours the addition of acetone 1:1 made the surface dry within 60 minutes.
3. Acetone lowered the surface tension of the water giving better wetting of the particles to be removed.

The final mixture used for the whole process was:
de-ionised water 50% V/V
acetone 50% V/V

In the final phase water alone was again tested to determine whether this would allow better removal of the residual overpaint. This proved not to be the case and the mixed solvents were employed throughout.

The method

The test program brought out one very important constraint on any paint removal technique and that was the sensitivity of the painting to abrasion. Not only would abrasive techniques be quite detrimental to the original but even slight abrasive action associated with solvent removal could damage the painting. The solvent in the tests had been applied with a cotton wool swab and this seemed to be the most desirable and efficient method to pick up the recent paint effectively.

Alternative methods were tested:

1. Applying the solvent in a poultice. This left the solvent on for longer periods and was not efficient in removing the overpaint. One poultice application would not have been sufficient to remove all of the overpaint in any test area.
2. Using the solvent in a swab through a protective tissue to protect the surface from abrasion. Similarly this technique required constant changes of tissue to complete the process.

The approach adopted was to use the swab in a rolling motion so that there was little lateral movement dragging across the surface. The cotton swab ranged from 3-20 mm in diameter depending on the surface condition. The larger swabs were used where the overpaint was applied directly onto granite and, in large part, over the original red. The advantage of this 'rolling poultice' was the control afforded.

1. The amount of pigment being collected could be seen and a fresh swab made to avoid re-depositing the pigment elsewhere.
2. Contact of the swab with the surface was very brief and could be adjusted to suit the condition. Some areas tolerated repeated rolling but the more delicate areas, such as the whites, could be permitted to dry completely between applications.

The painting was worked in sections corresponding to the divisions within the painting itself. The head, for example, was worked on as one separate unit, as were each partition of the body ornament. This enabled the process to be executed thoroughly and evenly over the whole painting without loosing track of how many applications had been made to each partition.

The process was repeated three times to every section over three successive days. This meant that the surface was wetted for no longer than one hour with 23 hours between applications. At the end of these three applications the painting was compared to the earlier photographs (Gunn 1988) and an assessment made of the appearance.

This assessment after day 3 showed that the majority of the paintings were very close to their previous appearance. The exceptions being the partition above the true left shoulder of Bunjil, the nose, mouth and eyes. These areas could be seen to be different to the photographs of the original painting (Gunn 1988).

Further work was undertaken on the remaining areas with some improvement. Very localised cleaning was undertaken under a stereo microscope of the nose and mouth, which were the most distorted, bring them closer to their true appearance.

Finally a point was reached where safe removal was no longer possible. The paintings were discussed with the Victorian Archaeological Survey and it was mutually agreed that the appearance was visually indistinguishable from the original painting and further paint removal would unduly threaten the pictographs.

Observations on overpaint removed

Charcoal (composition confirmed with Polarising Light Microscope): the charcoal particles were thickly encrusted on the surface. The charcoal had been applied almost exclusively to the outer perimeter of the paintings which had previously been left unpainted. Prior to wet removal the charcoal was partly removed by gentle brushing with a small bristle paint brush. This was to avoid smearing the bulk of loose charcoal into unpainted areas. Where the charcoal was applied over the red original the brushing was not carried out but repeated swab changes were necessary. The charcoal was easy to remove but left a discernible stain which could not be removed. This darkening of the surface was not considered to be disturbing as it lay outside the paintings themselves.

Yellow earth (confirmed with PLM): the yellow pigmented areas were much more readily removed than the orange.

Orange earth: mixture of yellow earth and hematite (PLM); the most difficult colour to remove.

ANALYSIS OF PIGMENTS

Although the question of the extent to which the paintings have been overpainted in the past still remains unresolved it was thought undesirable to sample the paintings as this has been done on several occasions already (McConnell n.d.). The results of these analyses have not been fully documented and this has diminished the understanding of the paintings. To better understand the behaviour of the overpaints and other surface phenomena associated with the paintings, samples were collected and analysed using Polarising Light Microscopy (PLM). The sites and results are described below. The results are not complete but raise some interesting points which may prompt further research.

Samples 01-08

01 yellow 1988 overpaint from Bunjil's face

Yellow earth, charcoal, iron red, silica; the yellow was a natural earth containing some haematite.

02 orange 1988 overpaint from Bunjil's face

Iron red, yellow earth, silica; the orange was a mixture of the yellow earth with haematite.

03 1988 black overpaint

Charcoal

04 red graffiti dated 25/1/11

Red brown earth, iron red, carbon black; the red graffiti dated 25/1/11 is composed of three pigments - haematite, red/brown earth and carbon black. A comparison with the red on Bunjil would substantiate the belief that the figure of Bunjil has been touched up.

05 black deposit from roof of alcove

Gypsum, silica, charcoal, red lead; the black on the roof of the alcove reveals an interesting mixture of particles. Of particular note are:

a. The black particles have a refractive index below 1.66 and are consistent microscopically with gypsum. It has been said that these black deposits are manganese but there is no manganese salt with a R.I. below 1.66.
b. Charcoal has been found in the roof of the alcove. This charcoal may be solely from the recent overpainting but bushfires or camp fires at any time are likely to deposit charcoal in the alcove. Charcoal is not the only black present but (a) above may be another product of combustion.
c. The trace of red lead is difficult to explain. It would seem unlikely to be a laboratory contamination as it is not found in other samples. Nor is it found in the red graffiti.

06 white salts deposit from outside alcove

Gypsum

07 black deposit from outside cave adjacent to salts

Gypsum

08 exfoliation within alcove with black crust and deposit behind

Gypsum, charcoal

Samples 06-08 all contain gypsum in different forms. These findings have been independently analysed (J. Clarke) and the deposits from which they were taken are typically those of gypsum formation. They manifest themselves in three bands:

1. The wash zone. The excess of water dissolves the gypsum and keeps it in solution.
2. White deposit. Immediately adjacent to the wash zone a narrow band of white-coloured gypsum forms.
3. Black deposit. A much broader band of black gypsum is formed as the outer band. This gypsum formation differs from the white deposit in the amount of water available to the calcium sulphate ions during crystallisation and thus determines the degree of hydration in the gypsum formed.

FUTURE PRESERVATION

It can be seen from the preceding sections that various environmental conditions prevail that may threaten the longevity of the pictographs at Bunjils Cave. These can be divided into three main areas:

1. Water movement around the site.
2. Exfoliation of the granite.
3. Lack of adhesion in the white pigment.

At this stage no treatments have been attempted as any intervention must be preceded by a sufficient period of research and cautious trials.

Summary of Recommendations

(not in order or urgency)

1. Photogrammetric and measured documentation to record the surface of the alcove. This will monitor any losses of granite.
2. Improved drainage of fissure B to prevent water spilling onto the alcove entrance.
3. Consolidation test program for both granite and paint.
4. Monitoring of microclimate. This should be specific to the problems identified in this report. Installing a thermohygrograph will be of limited benefit in isolation. It is necessary to be able to establish the relationship between the RH and air T° and the surface T° of the granite at selected relevant sites.
5. Reduce insolation to south eastern wall of alcove. This can be either natural vegetation planted some distance from the site or a modification of the existing barrier. Shade cloth or a louvre system should be considered despite their visitor impact. Appropriate design should minimise the impact which should in all events remain secondary to the preservation of the alcove itself.

REFERENCES

GUNN, R. G. 1983. *Bunjils Cave: Aboriginal rock art site.* Victorian Archaeological Survey Occasional Reports Series No. 13, Melbourne.

GUNN, R. G. 1988. *Bunjils Cave: Documentation of recent vandalism,* (photographic supplement), for Victoria Archaeological Survey, Melbourne.

HOUGH, D. J. 1986. *Grampians Conservation Project - Draft report.* Victoria Archaeological Survey Internal Working Document, Melbourne.

KOSINOVA, A. and THORN, A. 1988. Condition report: Grampians rock art sites, unpublished report for Victoria Archaeological Survey, Melbourne.

McCONNELL, A. n.d. *Pigment analysis for the authentication of the Aboriginal paintings at Bunjils Cave, western Victoria*, pp. 43-56.

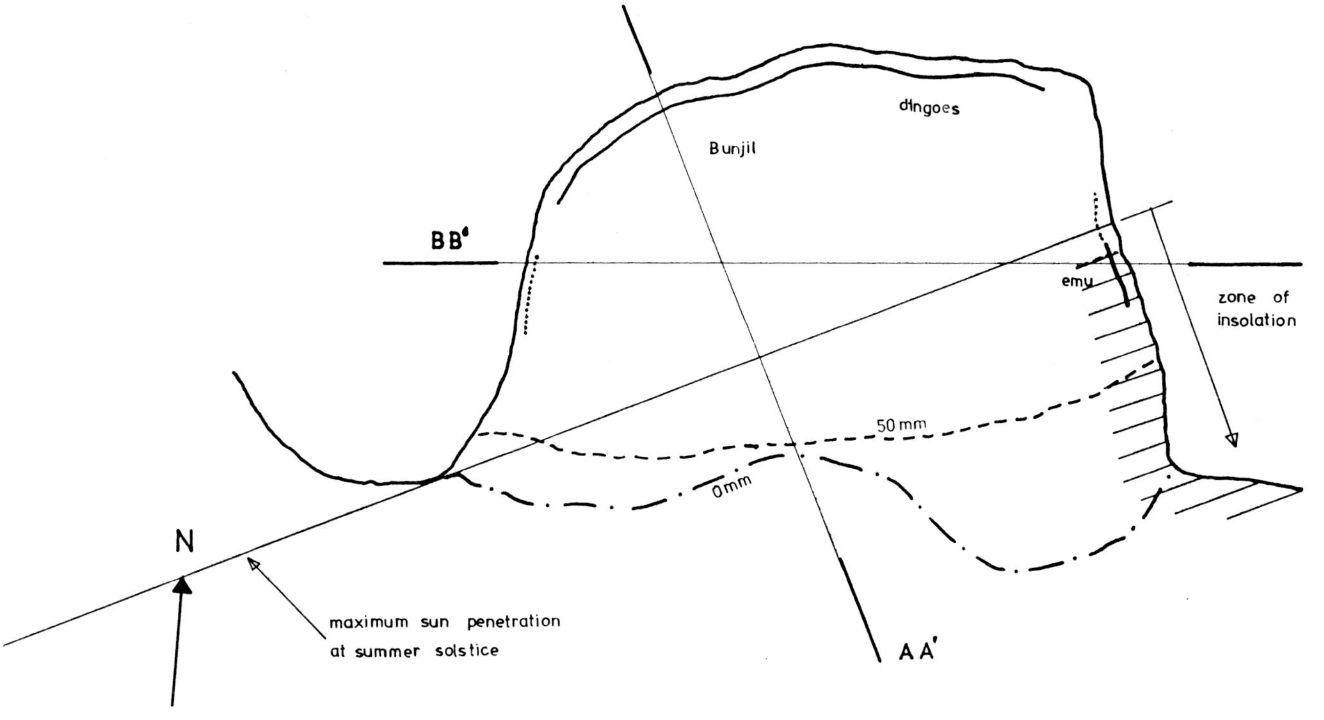

Figure 1. This plan view of the alcove illustrates the accumulation of water around the upper side of the alcove (right hand side of figure) together with the zone of insolation that penetrates into the alcove in the summer.

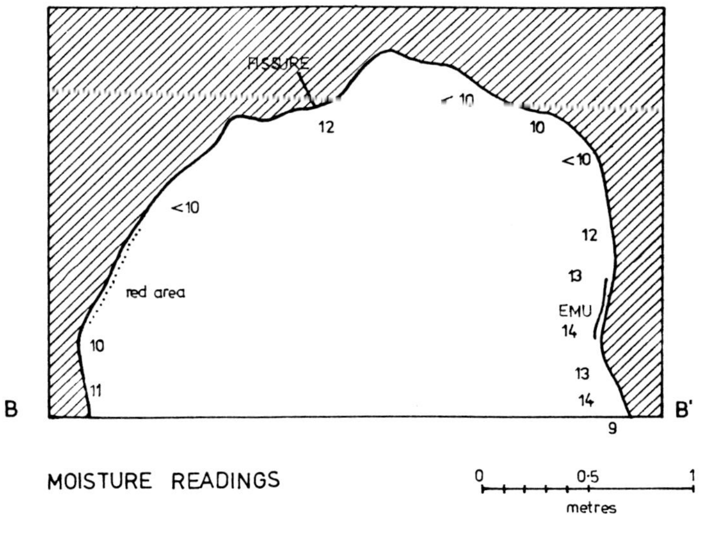

Figure 2. Section B–B' (refer Fig. 1 for section lines) shows the higher moisture levels of the upper side of the alcove with the slight increase associated with fissure A in the roof.

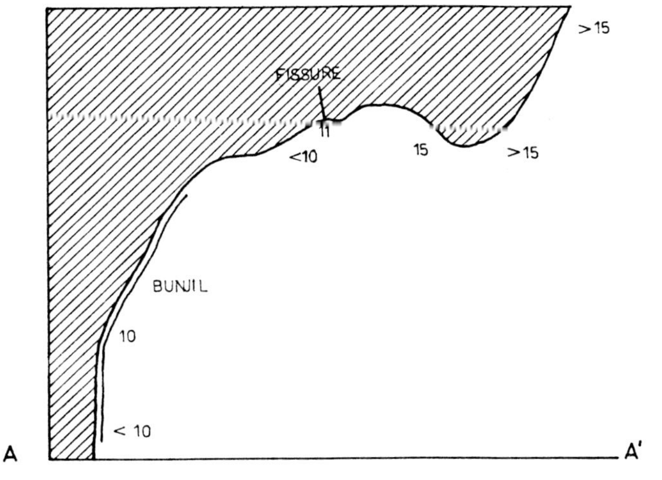

Figure 3. Section A–A' (refer Fig. 1 for section lines) indicates that the wall surrounding the painting is very dry compared to the natural dripline created in the arch of the alcove.

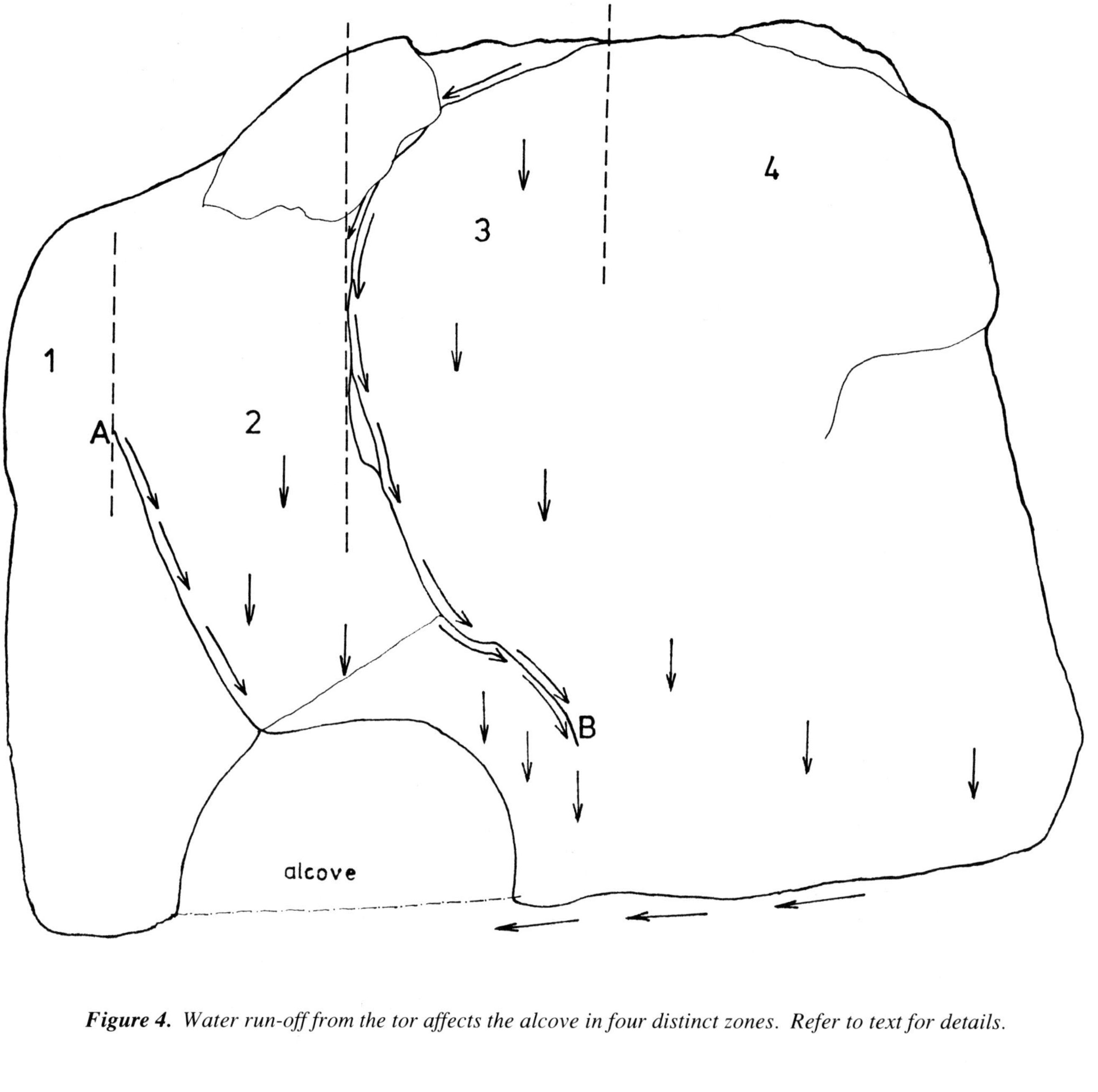

Figure 4. *Water run-off from the tor affects the alcove in four distinct zones. Refer to text for details.*

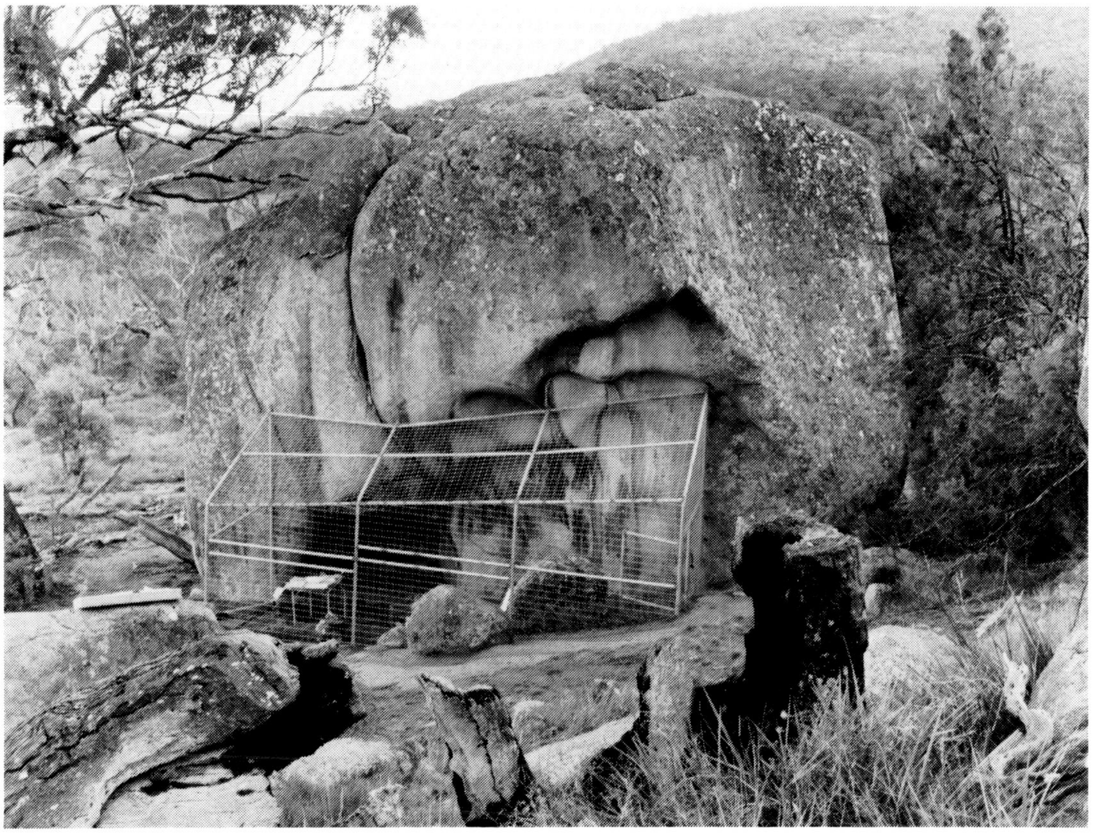

Plate 1. *General view of the alcove within a protective cage. The fissures can be seen running through the tor.*

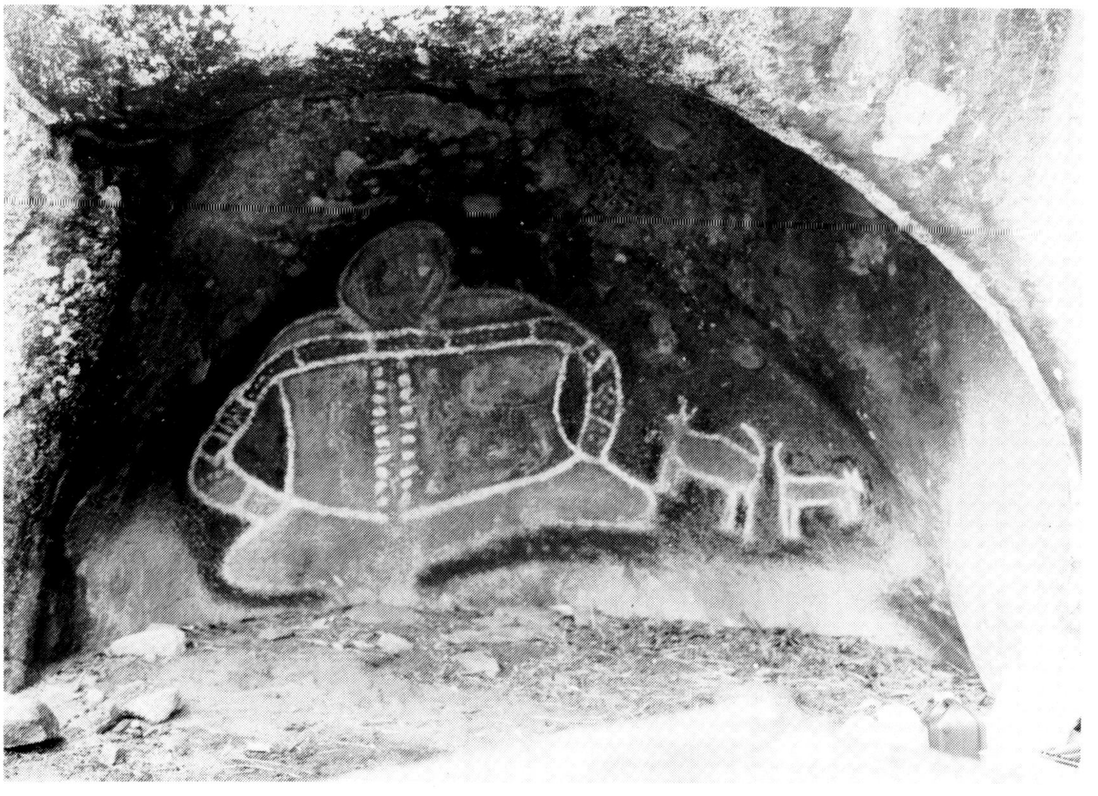

Plate 2. *View of Bunjil and the dingoes. The wash zones around the arch can be seen together with salts formation below fissure B.*

13

PIGMENT COMPOSITION OF POST-ESTUARINE ROCK ART IN KAKADU NATIONAL PARK

J. Clarke and N. North

Abstract. The results of a study into the chemical composition of pigments and paints used in post-estuarine period rock paintings in the Kakadu National Park (Northern Territory, Australia) are presented. Comparisons with contemporary Aboriginal pigments along with observations on deterioration and alteration processes are made. Natural mineral pigments as well as European materials have been identified. No evidence of organic binders or fixatives was found. The work formed the initial part of an ongoing conservation study. A companion paper presents details of the alteration products (soluble salts) present in the paintings and discusses the modes of formation.

Introduction

In March 1986 the authors undertook a consultancy for the Australian National Parks and Wildlife Service (ANPWS). The work formed the first phase of a research project on the conservation of post-estuarine period rock art in the Kakadu National Park. The scope of the first phase was to, in co-operation with the traditional Aboriginal owners and ANPWS staff:

(a) identify and describe the physical and chemical properties of the mineral paints used in the post-estuarine period rock art in Kakadu;
(b) identify and describe the physical and chemical properties of organic colours and mineral paints used by the Aboriginal population of Kakadu today;
(c) identify and describe the physical and chemical properties of naturally occurring fixatives in Kakadu.

The consultancy followed earlier work by Alan Watchman which presented the results of petrographic studies of rocks and mineralogical examination of salts associated with a wide range of rock art sites at Kakadu (Watchman 1983). The results of this earlier work and other conservation-related studies that have been undertaken at Kakadu have been reported in a special publication (Gillespie 1985).

The authors undertook a field trip to Kakadu between March 29 and April 7 in 1986. Under the direction of the ANPWS project officer, and following consultation with the relevant traditional Aboriginal owners, a range of samples was collected from rock art sites throughout the Park. These samples included:

(a) pigments used by Aboriginal artists in the Park from sources both in and outside the Park;
(b) paints prepared by Aboriginal artists;
(c) very small samples of paint and pigment from rock painting sites covering a range of colours and states of preservation;
(d) organic materials used as fixatives by Aboriginal artists;
(e) miscellaneous samples from rock art sites such as groundwater samples, salts, rock and soil.

All sample sites involving rock paintings were photographed and samples were collected in small glass vials. Every effort was made to prevent flake samples from breaking up so that internal structure could be examined.

The samples collected were returned to Perth where they were studied by a range of methods depending on the type, complexity and significance of the sample. The main methods used were:

(a) Optical microscopy
(b) Scanning electron microscopy (SEM)
(c) Energy dispersive x-ray analysis (EDXA)
(d) X-ray diffraction (XRD)
(e) Chemical analysis
(f) Infra-red spectroscopy

Details for each of these methods and limitations of each method, are described in detail in Appendix 1.

Pigments from contemporary Aboriginal and Park sources

Prior to the field trip Mr Ivan Haskovec, ANPWS Project Officer, had collected a number of pigment materials from sources within the Park and had them prepared as paints by an Aboriginal artist for an experimental rock painting site. Additional samples were obtained from Mr Talking Billy, an Aboriginal artist, who was using them to prepare paint for decorating artefacts. Several small lumps of possible pigment were also collected from eroded areas at the drip line of rock shelters.

Before discussing the types of pigments identified, it will be useful to consider some of the features of pigments and paints. Modern paints consist of an organic base (or medium) into which a very finely ground pigment or mixture of pigments is dispersed. In Aboriginal rock art paints there is little, if any, organic medium and the 'paint' is generally little more than the raw pigment mixed with water. The pigment particle size is also considerably larger than in European paints as only hand grinding procedures were available. Often pigments were simply dispersed into the water in the 'grinding' process and no physical size reduction occurred. In other cases simple settling processes were used to remove oversize particles from the paint.

White pigments

The majority of white pigments examined were clays of the kaolin type. These clays form by the weathering of many types of rock, usually rocks containing feldspars (silicate minerals). The samples examined all showed evidence of having been formed from the weathering of sedimentary rocks. A number of samples also contained mica (a group of silicate minerals midway between feldspar and true clays). In particular the mica, muscovite, was noted in samples and visible in some hand specimens. (Note: the mica, muscovite, has a very fine grained form known as sericite. Both have the same chemical composition and x-ray diffraction pattern and no distinction is made in this paper, the term muscovite being used irrespective of grain size.)

The presence of muscovite is often taken as an indication of immaturity in the clay formation process, being midway between feldspar and clay in generally accepted weathering (hydrolysis) series. However, more recent work has indicated that the processes involved are far more complex and that clays can also be converted to micas under conditions likely to occur in surface weathering.

Two samples contained well in excess of 50% silica sand grains and would have had to be subjected to some form of 'processing' to remove this much coarser material as no pigments from rock paintings contained any similar percentages of sand particles. One sample also contained considerable calcium and traces of magnesium indicating the presence of a carbonate mineral, such as dolomite, $(CaMg)CO_3$, in addition to the clay. Such mixed clay-carbonate pigments were also observed in the site samples. It is unusual to find a naturally occurring material containing clay and carbonate minerals in weathered rock profiles.

The two white pigment samples from Mr Talking Billy, reported to come from Murgonella, on the coast well north of the sites investigated, are of considerable interest. They consisted almost entirely of kaolin-type clay and had no sand grains present. Both samples showed significant minor amounts of iron and traces of titanium in EDXA. This feature was also noted in the vast majority of white clay-based pigments in the site samples. The SEM-EDXA results indicated that both the iron and titanium are evenly distributed throughout the clay mineral rather than being present as an accessory mineral as would normally be expected. The presence of iron and titanium at the levels indicated is most unusual and is believed to be one of the reasons that we have had difficulty in providing a precise identification of the clay mineral beyond being of the kaolin type. (See section on the Deaf Adder sites for further discussion on this feature.)

One sample was a very pure example of the mineral huntite, $(CaMg)CO_3$. This mineral, although known to be used extensively in northern Western Australia as a rock art pigment (Clarke 1976), was not identified in any of the site samples examined, while two other carbonate minerals, dolomite and calcite, were identified. These carbonates, while chemically very similar to huntite, have very different crystal structures and are easily differentiated.

Red pigments

Four red pigments were examined. Two were relatively pure examples of the mineral haematite (Fe_2O_3). This is a very distinctive mineral with considerable chemical stability under surface conditions. It it also known to have been widely used as a rock art pigment throughout Australia and displays considerable durability. It is easy to identify in hand specimens as it has a distinctive red streak (powder) when scratched. It was observed on many of the sites visited in the archaeological remains visible in the drip line of rock shelters. These lumps often display typical flat, striated faces due to grinding.

Although not always apparent in hand specimens, many haematites, when ground, produce a powder made up of very thin, mica-like plates. Both the samples examined displayed this platey structure as did some of the red site pigment samples. It is this platey structure that is responsible for the ability of haematites to adhere to rock surfaces, often under extreme conditions (Clarke 1976).

The remaining two samples consisted of mixtures of haematite, clay and mica. This type of mixture was also found in many of the red site pigment samples. This type of mineral assemblage is very common in the area and results from the weathering of rocks containing feldspar and a ferro-magnesium mineral. Numerous occurrences of both source rocks and material suitable for use as a pigment were observed.

Although none of the four samples examined displayed any distinctive trace element pattern, some of the site samples did show anomalous levels of elements such as nickel, chromium and titanium, that may prove useful in identifying a particular source.

Yellow and brown pigments

Three samples were examined. One consisted of a mixture of very angular silica sandgrains with minor clay minerals and iron oxide. The material as examined would have required considerable pre-treatment to remove the coarse sandy particles before its use as a pigment.

Two samples consisted of mixtures of clay, mica and an iron oxide (probably limonite) and as such were similar to many of the brown and yellow pigment samples from painting sites.

In the case of the red, brown and yellow pigments, the form of the iron mineral governs the colour. Where haematite is dominant a red colour is the result, while limonite produces a yellow, and mixtures of the two minerals produce the wide range of browns encountered. While the pure mineral haematite is used as a red pigment on its own, the addition of clay and mica can also produce red pigments. In the case of yellow and brown pigments, clays and/or mica were always present. The iron minerals are usually extremely fine-grained in such mixtures, and it is generally not possible to be specific as to precise mineral form or forms. It is often present as a very thin surface oxide coating on the surface of the other minerals present, resulting in some over-estimation when using the SEM-EDXA method.

Organic binders

In this work special emphasis was placed on detecting the presence of any organic materials present in the rock art samples. The main method was through infra-red spectroscopy where a very small amount of organic material will produce a significant characteristic absorption.

Samples of natural gums and resins collected by Kakadu Aborigines were found to give strong absorptions when mixed with large amounts of natural ochres. It was noted, however, that the supplied gums and resins had only limited binding ability and solutions of these could not be prepared by grinding and mixing with cold water. Modern Aboriginal artists in Kakadu area have used 'Aquadhere', a polyvinylacetate glue (PVA), in small quantities as a binder for natural ochres. Samples of pigment were collected from a rock art site which had been painted with such a mixture.

Although many of our collected samples were examined by infra-red spectroscopy there were very few that showed any indication of any organic material, and those which did showed such a low level that it could not be positively identified as to the type of organic material. In these cases the observed signals may have been due to fortuitous inclusion of other trace organic material such as bacteria, pollen, organic detritus material etc. The sample which was known to contain PVA showed the expected indication of organic material but again at too low a level for identification.

From these studies it is clear that the rock art pigments, as they exist in Kakadu rock art paintings at the present, do not contain sufficient organic material for it to have any significant effect on their physical or chemical properties. Considering the properties of the natural gums and resins it is unlikely that they were used in traditional rock painting pigments. However, their usage cannot be ruled out completely. If such organic materials had been used it is possible that, with time, they were attacked chemically and by micro-organisms and so removed from the applied pigments.

Rock painting site pigment samples

The pigments used in the post-estuarine rock paintings of Kakadu sampled in this study show that a wide range of natural minerals, mineral mixtures and modern industrial materials were used. Before discussing the particular pigment compositions of each of the site complexes sampled, some more general observations are warranted.

Of the samples examined, only one was a European-type paint containing an organic base with an aluminium oxide (possibly bauxite) pigment. While traces of organic material were detected in some samples, there were no other samples where an organic paint base had been used and had survived. This is not unexpected as organic material, although often reported to have been used in Aboriginal paints, would have a very short life expectancy under the conditions present at most sites. The residual decay products of any organic materials would be almost impossible to detect in the samples examined.

All the remaining pigment samples examined were either minerals or mixtures of minerals. Some, however, were not naturally occurring minerals, being either man-made or prepared in a grinding machine. For example the yellow pigment from a painting of a billy goat at 'David's old camp', East Alligator, was a chemically precipitated iron oxide produced by an industrial process. The white pigment from a painting of sexually active human figures at Yuwengayay was a finely ground mixture of quartz and feldspar, microcline, minerals that do not occur naturally in such small particle sizes, and would be too hard to have been ground by Aborigines using traditional methods.

In only one case, that of blue pigment taken from a fish figure at Blue Paintings, was the colour of the resulting paint believed to be due to an organic colouring agent. In this case the blue colour is not from the pigment, which is a carbonate (dolomite) similar to other pigments at the site, but results from the staining of the surface of the pigment grains with ferri-ferrocyanate (laundry blue-bag). In all other samples the colour could be adequately explained by the minerals present.

Each of the site complexes sampled displayed a distinctive suite of pigment types. Compositional features, such as trace elements, also tended to be site specific. These observations could be explained if pigments for each site complex were sourced locally, or the bulk of paintings were made in a short time period by the same artists using the same pigments. There is evidence to support both these concepts, and they will be discussed further in the following sections.

Discussion of site complexes

Nourlangie site complex

The pigments of the Nourlangie area are dominated by carbonate minerals, a feature not seen in any other of the site complexes examined. In the following table the composition of the various pigment colours is summarised. The table has been prepared by discounting elements and minerals resulting from salt accumulation on or in the pigment sample.

Pigment colour		Elements present	Minerals
White	9 samples	Ca with minor Si, Al, Mg	Dolomite, montmorillonite
	4 samples	Si, Al	Kaolin
	1 sample	Ca	Calcite
	1 sample	Ca, Si, Al	Calcite and clay mixture
Brown	3 samples	Si, Al, Fe	Iron oxide, kaolin
	1 sample	Fe, Ca, minor Al, Si, Mg	Dolomite, artificial iron oxide, and montmorillonite
Black	1 sample	Fe, Ca, minor Al, Si, Mg	Dolomite, artificial iron oxide and montmorillonite
Blue	1 sample	Fe, Ca, minor Al, Si, Mg	Dolomite, montmorillonite
Yellow	3 samples	Fe, Ca, minor Al, Si, Mg	Dolomite, montmorillonite, iron oxide
	1 sample	Al, Si, Fe	Kaolin, iron oxide
Red	3 samples	Fe	Haematite

Table 1. *Nourlangie pigment groups*

In some cases where the salt accumulation process has resulted in the alteration of the original pigment, it has been necessary to make some assumptions as to the original pigment material. At Nourlangie many of the carbonate minerals have been altered to gypsum. Details of this alteration process are given in the companion paper by the authors (Clarke and North this volume).

The most common pigment type and also the one presenting the most serious deterioration problems is the dolorite/montmorillonite mixture that has been widely used as a white pigment. It is also the basis of some of the other coloured pigments. This pigment is considered not to be a naturally occurring substance, and is probably of European origin. The dolomite is of the low magnesium type, typical of marine carbonate deposits. The grain size is very small and uniform, indicating a grinding and sizing process. The clay is of the montmorillonite type (often called ball clay) and this clay was not detected in any other Kakadu samples. It was, and still is, used to assist in increasing the viscosity of water-based suspensions. Its occurrence with carbonate minerals in nature is most unusual.

There was a range of interior house paints known as 'Kalcimine' used extensively up until the introduction of the present 'plastic' paints in the post-war years. The pigment may be of this type although we have been unable to obtain any comparative samples. The alternative is that the pigment was a proprietary whitewash to which the montmorillonite had been added to improve the suspension of the lime.

Ubirr site complex

While the sites in this complex generally showed reasonable states of preservation, detailed examination revealed that many of the better-preserved paintings are starting to show signs of deterioration. Cracks are present in the surface and flaking is starting in some cases. At the main gallery a very thick paint layer has accumulated. Examination of flake samples has shown that the individual paint layers have been homogenised through some mixing process. This has probably resulted from some form of surface preparation by artists prior to each painting phase, such as wetting the surface and mixing the previous pigments to provide a uniform surface upon which to paint. Some additional mixing may have also been caused by the alteration processes within the paint layer and the formation of soluble salts.

White pigments. Five different white pigments were identified at Ubirr. The most common was a clay of the kaolin type, containing high iron and traces of titanium (see earlier). This particular pigment has been identified at many other sites in Kakadu and corresponds to the raw pigment samples collected from Mr Talking Billy.

A sample taken from a fish painting at the south end of the main gallery, a multi-layered flake, contained a white pigment produced from the mineral calcite (calcium carbonate). While this was the only definite carbonate-based white pigment, many of the samples of clay-based white pigments showed high levels of gypsum as salt accumulations indicating that there may have originally been some calcium carbonate present in the pigment before alteration.

A sample from the painting of running figures known as the Mountford Frieze contained the kaolin-type clay described above, and it also included muscovite flakes and quartz, and had a very open porous structure. Without some form of consolidation treatment these paintings could not be expected to survive.

All of the white pigments at Ubirr appear to be of natural origin and no man-made materials were identified.

Red pigments. There were two types of red pigment identified. In a number of samples with red pigment on the surface of paint flakes the red paint was shown to be haematite. This was usually present as a very thin layer on the surface and showed a typical haematite platelet structure. The remaining red pigments were mixtures of haematite and kaolin-type clays.

Brown pigments. These were all identified as mixtures of iron oxide and clay minerals, while two samples also contained muscovite. All were of natural origin.

Yellow pigments. One yellow pigment sample from a turtle painting, where it was painted over an area of wasp nest and salt efflorescence, was shown to consist of a mixture of iron oxide and clay. However, this sample also showed very heavy salt accumulation (gypsum) and it is possible that it originally contained a calcium carbonate mineral.

All the pigment samples examined showed some degree of alteration and salt accumulation. Even in the case of very well preserved paint surfaces such as the tail area of a large fish at the main gallery, and a large fish painting at Mushroom Rock, where brush marks were still visible, there was still evidence of salt accumulation both on the surface and within the pigment layer.

Of particular interest was an accumulation of gypsum on the outer surface of the samples from the main gallery site. This accumulation is probably the explanation for the paintings losing some of the 'brightness'. The samples show that there has not been any significant loss of pigment from the surface but that the surface is being obscured by the formation of salts. Of more serious concern, however, is the considerable build-up of soluble salts within the thick porous pigment layers. These salts expand and contract in response to humidity changes and will lead to a break up of the pigment layer.

Cannon Hill area

A number of individual sites were sampled in the Cannon Hill area. One sample was from a painting known to have been painted in the early 1970s. It is of interest as a PVA emulsion (wood glue) was known to have been added to the pigment mix. Examination of this sample failed to positively identify any PVA, although a minor organic component was present in the infra-red spectra. This sample is also of interest as it did not show any salt accumulation, giving some indication of the slow rate of the build up of salts.

Two small sites were intensively sampled as they were to be used for experimental treatment trials and a complete pigment inventory was necessary.

White pigments. There were 3 types of white pigment identified in the samples collected. As with Ubirr the most common was the kaolin-type clay with iron and traces of titanium. Two samples contained the mineral calcite as the white pigment, while one was a mixture of kaolin-type clay and muscovite.

Brown pigments. All the brown pigments were mixtures of iron oxides and clay minerals with variable amounts of muscovite.

Red pigments. All red pigments were based on the mineral haematite, however, some samples also contained clay minerals. As with samples taken from other sites the red pigment layers were usually very thin and extremely well bonded to the substrate.

Yellow pigments. A total of 4 yellow pigments were identified, all were mixtures of iron oxides and clay minerals. A number showed considerable salt accumulation and may have originally also contained a carbonate mineral.

Cream pigment. A sample of pigment from an indeterminate shape at the edge of the Cannon Hill (vertical) site was of particular interest as it was a complex mixture of quartz, jarosite, dolomite and kaolin. The mineral jarosite ($FeKSO_4H_2O$) may be an alteration product although it is known to be produced in some rock weathering processes.

With the exception of the PVA sample described before, all samples contained secondary salt accumulations. Many of the samples showed very heavy salt accumulation making identification of the original pigment difficult. In several cases the salts (usually gypsum) had built up in a network of interlocking crystals and were in fact responsible for the survival of the paintings. The Cannon Hill sites showed a very wide range of salts, with potassium salts (e.g. syngenite) present in addition to the calcium salts gypsum and whewellite. At the site known as 'old camp site' containing a painting of a motor ship ('*Narlm*'), the pigment surface of a number of samples was completely covered by a thin transparent salt crust. Although it was not possible to confirm, it is believed that this salt was sodium sulphate ($Na_2SO_4H_2O$). This is a very soluble salt and appears to have been restricted to this particular site.

East Alligator area

A range of typical sites were sampled in the East Alligator area.

White pigments. Three types of white pigment were identified, again the most common was the kaolin-type clay with iron and a trace of titanium, similar to that at Ubirr and Cannon Hill. One calcium carbonate-based pigment also containing the kaolin-type clay and one very sandy (quartz) clay was also identified.

Red pigments. As with other areas, these were either pure haematite as a very thin paint layer of haematite/clay mixtures.

Yellow pigments. There were three types identified, one sample was a man-made iron oxide of high purity and small particle size. This sample came from one of a series of paintings reported to have been painted in 1967. The other two were naturally occurring mixtures of clay and iron oxides, and calcium carbonate (probably calcite), iron oxides and clay.

All samples showed evidence of salt accumulation and alteration. While gypsum was most common there was also whewellite and sveite present in a number of samples. In addition several samples also showed a build-up of potassium and phosphorous, although no specific minerals were identified.

Deaf Adder area

Only two of the larger sites in this area were sampled. The major site, Yuwengayay, was sampled in considerable detail. It is similar in many ways to the main gallery at Ubirr as a very thick paint layer has been built up on the rock surface and, like Ubirr, the pigments have been homogenised and, except in the surface layers, no individual pigment layers are visible. While the overall range of pigment types is similar to Ubirr the site also displayed some distinct compositional differences.

The site differs in that it has considerably less salt accumulation and very little evidence for alteration of pigment minerals. The calcium salts, gypsum and whewellite, that predominated at the previously described sites are present at much lower levels with minor potassium salts.

These lower overall salt levels are matched by very good states of preservation. Areas of poor preservation are limited to the extremities of the site and areas of groundwater seepage. Salt associated with these seepages was again sveite.

One white pigment sample from Yuwengayay was a mixture of the minerals quartz and microcline (a feldspar). These materials had been ground in some form of milling device, and are therefore not of traditional Aboriginal origin.

White pigment. The predominant white pigment was a kaolin-type clay and mica mixture containing titanium and iron at anomalous levels. While this pigment is very similar to the high iron and titanium clay common at many other sites the samples at Yuwengayay show much higher titanium levels. Chemical analysis confirmed the SEM-EDXA results that titanium was present at about the 1% level whereas in the earlier samples it was only present at trace levels.

The uniformity of the composition of this particular pigment (present in 12 samples) indicates that a particular pigment source may have been used for much of the painting at this site. The distinctive composition would enable potential sources to be easily identified.

Two samples, one a multi-layered flake from the right hand side of two female figures, the other a multi-layered flake from a kangaroo, in addition to titanium also contained magnesium at between 1 and 2% and it was not possible to specifically identify the clay mineral present.

Detailed x-ray diffraction studies of this clay mineral showed that the iron and titanium is present as ultra-small inclusions within the clay particles and as such are not chemically part of the clay as originally believed. This work also indicated that the 'clay' in fact showed characteristics of a mica rather than a true clay. These very distinctive features should enable the particular source of this pigment to be identified.

Only one sample, that of cream/white pigment from a painting of a kangaroo, showed evidence of carbonate minerals (probably calcite but altered partly to gypsum).

Brown pigments. All the brown pigments were mixtures of iron oxides (haematite and goethite) and clays. The clay mineral is often the same as the high iron and titanium clay present in the white pigment. This indicates that several materials may have been mixed to produce the required colour.

Red pigments. The red pigments examined were composed of the mineral haematite mixed with minor clay.

Yellow pigments. Only one sample of a yellow pigment was obtained and this was a mixture of an iron oxide with the typical high titanium clay.

As mentioned above, there has not been major alteration of the pigments at this site. A sample taken from an area damaged by seepage from a rock fissure shows evidence of the iron and silicon being leached from the pigment by the acidic ground water leaving behind an aluminium oxide. This phenomenon is, however, very localised.

The cream/white pigment from a painting of a kangaroo, mentioned in the white pigment section above, was the only carbonate-based pigment and shows some development of gypsum indicating that some sulphur dioxide must be present. However, the level of alteration is considerably less than that seen in the samples from the Nourlangie and Ubirr areas of the Park.

Ngarradj site complex

Two sites were sampled and overall the sites showed very variable states of preservation, while examination of the pigment samples showed considerable build-up of salts. Once again sveite was identified in a sample of disintegrated rock from an area of active weathering on the lower of the two sites visited. This site also produced the only example of a European-type paint found in the samples examined. The sample came from a poorly preserved white fish painting that did not appear to be physically different to other white paintings executed with natural pigments.

The site also produced the only example of gypsum having been used directly as a pigment. In all other cases it was established that the gypsum present in samples had formed as an alteration product of carbonate minerals, however, in a sample taken from a very poorly preserved white painting of a turtle, the uniform size and purity of the gypsum crystals indicate that it was the primary pigment and possibly of European origin.

White pigments. The most common white pigments are kaolin-type clays, with traces of iron and titanium present as noted at most other sites. Two samples also had high levels of calcium present indicating the presence of a carbonate mineral (probably calcite) in the original pigment. In both cases this has been largely altered to gypsum. In addition, as described above, there was one white European paint based on an aluminium oxide (bauxite) and one gypsum-based white pigment.

Red pigments. The red pigments were haematite clay mixtures and pure haematite.

Yellow pigments. Both yellow pigments examined were mixtures of iron oxides and clays.

Brown pigments. The only brown pigment examined was a mixture of iron oxides, clay and a calcium carbonate mineral.

As outlined above, the pigments showed high levels of alteration of carbonate minerals to gypsum, and the build-up of salts in some samples made it difficult to identify the original pigment.

Christmas Creek

Two sites at Christmas Creek in Stage 3 of the Park were sampled. The larger of the two sites had a very thick build-up of salts (gypsum and whewellite) on the roof of the shelter prior to the paintings having been made. The paint had been placed directly onto the black salt surface, but the overall preservation was good. A small lump of yellow pigment found in the drip line of the shelter matched well with a yellow paint from a painting of a large kangaroo on the roof of the shelter.

White pigments. Two samples were mixtures of mica and clay minerals while the other two were carbonate (calcite) based, one being mixed with clay minerals.

Pale yellow pigment. This material, found in both a painting and a floor deposit, was a mixture of very fine silica (sand), clay and iron oxide with minor mica.

Brown pigment. This was a mixture of haematite and clay.

All samples showed considerable salt build-up, with alteration of the carbonate-based white pigments. Both gypsum and whewellite are present in the pigments and on the rock surfaces. The source of the particularly heavy salt build-up is not known, but because of the large volume involved it may be a useful site for future study of salt deposition processes.

Namargon

A number of sites in this area were sampled. The range of minerals identified in these samples is the widest of any site examined in Kakadu. Although many of the pigments consist of known minerals, the particle size and composition indicate that a large number of pigments in these sites were not from naturally occurring sources, but were either man-made, or were at least prepared in a European grinding mill. Some of these prepared materials were subsequently mixed by the artist with naturally occurring pigments similar to those seen at other sites, thus making positive identification of the source difficult.

The use of these prepared pigments has resulted in very poor adhesion between the pigment particles in a number of cases and explains the highly dispersive properties noted in areas where water had entered the rock shelters. These adverse properties result from the absence of clay minerals and the even particle size distribution and equigranular shape of these European materials.

In contrast the man-made yellow iron oxide pigment used in three samples at Belly Button, a human type figure, a kangaroo and a fish, is extremely durable due to its very small particle size and flat tabular particle shape. Its properties are similar to the good quality haematite pigments described earlier.

White pigments. There were at least seven different combinations of minerals, both naturally occurring and artificially milled, used to produce white pigments. How such a strange collection of materials came to be used at the sites is extremely puzzling. A number of products are prepared for use in pottery glazes, but this is unlikely to have been a popular hobby in the area at the time the paintings were produced. The following mixtures and minerals were identified:

(1) Finely ground feldspar (microcline) and quartz. This is similar to a sample from sexually active figures at Yuwengayay.

(2) Coarsely ground enstatite (a magnesium iron aluminium silicate).

(3) Finely ground mica, probably muscovite, with minor clay.

(4) Kaolin clay and mica mixture (natural).

(5) Kaolin clay with high iron and traces of titanium (natural).

(6) A mixture of the above clay with finely ground quartz and jarosite (iron potassium sulphate).

(7) Finely ground quartz with a trace of clay.

Yellow pigments. The three samples previously described, taken from a human type figure, a kangaroo and a fish, were identical industrially made iron oxide of ultra-small particle size and very high purity. One sample was a natural iron oxide clay mixture.

Brown pigments. Only one brown pigment was examined and it was a natural iron oxide mixture with minor clay.

Red pigments. Two samples were haematite and clay mixtures, probably of natural origin, while one was haematite mixed with finely ground quartz and was not a naturally occurring material.

It is of interest to note that most samples showed minor salt (predominantly gypsum) accumulation. None of the man-made pigment materials is reactive with sulphur dioxide so it is not possible to make any estimate of salt accumulation rates. As the painting sites had relatively thick pigment layers it is possible that the salts seen in the samples had migrated from lower pigment layers. The salts would, however, be highly disruptive in the very weakly bonded materials used in many of the pigments.

Discussion

A very wide range of materials have been used as pigments in the post-estuarine period rock art in the Kakadu Park. While naturally occurring minerals and mixtures of minerals predominate, a significant number of man-made or prepared pigments of European origin have also been identified.

There was a strong association between particular site complexes and the individual range of pigments used. In some cases particular pigments were restricted to a site complex such as the dolomite-based white pigment at Nourlangie or the clay with very high iron and titanium levels at Yuwengayay.

While this would normally indicate that local pigment sources were being utilised, observation at several site complexes where European-sourced materials have been used raise the possibility that large numbers of paintings were done in a short period of time by individual or small groups of artists.

There are obviously differences in the durability of individual pigment materials. The white pigments are generally least durable, particularly those based on carbonate minerals such as calcite or dolomite. The clay-based white pigments show a greater range of durability, but this is due more to environmental factors rather than any physical property of the pigment.

The red pigments are clearly the most durable, especially those in which haematite predominates. This is the case on rock painting sites throughout Australia and some of the reasons have been discussed in the section dealing with red pigments.

We have found no evidence for the use of organic binders in the paint samples examined with the exception of the one obvious European paint sample. Neither the SEM examination nor the infra-red spectra showed any significant organic components. This observation is in line with similar work carried out by the authors on Aboriginal rock art pigments elsewhere in Australia.

The trace element chemistry of some of the pigments examined was of considerable interest and it may be possible to source particular pigments using trace element distribution. While such a study would be of interest in establishing trade patterns for pigments, it is not considered to be of relevance to the deterioration processes or conservation treatment.

Acknowledgments

The work covered by this paper was carried out for the Australian National Parks and Wildlife Service under a consultancy contract. The authors are very grateful to ANPWS for the support provided and to the rangers and staff at Kakadu National Park for assistance in this work. In particular the assistance provided by Ivan Haskovec was invaluable. The authors also gratefully acknowledge the assistance and permission from the traditional custodians in making this work possible.

APPENDIX 1

Analytical techniques used

The samples collected at Kakadu were analysed in the laboratory using a variety of techniques. Initially all samples were examined by scanning electron microscope which, through its x-ray analysis facility, also produced a semi-quantitative chemical analysis of the sample. In many cases the combined visual and chemical analysis provided by the SEM was sufficient to adequately identify the sample. However, some selected samples were subjected to further examination by other means. The samples so selected were either those for which confirmation of the SEM results was required or those where the SEM results did not adequately allow for identification. Additional analytical methods used were x-ray diffraction and/or infra-red spectroscopy and/or wet chemical analysis.

Each of the techniques applied provides information on a different aspect of the samples' structure or composition. As our final conclusions on structure and composition are based on one or more of the above analytical techniques, the following is an outline of each with particular emphasis on the limitations and the type of information given by each technique.

(1) Scanning electron microscope (SEM) with energy dispersive x-ray analysis (EDXA)

The SEM, like an optical microscope, gives an enlarged image of the sample being studied. The SEM used in this work provides a magnification of up to 30 000 times with resolution of individual particles down to 0.1 micron, although full power is very rarely required and most examination was done at magnifications of less than 1000 times. The SEM was a low pressure-type which did not require any pretreatment of the sample prior to examination.

The SEM operates by focusing a very small beam of electrons onto the sample and then collecting the back-scattered electrons. By moving the beam over the sample surface a picture is built up in much the same manner as with TV images. The intensity of the back-scattered electron beam at each point on the sample surface depends primarily on the atomic weight of the atoms at that point, hence minerals with high atomic weights appear bright and those of low atomic weight are dull. The resultant images are often strikingly different in contrast to the corresponding optical images.

When the electron beam strikes the sample surface some of the electrons interact with the sample atoms to produce x rays. By collecting these x rays and measuring their wavelength and energy an indication can be obtained of the sample composition (atoms of differing atomic number produce x rays of different wavelengths). The x rays from the sample are collected over a period of approximately 30 seconds and if, during this period the beam scans over the entire sample, a general analysis is obtained. However, optionally the beam can be immobilised at one point on the sample and this will give a point analysis. By locating the analysis point using the magnified SEM image, an analysis can be obtained of individual crystals or paint/pigment layers. As the electron beam does not penetrate far below the material surface the analysis obtained is essentially a surface analysis. If the surface being scanned is a natural surface, such as the outer surface of a painting, then the results obtained will show a large contribution due to surface salts and dust, and may not be representative of the bulk pigment. A better representation of the bulk pigment is obtained by either powdering the sample or by measuring at the edge fracture.

An important limitation of the EDXA facility is that low atomic weight elements do not give a detectable x-ray signal, and as a consequence elements with an atomic number of less than 24 are not detected. In practical terms this means that the following common elements cannot be detected by EDXA even when they are present in the sample in large amounts:

Cations: Sodium, lithium and ammonium

Anions: Carbonate, oxides, hydroxides, nitrates, oxalates and organic acid salts

Compounds: All organic compounds containing only carbon, hydrogen, nitrogen or oxygen

(2) X-ray diffraction (XRD)

In XRD a fine beam of x rays are fired into a powdered sample. As the x rays pass through the sample they are deflected by the crystal planes. The deflected x rays are collected and the angles of deflection and the intensity of the beams at each angle of deflection are recorded. Both the intensity and angle of deflection are related to the shape of the crystal structures present in the sample and to the types of atoms in the sample.

Sample identification is normally performed by comparing the observed x ray diffraction pattern with the known patterns of previously studied minerals. To assist this process tables are available which list the x ray diffraction patterns for known minerals.

The main limitation with XRD is that the unknown sample may contain mineral types which have not previously been recorded. This is quite common with rock art pigments as there has been comparatively little previous study done on rock surface mineralogy. In addition, if a sample contains several different mineral types then the resultant XRD pattern can be exceedingly complex. Some minerals, such as quartz, have a very intense x-ray pattern and they will generally show strongly even when they are present in relatively small amounts. By contrast amorphous substances have very weak patterns and can readily avoid detection.

(3) Infra-red spectroscopy (IR)

In infra-red spectroscopy a beam of light with a narrow frequency range is passed through a prepared specimen and the amount of light absorbed by that specimen is recorded. The frequency of the light beam is slowly changed from 2.5 to 50 microns and the absorption as a function of wavelength is continuously recorded. In this study all the samples were prepared by mechanically grinding to a fine powder and then incorporating this powder into a KBr disc.

In this region of the light spectrum all absorptions are due to covalent bonds within the sample molecules such as C-H, C-O, Si-O, H-O, Fe-O and similar bonds. Ionic bonds, such as in NaCl, do not absorb in this region of the spectrum. Consequently IR gives information only on the types of covalent bonds which exist in the sample. In compounds having a mixture of ionic and covalent bonds, such as calcium carbonate, only the covalent bonds (such as C-O) will show absorption peaks. However, the position of these peaks can be influenced by the ionic bonds present in the compound. A consequence of this selective absorption is that compounds having the same covalent bonds, but different ionic bonds, will show very similar IR spectra.

With IR spectroscopy those groups having sharp intense peaks can be readily identified even when present at quite low concentrations. In rock art pigments common compounds with these characteristics are oxalates, carbonates, organic materials and clays. Other compounds which have either no IR peaks (such as KCl, NaCl) or only broad peaks (such as FeOOH) will not produce distinctive spectra and consequently may not be detected at all by this method.

(4) Chemical analysis

A quantitative chemical analysis for certain elements was carried out on some rock art samples. Due to the small size of the collected samples there was not sufficient material for a complete analysis. Analysis was performed by dissolving the powdered samples in hydroflouric acid and then neutralising. The resultant liquid was analysed by inductively coupled plasma spectroscopy. The elements covered by this analytical method are boron, magnesium, aluminium, silicon, potassium, calcium, vanadium, chromium, titanium, manganese, iron, cobalt, nickel, copper, zinc, strontium, molybdenum, silver, barium, lead and some rare earth elements.

(5) Spot tests

Carbonates, oxalates and clays are three common minerals found in white pigments. These three minerals react differently with dilute acid and this provides a quick spot test for identification purposes. By spotting multi-layered samples with dilute hydrochloric acid, whilst the samples were under an optical microscope, it was possible to distinguish which compounds were present in which layers. The distinction used was:

Carbonates - React rapidly, dissolve and form copious gas bubbles.

Oxalates - React slowly with gradual gas formation.

Clays - No reaction but may disperse slowly. No dissolution.

REFERENCES

CLARKE, J. 1976. Two Aboriginal rock art pigments from Western Australia, their properties, use and durability. *Studies in Conservation* 21: 134-142.

GILLESPIE, D. (ed.) 1985. *The rock art sites of Kakadu National Park, some preliminary research findings for their conservation and management.* Special Publication No. 10, Australian National Parks and Wildlife Service.

WATCHMAN, A. 1983. Geological investigations into the formation of salts at Aboriginal rock art sites in Kakadu National Park, N.T. (unpublished).

14

CHEMISTRY OF DETERIORATION OF POST-ESTUARINE ROCK ART IN KAKADU NATIONAL PARK

J. Clarke and N. North

Abstract. A geochemical study of the materials present in the post-estuarine rock art of the Kakadu National Park (Northern Territory, Australia) showed that the major non-pigment minerals present were gypsum, sveite and whewellite. Minor amounts of other sulphates, chlorides and phosphates were also found. Modes of formation of these major alteration minerals are discussed. The significance of airborne sulphur compounds in the formation of gypsum is considered and preliminary results from a long-term gaseous sulphur monitoring program in Kakadu are presented.

Introduction

In 1986 a study was commenced, on behalf of the Australian National Parks and Wildlife Service, to investigate the materials which had been used in paintings of the post-estuarine period of Kakadu rock art; and to establish the physical and chemical changes which have occurred in these rock art paintings since they were painted. This work was undertaken in order to establish a basis for treatment methods to preserve these rock paintings. It follows earlier work by Watchman (1983). The results of this earlier work and other conservation-related studies have been presented in Gillespie (1985).

The pigment types found in these paintings are described in a companion paper elsewhere in this volume (Clarke and North 1991). In this paper we will cover the mineral and chemical changes which were observed in the rock art, and discuss the processes by which these changes occur and the effects they have on the stability of the rock art.

Deterioration of rock art

To appreciate the significance of the observed chemical changes it is best to look first at the various types of rock art deterioration which are readily visible in the post-estuarine rock art of Kakadu. This deterioration takes several forms and all forms are not necessarily present at every site. These are;

Rock weathering: This generally occurs in rock-shelters and overhangs, and is usually confined to the rocks which are closest to the ground level. In affected areas the surface of the rock has been converted to a crumbly sand-like material of very low mechanical strength. Frequently there are visible amounts of a white salt-like material mixed with the decaying rock, and the surface of the rock has a bitter taste.

The progressive decay of the rock often leads to a smooth sculptured appearance which is sometimes mistakenly attributed to wind erosion. The weathering, over a period of time, causes an undercutting of the rock wall producing a cavernous effect; hence the name cavernous weathering. The worst affected areas found in Kakadu were on overhangs at plains level on the base of rock outcrops, with massive weathering generally extending up to 1 m above ground level. However the same effect was found, although to a lesser extent, in most rockshelters examined in Kakadu.

Salt deposition and water seepage: During the wet season, water is frequently found slowly seeping from cracks and fissures in the rock surface. If this water runs down the rock face it generally completely removes any rock art over which it passes. The seepage water also carries dissolved salts and these can be deposited by evaporation at the edges of the damp area. In these deposition areas the rock art is not necessarily removed by the water, but it is often covered by the deposits from the water and associated biological growths.

Rainwater run-off: During the wet season rainwater may run down the rock face and then across sections of the rock art. This water is relatively pure and generally does not cause any chemical changes in the rock paintings but it does simply physically disrupt the porous pigment layer and wash the paintings off the rock surface.

Pigment flaking: This form of deterioration is characterised by the loss of paint flakes, sometimes up to several centimetres wide, from the surface of the painting, thus leading to a patchy appearance of the painting. This is most commonly found in those sites where there has been successive overpainting of earlier rock art. Overpainting is very common in the post-estuarine rock art of Kakadu, with several of the major sites consisting of twenty or more layers of pigment. In many cases the flaking has not yet produced major deterioration in the appearance of the paintings but, on closer examination, it is seen that the majority of the surface layer has disbonded from the underlying layers, is just barely held in place and can be easily dislodged by any physical disturbance.

Surface deposits: Many of the Kakadu rock art paintings appear to have become partially obscured over the years due to the build-up of extraneous material on the surface of the rock art. This build-up takes different forms at different sites, with the three following forms being the most common. Note that the terms used to describe each site are derived from the site appearance and do not describe the mode of formation of the deposits.

'Fire stains': These are matt black deposits which generally occur in patterns on the roofs and walls. In some cases these have formed over painted areas and in other cases the rock paintings have been made over the top of pre-existing black deposits. These are often incorrectly assumed to be soot deposits from campfires. In cases where the black deposit reaches areas affected by

water wash or seepage, a characteristic white salt deposit forms as a thin white margin.

'Dust' deposits: These are usually a brown-grey colour and look very similar to the dust and dirt found on the floors of the rockshelters. These deposits are usually firmly adherent to the rock face and cannot be removed by brushing or washing. The deposits are often quite hard and sometimes have a polished or glassy appearance. Some paintings have been obscured by the deposits and others have been painted over the deposits.

Organic materials: In Kakadu the most obvious are termite tracks and mud wasp nests. On closer examination, a mass of fine spider webs and small insects are found over most surfaces.

Mechanical damage: A common cause of damage is mechanical abrasion of the rock surface. In Kakadu this has been most commonly caused by cattle, wild pigs and water buffalo using the rocks as scratching pads. Similar damage can also be caused by movement of tree branches across the rock face and by people touching the rock art. Obviously if the painting is also suffering from flaking, then the damage caused by any contact with the surface is much greater. Prevention of mechanical damage is largely a matter of park management practices, and the extent of this damage has been greatly reduced in Kakadu in recent years (Gillespie 1983).

Chemical and mineralogical changes

A major factor in many of the damage types found with the Kakadu post-estuarine rock art, is chemical and mineralogical changes in the rock and/or the paint pigments since the time the art was originally painted. In many cases groundwater seepage is involved in these changes.

(a) Kakadu groundwater

Seepage of groundwater is a common feature at most rock art shelters in the wet season at Kakadu. In the dry season seepage, whilst much less frequent, can still be found at some sites. As several forms of chemical change in the rock art appear to be associated with groundwater seepage, samples of the groundwater seeping from rock surfaces were collected in both the dry and wet seasons for analysis. The following analysis is of water collected from a small permanent seepage at Nangalor.

Element	Dry Season (Aug '87)	Wet Season (Apr '86)
Calcium (mg/L Ca)	7.0	1.3
Magnesium (mg/L Mg)	4.0	0.3
Aluminium (mg/L Al)	0.8	<0.1
Silicon (mg/L Si)	12.0	8.6
Sodium (mg/L Na)	84.0	12.8
Potassium (mg/L K)	0.6	<0.1
Chloride (mg/L Cl)	110.0	17.0

Table 1. *Analysis of water sample collected at Nangalor*

The dry season seepage water contains a considerably higher load of dissolved minerals than does the wet season seepage, which is to be expected. However, the total mass of solids being moved in the wet season would be higher than in the dry due to the much larger volume of water involved. A significant factor of the wet season seepage is the very acidic pH of the water (4.6), which is an important factor in its ability to dissolve minerals. It should be noted that seepage of water from a rock face need not necessarily produce either visible water or a noticeably damp surface. When seepage rates are low, the water can evaporate in the surface layers of the rock thus producing no visible dampness but still leaving salts in the surface.

(b) Sveite

A white salt, subsequently identified as sveite, was first encountered in areas of active rock weathering at Ubirr-Kalabirri. The same salt was subsequently found at East Alligator (rear of Ranger's office), Jabiru Dreaming, Yuwengayay and Ngarradj. Confirmation that this was the same salt in all cases was obtained from x-ray diffraction (XRD) patterns.

In all occurrences, except at Jabiru Dreaming, the salts were found on rock surfaces undergoing active and rapid degradation. On this basis the occurrence at Jabiru Dreaming may be an indication of incipient weathering. Samples of this salt were collected during both the wet (March) and dry (August) seasons and it was noted that the salt deposits during the dry season were much harder and more brittle than those found in the wet season. This is a good indication of changes in degree of hydration of the salt with changes in local humidity. Such changes in salts have been found to be a major factor in determining their ability to fragment porous rocks. Both the wet and dry season salts showed the same XRD pattern when subsequently examined in Perth.

Scanning electron microscope (SEM) analysis showed the presence of only potassium, aluminium and chlorine in this salt. Subsequent chemical analysis showed that the salt contained:

Potassium approx. 4.0%
Aluminium approx. 20.0%
Chlorine approx. 7.0%

Initially the nature of this mineral could not be established as its XRD, infra-red patterns, and chemical analysis did not correspond with any known mineral. Subsequent work, by the Chemistry Centre of Western Australia, identified this mineral as sveite, a recently discovered mineral which was first reported in 1980 from samples collected in a Venezuelan cave (Martini 1980).

Sveite has the chemical formula of $KAl_7(NO_3)_4Cl_2(OH)_{16}.8H_2O$. This discovery at Kakadu was only the second known world occurrence of this mineral, but the geological conditions associated with its occurrence at both Venezuela and Kakadu suggest that it could be quite common in some tropical areas.

The significant properties of sveite are:

1. A very soft colourless mineral.
2. Readily soluble in water, with ability to absorb up to ten times its initial volume of water before dissolving.
3. Electron scanning micrographs show formation of aggregates of strangely distorted flakes. The texture is micaceous.

The Venezuelan specimens were collected from a series of caves located in a precipitous cliff face of a 850 m high pinnacle, the Cerro Autana, in southern Venezuela. This pinnacle is an outlier of the Pre-Cambrian pink quartzite of the Roraima group, forming a very thick horizontal strata resting disconformably on a granitic basement. On the walls of the cave, sveite forms white crusts and efflorescences which had been deposited by solutions trickling from a bedding plane close to the ceiling.

Martini (1980), in discussing the genesis of this material, states:

> It is classically believed that nitrate minerals in caves form from animal excreta such as bat guano. In the case of sveite, the origin from animal excreta meets some difficulty as ... the mineral has obviously been deposited from solutions trickling directly from the ceiling of the cave. As it is very probable that there are no other caves at a higher level, ... the nitrate and chloride must have been provided by rainwater and by the organic material contained in the soil on the top of Cerro Autana. The solutions, presumably very acidic and poorly mineralised, as must be the case in such a siliceous environment, dissolve aluminium and potassium from feldspar and mica contained in minor amounts in the quartzite. As the solutions are very dilute, the deposition of sveite can be achieved only after evaporation of many times its original volume.

There are close similarities in the occurrence of sveite in Kakadu and that described by Martini in Venezuela. His outline of the genesis is also in accord with our observations and water analysis at Kakadu. The mineral's property of absorbing large quantities of water before dissolving would also explain its highly destructive effect on the rock matrix. The high solubility of this mineral in water suggests it may be possible to remove the mineral from the rock face by judicious washing.

(c) Gypsum

Gypsum (calcium sulphate dihydrate, $CaSO_4.2H_2O$) is commonly found at rock art sites in Kakadu. It takes two general forms. Firstly, it appears as a major constituent, with whewellite, of the black 'soot-like' deposits which are commonly seen on the roofs and walls of rock art shelters. Secondly, it is found in the rock art pigments themselves.

Gypsum is a white mineral with a limited solubility in water (approximately 2.5 g/l). Once gypsum is formed the limited, but significant, solubility in water will promote the slow growth of larger crystals at the expense of the smaller ones. As a result clumps of gypsum crystals form over a period of time and they have the characteristic rosette morphology.

These gypsum rosettes grow out from the rock face. Due to their interlocking surface, the rosettes trap any incoming light and so appear as a black deposit.

Gypsum is also the most common alteration product found in the Kakadu paintings. This mineral is very common in the rock paintings of the northern sections of Kakadu where white pigments based on carbonates were widely used. The process of formation of gypsum in this case is thought to be through the action of acidic sulphur containing gases in the atmosphere with the basic carbonate pigments. The commonest sulphur containing gases are hydrogen sulphide (H_2S) and sulphur dioxide (SO_2). These gases can initially react with calcium carbonate in the following manner:

$$CaCO_3 + H_2S \longrightarrow CaS + H_2O + CO_2$$
$$CaCO_3 + SO_2 \longrightarrow CaSO_3 + CO_2$$

The products (CaS and $CaSO_3$) so formed are unstable in the presence of oxygen and water and will rapidly oxidise to form gypsum according to the reactions:

$$CaS + 2O_2 + H_2O \longrightarrow CaSO_4.2H_20$$

and

$$CaSO_3 + O_2 + 4H_2O \longrightarrow CaSO_4.2H_2O$$

Gypsum is appreciably more soluble in water than is the original carbonate. Consequently in damp weather, when a surface film of moisture can condense on the rock painting, the gypsum will dissolve slightly and can move for a short distance in the rock painting. The continued process of condensation and evaporation tends to draw the gypsum component towards the outer surface of the painting, and as the water evaporates the gypsum crystallises in the classical gypsum rosettes. The same cycle of wetting/drying will also cause the larger crystals to grow at the expense of the smaller ones.

The conversion of carbonate pigments to gypsum is very common in the Ubirr-East Alligator area, and to a lesser extent at Ngarradj and Nourlangie. Naturally for this process to occur there must first be present a carbonate mineral in the pigments. This carbonate need not be in the outermost layer of the painting but may be present in an earlier painting which has since been overpainted. The most probable source of the gaseous sulphur compounds is the wetlands biomass. The decay of organic materials at the end of the wet season will produce significant local concentrations of hydrogen sulphide (H_2S). Other contributions could arise from volcanic gases and engine fumes. A program was established in Kakadu to monitor atmospheric sulphur compounds, and this program is discussed in a later section.

The growth of gypsum crystals on the outer pigment surface has the effect of both obscuring the pigments and of acting as a dust trap. The net effect being a distinct loss of colour and life from the painting. However, in many paintings the gypsum does not form on the pigment surface but rather grows inside the outmost pigment layer where it forms clumps of crystals. As these crystals grow they force apart the pigment particles and thus reduce the bonding between the pigment particles. Their growth also distorts the outer surface of the pigment, changing the initially smooth and reflective surface into an uneven surface with numerous swellings of various sizes. These changes in the surface produce cracking of the pigment surface, with resultant adhesion loss, and a general dulling of the appearance due to light reflection losses.

(d) Whewellite

In many of the sites examined, there were large, and often thick, deposits of black soot-like material. Analysis of these deposits showed that they were predominantly a mixture of gypsum and whewellite (calcium oxalate dihydrate).

The whewellite mineral is probably formed through a reaction between calcium compounds and airborne oxalic acid. The oxalic acid would presumably be derived from the decay or burning of organic material and could readily be transported to the site by air movements (Bell 1987). In these deposits the origin of the calcium has not been clearly identified but it is thought to be derived from normal weathering of the feldspar minerals in the rock.

Analysis of seepage water in Kakadu rockshelters (see Table 1) showed that they are quite pure and, although the calcium content is a large fraction of the total dissolved salts, it is still small in absolute terms. This would imply that, if the calcium is supplied by the groundwater, then formation of the black deposits must be a slow process over many years. Similar black deposits are very common in rockshelters throughout Australia.

Whewellite is also found as a minor component amongst gypsum where the gypsum has formed as an alteration product of a carbonate-based pigment. The reaction in this case is presumed to be:

$$CaCO_3 + H_2C_2O_4 \longrightarrow CaC_2O_4 + H_2O + CO_2$$

In the scanning electron microscope the whewellite is frequently seen as a white 'fluffy' material between the gypsum plates. At high magnifications this 'fluffy' material sometimes resolves into extremely fine crystals.

(e) Minor salts

A range of other salts were found, in minor quantities, at various rock art locations throughout Kakadu. In most cases they formed only a minor constituent, and frequently there was not sufficient material present to provide an unequivocal identification. In the following list the salts which were positively identified by x-ray diffraction are noted as (XRD). The identification of the remaining salts is from scanning electron microscope analysis and is still tentative.

Jarosite: $(KFe_3(SO_4)_2(OH)_6$ (XRD). Found at Cannon Hill and Namargon. The iron may have been derived from dissolution of iron minerals in the painting pigments or through leaching of iron-containing minerals in the rock by the slightly acidic groundwater.

Sylvite: (KCl) (XRD). Found at Nourlangie and Ngarradj. Probable source is deposition from groundwater.

Syngenite: $K_2Ca(SO_4)_2.2H_2O$ (XRD). Found only at Cannon Hill where it occurred at a number of sites, generally in association with gypsum.

Sodium sulphate: Found only at 'old camp site', Cannon Hill.

Potassium sulphate: Found at Nourlangie, Ubirr and Yuwengayay as minor component only.

Phosphates of calcium, iron and potassium: Found at a number of sites, generally where the rock art had a dusty appearance. Appears to form as a surface layer from external deposition rather than water seepage. Presumably the phosphate is an offshoot of the biological cycle, and may be associated with animal excreta.

(f) Atmospheric sulphur compounds

The widespread occurrence of gypsum, as an alteration product of pigment carbonates in the absence of any indication of groundwater, suggested that atmospheric sulphur compounds, in particular hydrogen sulphide (H_2S) and sulphur dioxide (SO_2), were the cause of the carbonate to sulphate transformation. These sulphur compounds are normally associated with polluted industrial and urban environments, where their propensity to attack calcium carbonate is well known and documented (Winkler 1978). By contrast Kakadu is generally speaking a natural bush area, with no major industrial sources of sulphur compounds, either within its boundaries or close-by. Normally the levels of atmospheric sulphur compounds under these conditions would be expected to be negligible.

However, a distinctive feature of the Kakadu National Park is its extensive wetland areas, particularly in the lowlands sections of the park but also extending up to the escarpment. During the wet season there is massive biological growth in the wetlands, and as these dry out this growth withers, decays or burns. The amount of biological sulphur involved in the wetland biomass cycle would be extensive and, although most of this would return to the soil, a significant atmospheric release each year could also be anticipated. The wetlands biomass as a source of the atmospheric sulphur compounds was reinforced by examination of the distribution patterns for minor sulphate minerals, and the extent of conversion of pigment carbonates to gypsum. This showed high sulphate salt levels and extensive conversion in those sites, such as Cannon Hill and Ubirr, which are located near extensive wetlands. At drier sites, such as Yuwengayay, the minor sulphate minerals were rare and conversion was not as extensive.

To assess the local levels and to determine the distribution of atmospheric sulphur compounds in the Kakadu National Park, a long-term experimental monitoring program, devised in conjunction with the Chemistry Centre of Western Australia, was set up at various sites in Kakadu to measure levels of reactive atmospheric SO_2 and H_2S.

This was done by placement of sets of Petri dishes, containing powders selected so as to react with either H_2S or SO_2. A set consisted of six Petri dishes of each type. It is intended to remove one dish of each type from each location at half-yearly intervals, thus giving a total monitoring period of three years. To prevent contamination of the powder the open Petri dishes were covered with a nylon flywire mat. Every set was located in narrow deep fissures in the rock faces where they would not be disturbed by animals, and where they were reasonably protected from strong winds. The sites chosen were all close to rock art sites.

For the H_2S monitoring, the material used was lead carbonate ($PbCO_3$). This reacts with atmospheric H_2S as:

$$PbCO_3 + H_2S \longrightarrow PbS + CO_2 + H_2O$$

After retrieval the samples are analysed for hydrogen sulphide gravimetrically as barium sulphate, after dissolution of the sample in nitric-bromine mixture and separation of the lead with ammonia and ammonia carbonate solution.

For the SO_2 monitoring, the material used was manganese dioxide (MnO_2). This reacts as:

$$MnO_2 + SO_2 \longrightarrow MnSO_4$$

After retrieval the manganese dioxide samples were analysed for sulphur dioxide gravimetrically as barium sulphate after dissolution in dilute hydrochloric acid.

This monitoring was started in August 1987, and at the time of preparing this paper only the results for the first six months, covering the period of August 1987 to February 1988, are available. These were found to be as follows;

Site	Sulphur dioxide uptake (mg)	Hydrogen sulphide uptake (mg)
Christmas Creek	0.2	0.1
Namargon	0.1	0.3
Nourlangie	<0.1	<0.1
Cannon Hill	0.1	0.4
Ubirr	<0.1	0.4
Ngarradj	0.2	0.3

Table 2. *Results of SO_2 and H_2S sampling at Kakadu rock art sites.*

If, as postulated, the primary source of the gaseous sulphur compounds is from the decay of wetlands biomass at the end of the wet season, then the next set of samples (due for removal in August 1988) should show a substantial increase in sulphur dioxide and hydrogen sulphide levels. (Authors' note: the August 1988 samples did in fact show a substantial increase in both SO_2 and H_2S uptake.)

The levels observed to date do demonstrate that, even in a remote area such as Kakadu, there are detectable levels of gaseous sulphur compounds in the atmosphere.

Effects of salts on Kakadu rock art

In comparison with Aboriginal rock painting pigments from elsewhere in Australia, the Kakadu samples did show one significant difference. Virtually all the samples examined contained accumulations of salts, both on the surface of the painting and within the paint layers.

These accumulations of salt are responsible for much of the deterioration of the painted surfaces. There are a number of mechanisms by which the salts cause deterioration, including:

- chemical reaction with the pigment;
- physical disruption of the paint surface resulting from crystal growth;
- expansion and contraction of the salt crystals due to cyclic changes in hydration with changes in humidity;
- solubility of the salt causing the collapse of the paint layer;
- ionic exchange causing failure in clay-based pigments,
- accumulations physically obscuring the paint surface.

The nature of the salts and the site conditions showed that the origin of the salts was often atmospheric rather than the more usually observed source in the groundwater within the rock mass behind the paintings. Initial studies have confirmed that gaseous sulphur compounds are present in detectable levels in the Kakadu park area despite its natural bushland environment.

The salts occurring in the rock art of Kakadu are generally water soluble and are present at such high levels now that they effectively form one of the main binders holding the art pigments to each other and to the rock. In the case of carbonate pigments, these have frequently been completely converted to sulphate salt. As most of the salts are water soluble, they could be removed from the rock surface with water but their removal would also destroy the rock paintings. Consequently any approach to the stabilisation of the Kakadu rock art must be based on preservation of both the pigments and the salts. Trials are currently underway on such methods of stabilisation for the Kakadu paintings, and the results will be published in a later paper.

Acknowledgments

The work covered by this paper was carried out for the Australian National Parks and Wildlife Service under a consultancy contract. The authors are very grateful to ANPWS for the support provided and to the rangers and staff at Kakadu National Park for assistance in this work. The authors also gratefully acknowledge the assistance and permission from the traditional custodians in making this work possible. The technical support of the Chemistry Centre of Western Australia in identifying the sveite and developing the atmospheric sulphur monitoring is also acknowledged.

Disclaimer

The opinions and views expressed in this paper are solely those of the authors and are not necessarily the views of the Australian National Parks and Wildlife Service or the staff thereof.

REFERENCES

BELL, A. 1987. It rains formic acid in the Top End. *Ecos* 50: 28-29.

CLARKE, J. and N. NORTH 1991. Pigment composition of post-estuarine rock art in Kakadu National Park. This volume.

GILLESPIE, D. (ed.) 1985. *The rock art sites of Kakadu National Park, some preliminary research findings for their conservation and management.* Special Publication No. 10, Australian National Parks and Wildlife Service.

MARTINI, J. E. J. 1980. Sveite, a new mineral from Autana Cave, Territorio Federal Amazonas, Venezuela. *Transcripts of the Geological Society of South Africa* 83: 239-241.

WATCHMAN, A. 1983. Geological investigations into the formation of salts at Aboriginal rock art sites in Kakadu National Park, N.T. Unpublished report lodged with ANPWS, Canberra.

WINKLER, E. M. 1978. Stone decay in urban atmospheres, *Engineering Geology Case Histories Number 2*, pp. 53-58. Geological Society of America.

15

YIWALALAY:

An assessment of the current state of preservation and recommendations for future conservation.

J. Clarke, D. Lambert, P. Haydock and J. Flood

Abstract. For many years, the deterioration of the Lightning Brothers Site known as Yiwalalay in the Victoria River District has been the subject of considerable concern. The gradual accumulation of salts which covers a significant area of the painted surface has led to the deterioration of the inter-granular bonds in the sandstone substrate. This in turn has led to the lifting of sections of the painted surface. It is proposed that management strategies be implemented to enhance the preservation of this site. Part of the multidisciplinary approach to the problem will include field trials on the effectiveness of consolidants in countering such problems.

INTRODUCTION

Yiwalalay - the site of the Lightning Brothers

Yiwalalay is situated approximately 14 km west-southwest of the new Delamere Station and is arguably the most well known site in the region for its depiction of the Lightning Brothers — Yabiringgi and Yabjagbula. It is located amongst a number of eroded sandstone outcrops and ridges in a savannah woodland (Chaloupka 1978). The site itself is a large sandstone outcrop and rises to a height of over 10 m, with the major shelter, Yanginjanawuya, located with a western aspect. Another two painted shelters exist on both the eastern and southern sides. The Yanginjanawuya shelter is well protected from the elements with its substantial overhang, however it is this shelter in which the paintings are the subject of considerable concern.

The subject of attention for some time has been the condition of the rock paintings. The site has been the subject of recording and interpretation at various stages since 1930 (Chaloupka 1978), and was placed on the National Estate Register in 1979. However, only in recent years has the site been studied from a conservation view point.

Chaloupka (1978) and Clarke (1978a) have both described the deteriorating condition of the site, and previously stressed the need for recording and conservation action. In this field tip, as well as the photogrammetric recording of the most significant part of the site, preliminary analytical work has been concluded with a series of conservation recommendations and strategies outlined in this paper.

For the Wardaman people, the site still has immense traditional significance. With the active role of this site within their culture and belief system, it is necessary that the issues addressed within this study, be considered with an appropriate degree of sensitivity and respect.

Geology

Sweet (1972) describes the physiography of this area as the Delamere Plains and Benches, made up of low round and terraced hills. These terraces have been moulded by resistant basalt layers and chert bands of underlying volcanic rock. Between Delamere and Aroona Creek, a number of north-westerly trending ridges are formed by sandstones interbedded within the volcanic sequence. Barclay and Hays (1965) have suggested that the new weathered sandstone mounds and ridges within this area are of aeolian origin and that the deposits were originally longitudinal dunes. Thin section and Scanning Electron Microscope (SEM) examination here by Clarke has confirmed the aeolian origin of the sandstone. The rock is composed of very even sized, well rounded grains of silica showing micro pitting, typical of wind blown material. The grains have very thin haematite (ferric oxide) and clay coatings giving the rock its typical reddish colours. The rock is very porous with no evidence of cementation between the grains.

Deterioration - western shelter

Salt accumulation is causing structural failures within the rock surface, particularly over the upper body to lower face area of the figure known as Yabiringgi. These salts are active both within the rock and particularly within the immediate surface. There are three major mechanisms by which this deterioration takes place :

(i) granular disintegration of the sandstone caused by salt crystals within the pore space and undergoing dimensional change with corresponding moisture changes;

(ii) salt accumulation along fractures and joint planes resulting in accelerated weathering, which may lead ultimately to the detachment of large blocks of rock;

(iii) specific salt-induced weathering associated with paintings, especially Yabiringgi, where the weathering of the sandstone is occurring immediately behind the paint layers.

A scanning electron microscope (SEM) examination of a red-coloured detached 'paint' flake from the upper part of Yabiringgi showed that what appeared to be simply a paint flake was in fact thin layers of pigment over disintegrated sandstone. Examination of the paint layer proper showed evidence of layering typical of overpainting. However, there was no significant variation in the composition of the individual layers.

Energy Dispersive X-Ray Analysis (EDXA) of the above sample showed that the major elements present were calcium and iron with minor potassium, chlorine, silicon and aluminium, with a trace of magnesium. This analysis, combined with the visible structure of the paint layers, indicates that the pigment used was a mixture of the mineral huntite (magnesium calcium carbonate), iron oxide and clay. The paint layers were porous but the very fine particle size of the huntite resulted in equally small pore spaces (in the order of several microns). The individual paint layers were relatively thin (10-15 microns) but showed considerable variation resulting from irregularities in the rock surface.

The balance of the sample 'flake' was composed of well-rounded quartz grains bonded together by the surface paint layer and an interlocking mass of salt crystals between the grains. EDXA confirmed the presence of silicon, calcium and sulphur as the major components, with minor amounts of potassium and aluminium. It may therefore be assumed that the predominant salt is gypsum (calcium sulphate dihydrate). The individual sand grains were highly etched which is typical of quartz that has undergone salt attack, and there was no evidence of any bonding or cementation between the grains other than the salt described above.

A salt sample taken from the rock surface adjacent to the Yabiringgi was examined in the SEM. These tests showed that it was composed of a porous but non-crystalline mass with a smooth irregular surface suggesting possible solution and redeposition cycles. EDXA indicated that the major element was potassium with minor sulphur, silicon and aluminium (EDXA cannot detect either nitrogen or carbon). Based on previous analytical work by Clarke (1978), this analysis is interpreted as representing predominantly potassium nitrate with minor amounts of potassium sulphate and clay or dust.

A second salt sample taken from the northern end of the same shelter when examined in the SEM consisted of a mixture of sand grains and salt crystals. EDXA indicated the major constituents to be silicon, sulphur and calcium with minor potassium and aluminium. This is interpreted as representing sand grains (silica) from the disintegrating rock mixed with gypsum and whewellite (calcium oxalate dihydrate) with minor potassium nitrate and clay.

It is of interest to note that SEM examination of a white pigment used by Mr Bill Harney in 1986 in preparing ceremonial dress established that it was also the mineral huntite (cf. Clarke 1978b). The use of this mineral, previously considered to be rare, appears to be fairly common in Aboriginal art in Northern Australia.

A sandstone sample from the exterior surface of the outcrop near Yanginjanawuya shelter was confirmed to be a typical desert varnish (Dragovich 1984; Clarke [unpubl.]). Its presence on the outer rock surface combined with a surface silicification is evidence of the long-term stability and resistance to erosion of the exposed sandstone.

Salt sources

Clarke (1978) and Lambert (1980) have reported on the accumulation of soluble salts and their damaging effects on rock art throughout Australia. There is some debate as to the origins of these salts, but it is generally accepted that they are derived from:

(i) The atmosphere:
North and Clarke (1987; 1988) have shown that at Kakadu, soluble salts, particularly gypsum and whewellite, are accumulated either directly on to the surfaces within rock shelters from the atmosphere, or by surface relations with atmospheric gases.

(ii) Rainfall:
It has been discussed by Bell (1986/87) how CSIRO research has shown that rainwater in Northern Australia can contain a wide range of cations and anions. In situations where rainwater can infiltrate through a rock mass and evaporate from dry areas (such as a rock shelter), these ions, plus any reaction products from the rock mass, will accumulate as soluble salts at the evaporation surface.

(iii) Groundwater:
Most groundwaters contain a range of ions resulting from rock weathering products, residual geological formation water etc. Again these will accumulate at evaporation surfaces in dry areas as salts.

(iv) Local sources:
The decay of organic materials, the action of biological processes (lichen, fungi, algae, plants etc.), the leaching of soils and the weathering of some rock types can all produce soluble salts that can be transported by rain or groundwater to evaporation surfaces.

At this stage there is insufficient evidence to be definitive about the exact mechanisms involved at Yiwalalay. The SEM work clearly shows that there are wide variations in the composition of the salts throughout the site. The presence of gypsum and whewellite is more typical of atmospheric sources while the nitrate is more typical of a rainwater source. It is therefore possible that more than one of the above processes is operating at the site, and that local sources as well as water movement within the rock mass will determine the mount and composition of salts at a particular location. This is supported by observations at the site where the major salt accumulation is associated with the head and body of Yabiringgi, whereas the adjacent smaller figure of Yagjagbula on the same rock surface is largely unaffected. The salts in the eastern shelter are composed of gypsum and whewellite and nitrates are absent.

There has been some discussion regarding the source of nitrate within the shelter. It has been suggested that the nitrate may have originated from the breakdown of organic materials in the floor deposit. However, Bell (1986/87) and particularly Johnson (1978) have discussed the nitrate levels in rainfall. Johnson has measured high levels of nitrates in Sydney rainfall apparently derived from endemic Australian plant species.

In addition, Martini (1980), in discussing a rare mineral discovered in Venezuelan caves called sveite - a potassium aluminium nitrate chloride, believes it is most probable that the source for the nitrate and chloride is rainwater and that the aluminium and potassium are derived from feldspar and mica as the rainwater percolates through the rock. North and Clarke (1988) have identified sveite within rock shelters in Kakadu where it is causing damage to both paintings and underlying rocks. They also believe that the nitrate has been introduced in rainfall entering the rock mass.

It is considered that the most probable source of potassium nitrate at Yiwalalay results from rainwater with elevated nitrate levels reacting with clays within the rock thereby releasing potassium. We do not believe that the floor deposit is contributing significantly to the salt accumulation and therefore do not recommend its removal for conservation purposes.

RECOMMENDATIONS AND CONSERVATION STRATEGIES

Due to the seriousness of the problem of deterioration, it has been deemed necessary to introduce recommendations on two levels: non-interventionist and interventionist. They are as follows:

A. *Non-interventionist recommendations*

It is perceived that water may be seeping down the painted rock wall from an extended horizontal fissure and that this is contributing to the salt build-up and demise of the art. It is necessary therefore to divert this water by the placement of a dripline below this fissure. During this operation some fissure caulking may be considered necessary, particularly on the vertical face located on the southern wall of the site outside the shelter. It is considered that this fissure may be diverting water into the site.

- The site should be treated by diverting surface water from the art by placing a dripline below the main horizontal fissure located above the large figure (Yabiringgi) and running the entire length of the site.

The detrimental action of the salt is clearly amplified by its consolidation in large deposits over long periods of time. By dry brush and wet poultice removal of much of this salt, the rate of deterioration may be slowed in the short term, and some quantitative study may be made regarding the rate at which the salt is delivered to the rock face.

- Visible salt present on the surface of the site should be removed by dry brushing and wet poulticing where this is practical so as not to result in figure damage.

Salt taken off the rock surface will be collected and removed from site. It is considered that salt left on the site may be redistributed by way of either water seepage or condensation.

- The site should be subjectively monitored both visually and photographically after 12 months for salt re-crystallisation.

This work can be carried out during other scheduled rock art recording work planned for the area.

If salt re-crystallisation is evident, then an alternative (and probably more expensive) strategy will need to be formulated. It is considered that the strategy outlined above for salt treatment is the most likely to succeed; low cost, practical, and will have the least visual impact on the site. If it fails, the dripline can be either modified or removed, and the first deterioration model will have been eliminated.

During 1987 Mr Bill Harney constructed a fence to protect the site from stock. This fence was not made with a gate or formal entrance point, which may lead to excessive vehicular movements around the site by visitors looking for an entrance. This would generate considerable dust and cause parking in certain sensitive areas. To build a stile over the fence would be to formalise an entrance, and a simply constructed barrier near this should keep vehicles from driving around the site, enhancing the management of visitor movements around the site.

- An entrance into the enclosure should be formalised by way of a stile or gate.

A barrier should be erected in a suitable and adjacent area, to act as a boundary for a car parking area.

With increased visitation to the site, the possibility of damage to the site from vandalism becomes a very real threat. Already in the eastern shelter there are scratched initials (which subsequently, were partially removed by Dave Lambert). To deal with such incidences, the usual practice has been to place a visitors book and make specific site information available to the visitor, so as to foster greater awareness of the site's importance.

- An official sign and visitors book should be placed appropriately.

Before implementation however, these matters should be discussed with the Wardaman people, in the event that this type of development may not be in accord with their wishes or traditional use of the site.

The shelter that the site offers to stock and native animals has ensured that whilst the site was not fenced, the area surrounding the outcrop has been preferentially grazed. With a secure fence now in place, excessive vegetation growth due to the exclusion of stock could at some stage prevent a fire hazard.

- The Yiwalalay sites should be monitored for excessive vegetation growth on a regular basis, and cleared by hand if necessary to eradicate a possible fire hazard.

The vegetation should be cut by hand and removed from the site, reducing the threat of fire damage. No such problem was evident at the time of inspection in September 1987, as the fence had only recently been installed.

In the southern shelter, wasp mudnests pose some threat to the paintings present.

- The wasp mudnests should be removed from the southern shelter.

B. *Interventionist Recommendations*

On the basis of the SEM examination it is clear that the deterioration of Yabiringgi is caused by a salt-induced failure of the rock substrate immediately beneath the paint layer. In fact, the lower porosity and permeability of the paint layer may be at least partly responsible for this deterioration. On the areas without paint layers the salts have been able to escape from the rock and accumulate on the surface with little apparent damage to the rock mass. However, where the paint layer has prevented the salts from reaching the surface, serious granular disintegration has occurred.

Irrespective of the success or otherwise of the non-intervention procedures, we are left with a serious problem in those areas that have already been disrupted by the salt accumulation. Without a consolidation treatment these areas will continue to deteriorate and in the longer term up to half the figure of Yabiringgi may disappear.

North and Clarke (1987) have encountered similar

problems at Kakadu where salts have accumulated within the relatively thick paint layers of much of the post-estuarine paintings. A series of trials to examine the effectiveness of various synthetic consolidants and surface treatments have now been established.

- It is recommended that trials similar to those currently established in Kakadu, using the most promising consolidant products, be established on suitably salt-affected painted rock surfaces around Delamere.

Preferably, a nearby site with similar salt deterioration problems should be found so that Yiwalalay can be avoided as the trial site.

These trials would of course need to be fully discussed with the traditional custodians and their approval of both the concept and of individual trial sites would need to be obtained.

It has been suggested that the placement of a roof over the site might stop the entry of water and hence positively alter the salt deterioration pattern. Alternatively, a screen could be placed so as to shelter the afternoon sun from the painted wall, limiting the evaporation regime and creating more consistent microenvironmental conditions. However, in addition to such a dilemma as changing the fabric of the site and the subsequent effects on the site's integrity, there are problems of visual impact, cost, logistics and maintenance. Even with the construction of a roof, water may still enter the shelter via fissures outside the roof area, or alternatively if the site did dry out there could be deleterious side effects that are as yet unknown.

Bearing all this in mind, it is recommended that at this stage the question of a roof or screen construction be deferred pending the results of the above-mentioned treatments.

Conclusion

Chaloupka (1978) describes Yiwalalay as an important link between the Kimberleys and the Arnhem Land Plateau in the depiction of figures associated with rain and lightning. On top of this, he adds, it is a very important stylistic group located between these two regions.

At the present rate of deterioration, if a conservation program is not implemented quickly, a considerable portion of the Lightning Brothers could well disappear. This would be an irreplaceable loss.

There have been numerous discussions with the Wardaman people, at which they have expressed concern for the site and its current deteriorating state. However, in our attempts to examine the major problems of this site we have simply looked through a European perspective at conservation practices involving conventional techniques. At some stage the Wardaman people may take it upon themselves to consider the regular repainting of the site. Whatever is to happen in the long term, it is clear that the final decision should lie with the Wardaman people themselves.

REFERENCES

BARCLAY, J. and HAYS, J. 1965. Minor investigations by the Northern Territory Geological Section: water investigations at Victoria River Downs and Wave Hill 1963. Bur. Mines. Aust. Rec. (unpubl.)

BELL, A. 1986/87. It rains formic acid in the Top End. *Ecos* 50. CSIRO Summer 1986/87, pp. 28-29.

CHALOUPKA, G. 1978. Rock art deterioration and conservation in the 'Top End' of the Northern Territory. In *Conservation of Rock Art*, C. Pearson (ed.), ICCM.

CLARKE, J. D. n. d. Rock patination and the age of Aboriginal engravings at Dampier, W. A. (unpubl.).

CLARKE, J. D. 1978a. Rock art conservation, W. A. Museum (unpubl.).

CLARKE, J. D. 1978b. Deterioration analysis of rock art sites. In *Conservation of Rock Art*, C. Pearson (ed.), ICCM.

DRAGOVICH, D. 1984. Varnished engravings and rock weathering near Broken Hill, western N.S.W. *Australian Archaeology* 18: 55-62.

JOHNSON, M. 1978. Salt aerosols and rain in Sydney. In *Maintaining and Restoring Masonry Walls*, G. S. Gibbons (ed.), pp. 72-79. National Trust, N. S. W.

LAMBERT, D. 1980. The influence of groundwater salts in the formation of sandstone shelters near Gosford, N. S. W. *ICCM Bulletin* 6 (3, 4): 29-34.

MARTINI, J. E. J. 1980. Sveite, a new mineral from Autana Cave, Territorio Federal Amazonas, Venezuela. *Trans. Geol. Soc. S. Afr.* 83: 239-41.

NORTH, N. and CLARKE, J. D. 1987. Conservation of Post-Estuarine Period rock art in Kakadu National Park. Report on phase 1 study: Pigment Identification, ANPWS (unpubl.).

NORTH, N. and CLARKE, J. D. 1988. Conservation of Post-Estuarine Period rock art in Kakadu National Park. Final report on phase 2 study: Conservation Treatment Trials, ANPWS (unpubl.)

SWEET, I. P. 1972. Delamere-Northern Territory - 1:2,500,000 Geological Series. Bur. Mines. Res. Aust. Explan. Notes SD/52/16.

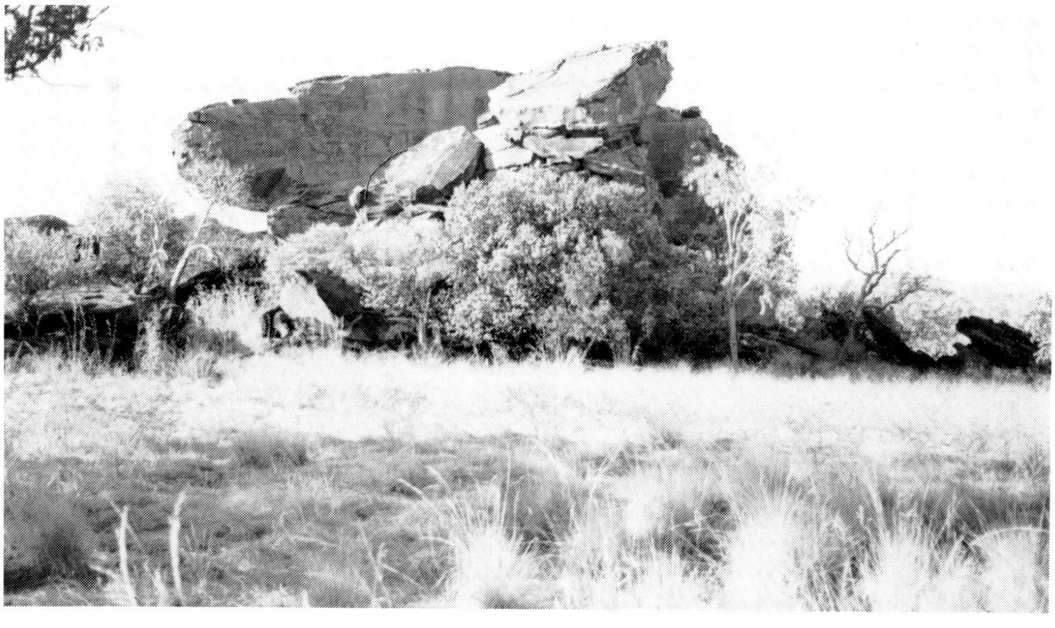

Figure 1. ***Yiwalalay, Northern Territory. (Photograph by J. Clarke)***

16

ON SOME NON-TECHNICAL ISSUES OF CONSERVATION

Ivan Pavel Haskovec

Abstract. This paper discusses conservation as a social phenomena in relation to the preservation of rock art. Any decision as regards conservation treatments must take into account not only conservation ethics but also, and more importantly, the views and wishes of the owners of the culture and the art.

INTRODUCTION

This paper is not a paper on technical aspects of preservation or a description of conservation practices. In this paper I shall attempt to discuss two issues:

1) some ethical issues associated with conservation; and
2) criteria for evaluation of some conservation experiments.

ETHICAL ISSUES ASSOCIATED WITH CONSERVATION

Conservation as a social phenomenon

I find the need to discuss conservation as a social phenomenon since we all work on the understanding that conservation is good, also on the understanding that by prolonging life of material culture we derive greater pleasure from the objects and that by exercising conservation practices, we in a sense exercise humanity itself.

The AICCM Code of Ethics and Practice, and the Burra Charter as an Australian chapter of the ICOMOS code are in agreement about the issue of conservation and are interpreted by all practitioners as good codes, articulating the good act of conservation. On the almost Aristotelian assumption that by conserving we are creating greater good, we are ever increasing the scope of conservation and find a need to conserve an ever greater variety of objects including prehistoric rock art.

Of course as the number of objects increases and the methodology of conservation becomes ever more complete, demands on resources, human and financial, is ever greater. I often think that this is a social entropy in action, as an ever greater proportion of social energy is required to maintain and conserve society and social institutions. Consequently, an ever smaller proportion of social energy is left to break new grounds of humanitarian growth. This is a serious issue, since the available human resources are not growing at the rate of the growth of perceived conservation needs. It appears that we, the conservators have to take some responsibility for the situation and use human and financial resources in the most efficient way.

My personal attitude is that studies and experiments should be conducted with clear research design and practical ends in mind, and not as an elegant exercise of complex methodology without any clear goals. Such activity may be not only futile but also a wasteful exercise. In some cases this may endanger the integrity of objects themselves. I shall refrain from citing examples, but there are a number of them in the literature and in living memory. The waste of an intellect and its energy in the process is alarming.

Conservation as an ossification agent

A main problem with 'goodness' of rock art conservation lies, it appears, in it's ability to discourage and suppress creativity. The issue may not be as important within national states as seen in Europe (i.e. France, Germany, Spain, Czechoslovakia) where the objects are conserved within the shared attitude to material culture. The issue is very important in multinational and multicultural states such as USSR, India, South American and North American states, Australia, New Zealand, etc., where the centralised institutions are applying the 'conservation-is-good' measure to all material culture of all ethnic groups equally, and are in the process of limiting cultural activities and creativity of the ethnic groups, particularly when these can interfere with the perceived 'goodness' of conservation.

I am referring specifically to the Australia Aboriginal/non-Aboriginal interaction where it is the increasingly non-Aboriginal attitude to Aboriginal art and 'what-really-is-Aboriginal', that commands the Aboriginal production of artefacts such as rock art, and selection criteria for their conservation. The following example from Kimberley (West Australia) is a perfect case study.

The Commonwealth of Australia has a job-creation scheme under which many things can be done. In this case, a rock painting has been decided upon by a European adviser and some Aboriginal people. After some consultations with elders, young Aboriginal people overpainted some older paintings, and painted new paintings at some sites. This caused commotion in Aboriginal and non-Aboriginal communities and many different opinions have been expressed, ranging from unconditional support to total rejection of the action. The paintings were not executed in any of the older styles, although the mineral paints were traditional. Some people questioned the appropriateness of the job-creation programs, of the European involvement, painting style, kind and type of consultations, 'traditionality', quality of art, etc. Reports were written about the paintings by non-Aborigines. TV shows reported on the paintings, judgement was made by some non-Aborigines as well as some Aborigines, and as a result some of the paintings were removed by non-Aborigines.

Now, in some Aboriginal traditions, paintings are

painted, repainted, retouched, painted over, covered up or left alone. We would not have a good 60% of the paintings had Aboriginal people adopted our conservation ethics. One has to wonder what impact this advisory and conservation attitude of one part of the population has on the cultural life of the other part of the population.

The Kimberley's example is not a unique one. When I invited a local Aboriginal artist to paint a painting in Kakadu, his first reaction, to my horror, was to repaint some of the older ones. I managed to dissuade him from it, but I was particularly concerned that I had initiated the painting for a planned treatment experiment and the action was not spontaneous.

So what in these two examples of practical application of the 'conservation-is-good' attitude have we done? We can say that we acted within the ethical guide-lines and conserved prehistoric rock art. What we also did was to impose our well reasoned and considered conservation ethics and views about our advisory role on people who do not necessarily share these views or ethics. Taking the line of some British, French and other institutions which hold a number of Egyptian and Greek artefacts, we can say that by conserving we are assuring survival of the artefacts. But unlike conservation of Egyptian and Greek artefacts which does not stop or suppress activity of today's Greeks and Egyptians, our activities in the field of art practice and conservation may in fact interfere with the cultural life of people who do not share our 'conservation-is-good' attitude, and who may see the perpetuation or change of their culture in different ways.

The issue of rights of use is returning with a frequency in various versions, be it as the right to use Stonehenge by Druids, or conferences on who owns the history etc. I do not claim to know the answers, however, I would like to make us all aware of the fact that side-effects of interference with rock art are not restricted to physical and chemical changes in the rock art. I also like to think that the issue of rock art conservation as a social phenomenon will be taken up by the Australian Institute for the Conservation of Cultural Material (AICCM).

CRITERIA FOR EVALUATION OF EXPERIMENTS

The second half of this discussion paper is focused on conservation standards, specifically on the articulation of criteria for evaluation of conservation experiments and procedures, proposing some criteria for the evaluation of conservation experiments in consolidation and stabilisation of rock art.

In 1988, I went to the ICOM Committee for Conservation Triennial Meeting in Sydney, where I attended sessions on conservation of gallery placed art, oil paintings, statues, etc. I left with an impression that even in the controlled environment of galleries where special lights, humidity and temperature manipulation are standard procedures, some aesthetical changes associated with conservation are not only inevitable but accepted and catered for.

The issue of standards and criteria came to a watershed when a chemical company wrote to me asking what results are we expecting from various treatments and consolidants. I had to then start to think what would be acceptable results in the conservation of rock art. Oil paintings and organic objects stored in museums and galleries are, for example, expected to undergo some colour shift due to inherent instability of conservation materials used. Or, highly polished or glazed surfaces seem to be a desired property, which improves optical sharpness of objects and their details.

I have searched rock art literature, but did not find any discussion of acceptable or desirable criteria of conservation procedures, although everybody appears to work towards these standards within unstated criteria.

The AICCM code of ethics states that '*all actions of the conservator must be governed by an unswerving respect for the* physical, historical and aesthetical integrity *of the objects*'. (AICCM 1986: 4).

Obviously, without interpretation and some discussion of expected standards of '*physical, historical and aesthetical integrity*' these words have no meaning.

To start with, we must accept the fact that rock art is undergoing a change regardless of conservation intervention, and that conservation procedures will affect these natural processes and that whatever we do will alter the cast history of a rock art site. Now, accepting that '*physical, historical and aesthetical integrity*' of rock art must be respected, what changes are acceptable and what criteria are we going to have and to observe?

In the case of Kakadu National Park we must observe first of all the Aboriginal wishes in regard to rock art and only then try to incorporate our views into the Aboriginal perception of '*physical, historical and aesthetical integrity*' of rock art.

These views luckily for us to a large degree coincide. On occasions Aboriginal people request that the surrounding of an art site be maintained as a living area, since the immediate environment of an art site is seen as an integral part of the site. On other occasions we may be asked to carry out preservation procedures as a matter of priority, or not to approach a site at all. Or we may undergo a small ceremony prior to entering a site, not to disturb its cultural integrity. In this way Aboriginal people can and do assess and maintain the *historical and cultural integrity* of the rock art.

We have in our experimental work decided to concentrate on the development of suitable treatments and stabilisation procedures. Consequently, some criteria for acceptance of experiments have been designed and are monitored for. These criteria are designed in regard to the aesthetical but more importantly in regard to *physical integrity*. The historical and aesthetical integrity will be addressed later, when we will be ready to implement the program of stabilisation treatment.

(Perhaps I should first clarify that since we did not find any binders in the mineral paints used by Aboriginals during the analytical stage of the work, experiments with binders were not included. However, in the first stage of the research we identified rain and ground water, salts and atmospheric sulphur to be the major causes of rock art deterioration in the region, we therefore concentrated on experiments with treatments counteracting or minimising the effects of these causes.) We have treated an experimental, newly painted site, as well as a number of derelict art sites in an attempt to provide a broad representative sample. The work was carried out with agreement and under direct supervision of Aboriginal owners. The sites are now monitored for any changes and processes affecting the site.

The criteria for evaluation of a consolidation treatment are as follows:

a) One criteria is related to the *visual quality* of the treatment, where a matt, non-reflective clear appearance of the rock art surface is a desirable and necessary quality. Glazed surfaces such as seen on photographs or oil paintings may improve the optical quality of the painting, where the light would be reflected uniformly, thus deepening the hue of colours and optically sharpening the divisions between colours. However, it would interfere with the aesthetical qualities of the essentially dry, dusty appearance of most paintings. Some older paintings covered by natural silicate skin have the deeper hue. These are however, not in a need of conservation. (A temporary deepening of colour used to be achieved by wetting the painted area with water for photography. I have just discovered that this harmful practice has not been discontinued at some places.)
b) Another criteria is *colour stability*. It appears that a small degree of tolerance exists for these reasons:
   1) We do not know the original colour scheme of 98 per cent of the paintings and therefore cannot attempt to approximate this scheme.
   2) The scheme of polychrome paintings is based on inter-related positions of red, white, yellow, black and blue colours within a painting. We have not observed a painting where gradual changes in hue of colours would be employed as a part of the composition of a painting.
   3) Daylight, unlike controlled artificial light changes its colour temperature, depending on the time of day, cloud cover, season, surrounding vegetation, etc. The perception of colours in rock art exposed to natural light is therefore constantly changing.

The difficulty may arise with a *selective shift* of colour hue in some of the colours in the painting, where due to different chemical composition and size of grains, some colour(s) such as red (iron oxide) may selectively react with consolidants and upset the balance of the colour scheme.

c) *UV Stability* in our subtropical climate, with almost uninterrupted sunshine for eight months a year, is of paramount importance. The sunlight intensity is extremely high and this in combination with reflected light, affects physical and chemical stability of all materials.
d) *Chemical stability* is also very important, since some amount of salts, particularly the sulphur-based salts, are deposited on rocks. Early wet season rains are quite acidic which require consolidants to be acid-resistant.
e) *Surface hardness*. Some consolidants may soften slightly by heat and trap and absorb dust particles, thus progressively obscuring the painted surface.
f) *Depth of penetration* is an important criteria, which is difficult to evaluate. Some rocks are very porous, some non-porous; porosity varies even within the same rock, and to test for absorption under paintings is impossible. However, it is clear that minimal penetration requirement would include the painted layer and the weathered layer of the rock surface.
g) *Water permeability* is necessary to allow the water inside the rock reaching the surface and not to build up pressure behind the art which could exfoliate a whole area.
h) *Flexibility*. A degree of flexibility is necessary for consolidants to expand and contract with the rock.

These are the criteria for the physical evaluation of our experiments.

Often the point of *reversibility* is raised as a most important aspect of conservation treatment. Since conservation is an interventionist process, the theoretical view that a process can be reversed is not attainable. Also, if it can be reversed, it will most likely be reversed by natural processes and therefore inefficient.

A second point that the procedure should *not interfere* with possible future treatments is also partially unattainable. It demands a knowledge of possible future processes, which are as yet unknown.

We therefore focused on treatments by such materials, eventual *decomposition* of which would not cause or accelerate deterioration of the rock art.

The object of this exercise is to select a small number of technically suitable materials. However, the ultimate selection of the treatment or materials (if any) will be a choice made by Aboriginal owners. Our role in this exercise is to prepare suitable *technical options* and execute one treatment(s) selected by Aboriginal owners, if they wish to proceed with a treatment.

It seems to me, that the Aboriginal owners are, within the Aboriginal tradition, better able than we could ever be to judge whether an offered treatment affects the historical and aesthetical integrity of the rock art. In view of this issue our role as conservators is restricted to preparation of several suitable options of physical treatment.

CONCLUSION

The two discussed issues, (1) conservation as a social phenomenon and (2) criteria for selection of conservation treatments, appear to be disjointed aspects of conservation, however, in both issues I seem to have reverted to the owners of the culture and the art as the final arbiters of the issues.

It may be that while we constantly search for technical excellence, we may not be developing the philosophical and ethical issues of conservation fast enough to keep up with the technical development of conservation.

REFERENCES

AICCM Australia (1986) *Code of Ethics and Guidance for Conservation Practice.*

ICOMOS Australia (1981) *The Australia ICOMOS Charter for the Conservation of Places of Cultural Significance* (*The Burra Charter*).

17

TRAINING FOR ROCK ART CONSERVATION:

Recent initiatives of the Getty Conservation Institute

Nicholas P. Stanley Price

Abstract. The deterioration of rock art is a world-wide phenomenon but there exist very few conservators qualified to treat it. The Getty Conservation Institute has attempted to tackle this problem by two different routes: by helping to establish a year-long specialised course in rock art conservation, and by encouraging already qualified conservators to extend their expertise to include rock art. The paper describes the conception and implementation of a two-week course designed to achieve the second aim, and places current initiatives in the context of long-term needs in rock art conservation.

Introduction

It is generally recognised that there is great shortage of people qualified to carry out the conservation of rock art. Here I mean 'conservation' in a technical sense. Often the term 'conservation' is used for policies or practices that I think of more in terms of the protection or management of rock art sites - itself an important objective. I am using 'conservation' to refer to a technical discipline based on a combination of humanistic and scientific knowledge that contributes to prolonging the life of works of art.

Considering the range of humanistic and scientific disciplines on which the conservation of rock art must draw, it is not surprising that few specialists exist and that no formal training opportunities have been created. A rock art conservator requires not only a sound training in conservation theory and procedures, but also a good understanding of such sciences as geology, petrology, hydrology and chemistry.

The lack of development in this area is sometimes attributed in part to the slow acceptance of rock art studies as a serious contribution to archaeological research. However, even where there exists a tradition of serious rock art studies, conservation has not developed in parallel fashion (France constitutes an exception in this respect). The reason must lie rather in the failure of the disciplines of anthropology/archaeology and conservation to communicate with regard to rock art.

Recognising that rock art is one of the under-developed areas of conservation, the Training Program of the Getty Conservation Institute is attempting to increase the number of qualified rock art conservators. By drawing on the expertise that does exist to provide training opportunities, it is hoped that these will stimulate other institutions to incorporate rock art conservation in their syllabi or to establish their own specialist courses.

Current initiatives by the Getty Conservation Institute

The GCI is approaching this task by following two parallel paths. One is by establishing professional training opportunities for potential rock art conservators. In cooperation with the Canberra College of Advanced Education in Australia, it has established a one-year Graduate Diploma in the Conservation of Rock Art, which began in February 1989.

This Graduate Diploma was advertised internationally. Of the fifteen participants selected to follow the course, seven are from outside Australia. Unfortunately one overseas participant had to withdraw at the last moment. Once qualified and working in rock art conservation in their own countries, these should provide a nucleus for organising regional training opportunities in this field. (The Graduate Diploma course is the subject of a separate paper in this volume by Colin Pearson.)

The other approach being followed by the GCI is to encourage professional conservators to extend their area of expertise to include rock art. If the audience consists of conservators already trained in the theory and practice of conservation, a short course concentrating on the particular characteristics of rock art conservation can afford to be specialised in its content.

It was with this aim that the GCI organised a course that took place in April, 1988, in California. Of the twelve participants selected for the course, ten had formal training in conservation, several of them in the conservation of stone and mural paintings. Five came from outside the United States - from Argentina, Australia, Malaysia, New Zealand and Sri Lanka. As with the Canberra Graduate Diploma course, it is hoped that these will help develop local training schemes in their own countries.

While efforts are being concentrated on training conservators, the management and protection of rock art sites has not been overlooked. This task tends to be in the hands of archaeologists employed by national and state agencies responsible for the management of archaeological sites, of which rock art sites are merely a special case. Another element of the potential audience for training in this area are those rock art researchers who become concerned about the preservation of those sites that are the object of their study.

In order to reach this audience, the GCI organised a one-week course on the management and protection of rock art sites in April, 1989. The course concentrated on site and visitor management and interpretation techniques. The principal instructor was Sharon Sullivan of the National Parks and Wildlife Service of New South Wales in Australia. Twenty-five participants, either archaeologists from Federal and State agencies or rock art researchers, were selected from twice that number of applicants.

The GCI course for conservators in April, 1988

With this summary of current initiatives by the GCI, the rest of this paper describes in more detail one only of them, namely the course designed for conservators held in April, 1988. An earlier course organised at Valltorta in Spain by the GCI in cooperation with the Generalitat de Catalunya in April 1987 had helped to defined the appropriate subject matter for a course of this kind. The Valltorta course was aimed primarily at French and Spanish rock art specialists and drew upon the work on conservation of rock art in caves and shelters in those countries.

The California course, on the other hand, gave emphasis to open-air rock art. It was planned as a program of two weeks, the first being devoted to teaching in the lecture room and laboratory, and the second to practical work at a rock art site. It was designed to introduce the cultural context and ethics of rock art conservation, and then to consider in turn mapping, geology and hydrology, stone deterioration, analytical techniques, documentation and intervention. In the second week the lessons learned from the formal teaching were applied to the rock art site under study.

The week's fieldwork took place at the site of Painted Rock, a well-known site of the Chumash Indians in central southern California, some 300 km from Los Angeles (see Plate 1). The site was selected because it exhibits a range of deterioration problems suitable for study and analysis, and because it was accessible from Los Angeles.

Since the course was to be advertised internationally, the instructors were chosen so as to provide experience of rock art conservation from different parts of the world. Andrée Rosenfeld of the Australian National University introduced some of the ethical issues of rock art conservation, drawing on recent work on Australian Aboriginal rock art. She also reviewed the historical development of rock art techniques and materials since the Palaeolithic.

Jean Vouvé from the Université of Bordeaux taught the sections devoted to mapping, geology and hydrology, drawing on his extensive experience of the French Palaeolithic sites, and Isabelle Dangas described the documentation and intervention techniques that she uses in her work in France. Connie Silver, one of the few conservators working on rock art in the United States, also spoke on documentation and intervention. Stone deterioration and conservation were covered by Norman Weiss (Columbia University), who showed how much of the recent work in this field is relevant to rock art preservation.

In the first week's program, practical exercises and visits complemented the formal teaching. A visit to the Scientific Laboratories of the GCI illustrated the analytical techniques previously described by Norman Weiss, and a visit to the Rock Art Archive in UCLA emphasised the importance of documentation and maintenance of records.

The practical exercises consisted of examination of stereoscopic aerial photographs and maps; microscopic and thin-section identification of rocks; and water-absorption tests on samples of the sandstone from the site under study. Air photos and a satellite photograph of the area of Painted Rock were available, and were subjected to interpretation as a preliminary to the fieldwork to be carried out on the site during the following week.

A further exercise was designed to simulate the problems of graffiti removal from a vandalised site such as Painted Rock. Each participant was given a sample of the sandstone from the vicinity of Painted Rock, and invited to paint or engrave it. The paints available were of the type that had been available to the indigenous inhabitants of the area - yellow and red ochres, charcoal and a white diatomaceous earth. The diatomaceous earth proved more difficult to apply than a commercial kaolin that was also provided for comparative purposes. Participants could choose from different binders - water, animal glue, carrageenan, plant gum and vinegar.

The resulting works of art, as some of them proved to be, were then sprayed with ethyl silicate to simulate mineralisation over time. Three days later, the decorated rocks were deliberately vandalised, using a variety of scratching and painting techniques, including house paint, marker pen, lipstick, chalk and Krylon spray paints. The second exercise consisted in attempting to remove the graffiti and paint by testing different solvents. The exercise proved how variable success can be whatever methods are employed.

In preparation for the second week of the course in the field, participants were introduced to the cultural content of Chumash rock art. Georgia Lee lectured on Chumash culture and rock art, and participants visited another Chumash rock art site at Painted Cave near Santa Barbara, and the gallery of Chumash culture in the Santa Barbara Museum of Natural History.

Field studies on site at Painted Rock

The site of Painted Rock consists of a prominent sandstone outcrop of irregular circular outline c. 90 m in diameter. It encloses a central hollow open to the north (see Plate 1). Much of the surviving art is on the walls of this central hollow. The site lies at the juncture of open grasslands and scrub-covered hills in an area notable for its wildlife - there are nine endangered and/or threatened wildlife species on the plains north and east of the site, the highest concentration in North America.

The rock art is attributed to the Chumash Indians who inhabited this area and is considered to have been executed within the last five hundred years, though no direct dating has yet been possible. The paintings have much in common with Chumash art styles on other sites nearer the Californian coast, and make use of red, yellow, black and white colours.

The paintings have been very badly vandalised with paint, chalk, engraving and other forms of graffiti. Much of this vandalism took place in the first third of this century, although the earliest dated graffito is in the 1880s. In the opinion of Campbell Grant, doyen of Chumash rock art studies, the paintings must once have been among the most spectacular in North America.

The course's practical work concentrated on a condition survey of selected panels. Each group of three participants worked on an area of c. 4 square metres (see Plate 2). It had to produce a base drawing using survey techniques, on which were recorded the main features of the rock and of the art represented upon it. On the first transparent overlay they recorded surface condition features such as exfoliation, loss of adhesion, loss of cohesion, abrasion and incised graffiti. On a second overlay, they recorded superficial phenomena, including lichen and bird guano, painted graffiti and accretion deposits. The nature and occurrence of lichen at Painted Rock were described by Charis Bratt of the Santa Barbara Museum of Natural History.

The aim of this condition survey exercise was to teach participants to observe closely a decorated rock art surface and to record visible features, whether natural or cultural in origin. Only after this could analysis of the causes of alteration be approached. This exercise took more than three days to complete but it resulted in valuable documentation drawings. At the end of every day, each group had to summarise their documentation work and explain their drawings.

Parallel with the exercise Jean Vouvé conducted investigations of the geology, hydrology and microclimate of the site, and summarised his preliminary observations in periodic talks to the group. He stressed a methodology to adopt when approaching any rock art site, one that depends on very simple techniques of measurement to produce information of value. In this way, instruction in the field alternated between the macroscopic view of the total site in its environmental context, and the smaller-scale approach required by the detailed condition survey drawings. This proved to be an effective teaching procedure.

On the final day and a half of fieldwork, a methodology and some techniques of intervention were demonstrated concentrating on the consolidation of sandstone and the removal of undesirable accretions such as bird guano, mud and painted graffiti. A limited number of tests were carried out by the conservators in the group on areas of rock that did not contain original art.

A large area surrounding the site of Painted Rock had been acquired shortly before the course by a consortium of interests, led by the Nature Conservancy and the Bureau of Land Management. With the prospect of its future protection thereby much improved, management needs for the site were the subject of on-site discussion by the participants. Andrée Rosenfeld described various options, based on experience with site and visitor management in Australia. These were in turn presented to a larger audience on the closing day of the course. A number of people who have long been concerned about the preservation of Painted Rock were invited to attend the closing day to have explained the aim of the course and to contribute their suggestions.

Retrospect

In conclusion, this two week course succeeded in introducing a group of conservators to the particular problems of rock art conservation and to a methodology for approaching them. Two weeks of specialised training do not qualify them to be rock art conservators; for that more training and much practical experience are necessary. But they are in a position now to give advice on the conservation of rock art sites, not least to say what solutions are not to be recommended.

This represents a notable advance, given the shortage of specialists in this field and given the number of well-intentioned but disastrous interventions made by non-specialists to remove the results of vandalism at rock art sites.

The particular case of Painted Rock has served well as a venue for field training both for conservators - as on the course described above - and for site managers: it provided the principle case-study for the recent course on rock art site protection and management held at the GCI. In the meantime, measures to improve access and management of the site have already been implemented by the Nature Conservancy, and these will be further developed by the Bureau of Land Management which has taken over responsibility for the site.

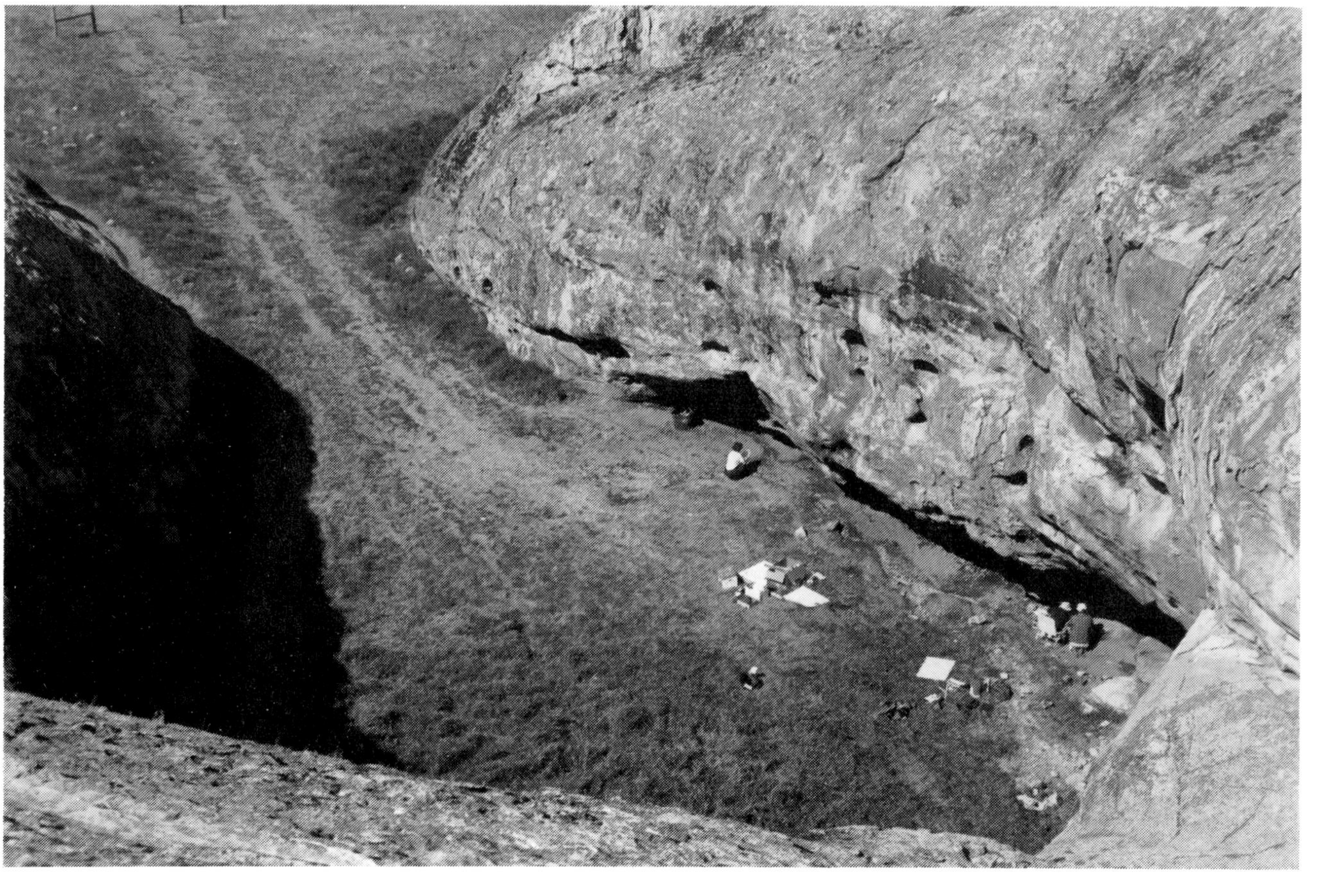

Plate 1. The central area of Painted Rock site, California.

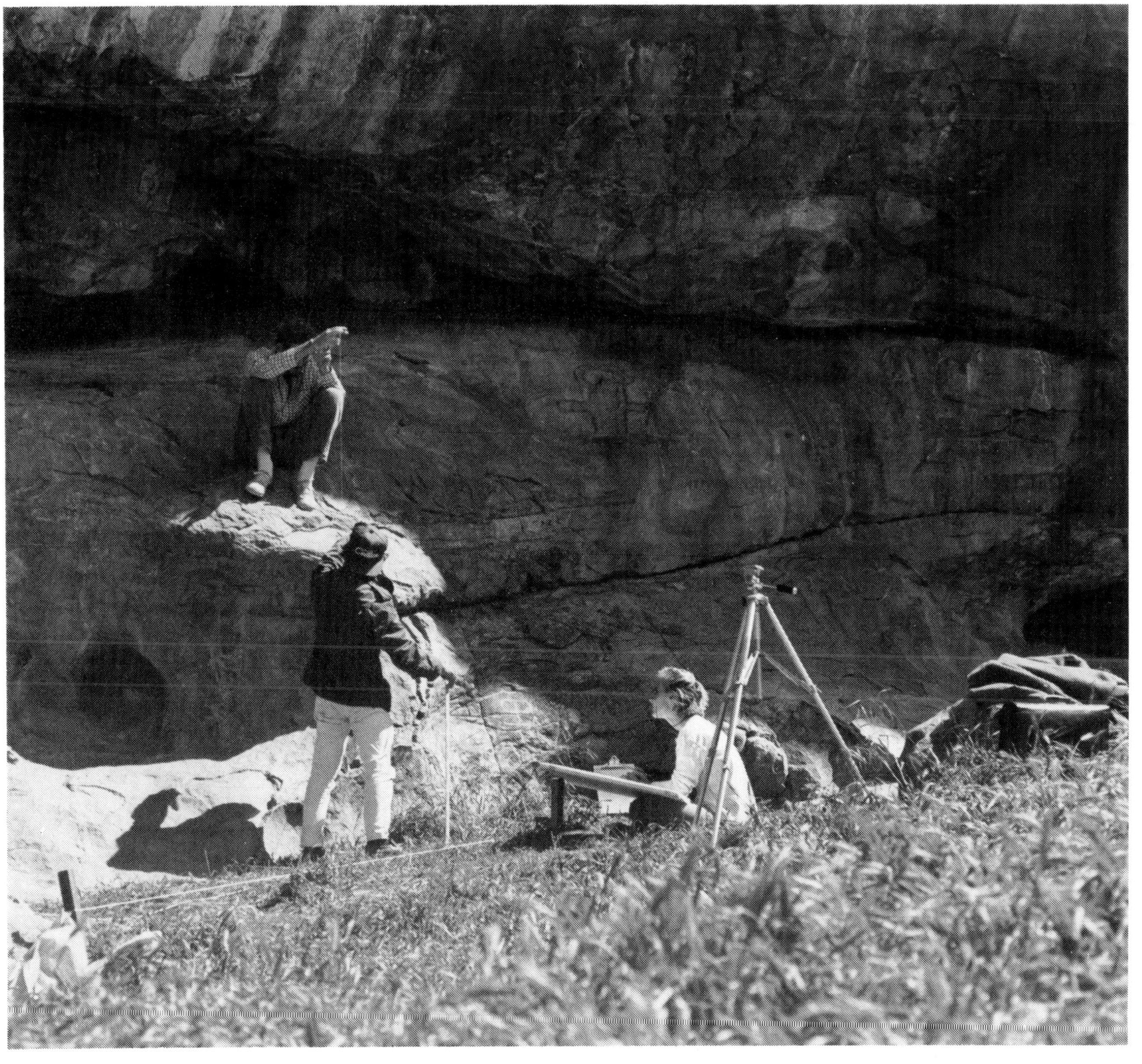

Plate 2. *Course participants preparing condition documentation of rock art panel.*

18

GRADUATE DIPLOMA IN CONSERVATION OF ROCK ART

A Joint Project between the
Canberra College of Advanced Education and The Getty Conservation Institute

Colin Pearson

Abstract. Apart from short courses there is no formal training available in the conservation of rock art. In 1989, a graduate diploma in the conservation of rock art is being offered jointly by the Canberra College of Advanced Education and The Getty Conservation Institute, and will be held at the CCAE in Canberra. This paper discusses the aims and objectives of the course, with details of the units being offered. Although based on Australian Aboriginal rock art, the course will cover rock art on a global basis, and is intended to attract overseas participants.

(1) *Introduction*

Rock art is one of the world's richest cultural resources, depicting as it does some of man's earliest expressions of beliefs and ideas. It combines art often of the highest quality, with significant scientific and cultural values, particularly in countries such as Australia where this art is part of the living cultural heritage of its Aboriginal people. This invaluable resource is at continuing risk everywhere because of the natural agencies of wind, water, insects and vegetation, and because of human activities associated with farming, industry, tourism and development. A professional-level course to train persons to understand the causes of deterioration of rock art, and to conserve, preserve and manage it, has long been overdue. There is an urgent need to train rock art conservators to preserve rock art on a world-wide basis.

Until this course there was no training course at tertiary level in the conservation of rock art at any institution inside or outside Australia. The demand for conservators on a world-wide basis has been recognised by The Getty Conservation Institute (GCI); and UNESCO has indicated a need for training in conservation of rock art in accordance with the priorities given by the Organisation to the preservation of world heritage.

The Australian Joint Academies set up a Joint Consultative Committee for the Preservation of Prehistoric Places; and in conjunction with a consequent proposal for the establishment of a Conservation Resource Unit for Aboriginal Rock Art, made a request to the Canberra College of Advanced Education (CCAE), to introduce a Graduate Diploma in Conservation of Rock Art.

In 1988, following negotiations between the GCI and the CCAE, an agreement was signed to offer a joint course at Postgraduate diploma level (i.e. postgraduate in time, standard and content), in the conservation of rock art. The course offered in 1989 is of one years duration (February to November) of full-time study and is located in the National Centre for Cultural Heritage Science Studies (NCCHSS).

Initially an Australian Course Advisory Committee was established to prepare a draft of the outline and contents of the course. The Committee had a membership of 17, representing all aspects of management, recording and conservation of rock art activities in Australia. There was also involvement from Aboriginal people. A Board of Studies was then established with representation from the GCI and CCAE, plus two Australian and two overseas members (the latter being from the USA and France). The Board of Studies finalised the Course Accreditation document and is over-seeing the progress of the course.

(2) *Objectives of the course*

The aim of the course is to draw on existing knowledge and experience to provide specialised training in the conservation of rock art.

Its particular objectives are to:

- develop standards of professional practice and conservation ethics, taking into account the views and concerns of traditional landowners and custodians;
- provide an understanding of rock art as material culture on a world-wide basis;
- provide an understanding of the methods of producing rock art and the causes of its deterioration;
- provide training in the theory and practice of recording and conserving rock art;
- develop an awareness of the management requirements of rock art sites, and
- develop an ability to plan applied research in the field of rock art conservation.

(3) *Entry requirements*

For entrance into the graduate diploma applicants require a degree or equivalent in a relevant field such as earth sciences, prehistory, anthropological or social sciences, conservation, rock art studies, or cultural or natural heritage management. Since an understanding of materials science is basic to conservation, applicants need to have passed chemistry at first-year university level or equivalent. There are special admission provisions for indigenous people who do not meet the normal College entrance requirements.

Applicants are required to provide details of their training and experience and explain why they wish to study rock art conservation. Their application has to be supported by two referees. Where possible, all applicants who meet the entry requirements are interviewed for admission to the course. The language of the course is English, and overseas students have to show a proficiency in English to be admitted.

(4) *Course structure*

The academic requirements for the course are 24 credit points from 8 postgraduate units. Students will normally be expected to take all 8 units, however, depending on previous training/qualifications, etc., alternative units at the appropriate level may be substituted.

The study program for the course is:

Semester 1:
Resource Management (summer school)
Archaeology and Anthropology of Rock Art
Rock Weathering
Conservation Procedures (half unit)
Rock Art Recording (half unit)

Semester 2:
Field Conservation (winter school)
Analytical Techniques
Research Planning,
Conservation Procedures (half unit)
Rock Art Recording (half unit)

Details of these units are given in Appendix 1.

The course initially provides an introduction to Resource Management in the form of a month-long summer school held in February before the commencement of the course proper in March.

The units that comprise the remainder of first semester studies are two semester-long units, Archaeology and Anthropology of Rock Art, and Rock Weathering, plus the first half of two year-long units, Conservation Procedures and Rock Art Recording. By the use of these units, students are introduced to the identification and deterioration of rock forms, basic recording techniques, sampling techniques for analysis, and conservation and restoration processes, all of which are applied in the winter school of three weeks duration (Field Conservation) in July.

In semester two, the materials and techniques used in the field, along with samples taken, are analysed and evaluated in the semester-long units, Analytical Techniques and the second half of the two year-long units, Conservation Procedures and Rock Art Recording, For example, measurements and photographs are plotted, rock and pigment samples analysed, and conservation and restoration techniques evaluated with input from Aboriginal people. The final semester-long unit, Research Planning, leads to a research project on some aspect of deterioration or conservation of rock art.

(5) *Progress to-date*

In 1989 there was an intake of 14 students (8 male, 6 female), eight being from Australia and the others, three from the U.S.A. and one each from South Africa, Sri Lanka and Tanzania.

The senior lecturer responsible for directing the course is Alan Watchman, with technical and clerical support from the NCCHSS. There has been significant lecturing support from staff of the School of Applied Science, the Regolith Studies Centre and from the Australian National University, particularly in the units Rock Weathering and Rock Art Recording. Dr Andrée Rosenfeld from the Australian National university has taught the unit, Archaeology and Anthropology of Rock Art, aided by visiting lecturers, and in the unit Conservation Procedures there has been input from well known specialists including Jacques Brunet from France, and from Australia, Sharon Sullivan, Dave Lambert, John Clarke and Grahame Walsh.

The Board of Studies will evaluate the course at its conclusion and provide recommendations for future courses.

APPENDIX 1

UNIT OUTLINES

ANALYTICAL TECHNIQUES

Unit Value	:	3 credit points
Prerequisites	:	Graduate entry
Class Contact	:	1 hour lecture and 1 2-hour practical/tutorial per week.

Aims and objectives
The aim of the unit is to provide students with an understanding of the different analytical techniques used for examining, dating and recording rock art sites, and also the materials used in conservation practice, emphasising their limitations and capabilities.

Syllabus
Emphasis will be placed on the application and limitations of analytical techniques, including sampling, sample size, and preparation, rather than on in-depth training in analytical chemistry.

Recommended Reading

PEARSON, C. (ed.) 1978. *Conservation of rock art*. Proceedings of the International Workshop on the Conservation of Rock Art, Perth, September, 1977. ICCM, Sydney.
ROSENFELD, A. 1988. *Rock art conservation in Australia*. Special Australian Heritage Publication No. 2. 2nd Edition, Australian Government Publishing Service.
PECSOK, R. L., SHIELDS, L. D., CAIRNS, T. and McWILLIAM, I. G. 1976. *Modern methods of chemical analysis*, 2nd Edition. John Wiley and Sons, Sydney.

ARCHAEOLOGY AND ANTHROPOLOGY OF ROCK ART

Unit Value	:	3 credit points
Prerequisites	:	Graduate entry
Class Contact	:	4 hours per week

Aims and Objectives
The aim of the unit is to provide students with an understanding of the reasons for the creation of rock art, the different methods of production, and an awareness of the significance of art to indigenous peoples.

Syllabus
The unit is divided into the following sections.

1. *Rock art on a global basis*
   Students will be provided with an introduction to rock art from a global perspective, but emphasis and case studies will

focus mainly on Australian Aboriginal rock art.

2. *Anthropology of rock art*
   Rock art will be examined as a special case of artistic expression in non-literate societies, emphasis being given to its function as a social phenomenon rather than to aesthetic considerations. The meaning of rock art to Aboriginal people will be considered in terms of traditional and contemporary values.

3. *Archaeology of rock art*
   The archaeology of rock art will be outlined with emphasis on methods of relative and absolute dating, materials and techniques, and archaeological associations.

Recommended Reading

ANDERSON, R. L. 1979. *Art in primitive societies*, Prentice Hall.
FLOOD, J. 1983. *Archaeology of the Dreamtime*, William Collins, Sydney.
LAYTON, R. 1981. *The anthropology of art*, Granada Publishing.
LAYTON, R. 1986. *Uluru. An Aboriginal history of Ayers Rock*, AIAS.
LEWIS, D. and Rose, D., 1987. *The shape of the Dreaming*, AIAS, Canberra.
McBRYDE, I. 1985. *Who owns the past?* Oxford University Press.
McCARTHY, F. D. 1979. *Australian Aboriginal rock art*, Australian Museum.
NEIDJE, B. 1985. *Kakadu Man.*
ROSENFELD, A. 1988. *Rock art conservation in Australia.* 2nd Edition. Australian Heritage Commission, Canberra.
SIEVEKING, A. 1979. *The cave artists.* Thames and Hudson.
UCKO, P. J. 1978. *Form in indigenous art: Schematisation in the art of Aboriginal Australia and prehistoric Europe.* AIAS, Canberra.

CONSERVATION PROCEDURES

Unit Value	:	3 credit points
Prerequisites	:	Graduate entry
Class Contact	:	1 hour lecture, 3 hour practical/tutorial per week over two semesters.

Aims and Objectives
The aim of the unit is to provide students with an understanding of the reasons for the intervention and non-intervention of conserving rock art sites, and the skills and techniques necessary for consolidating rocks, controlling water and the movement of people, and removing graffiti, salts, biological matter and other harmful agents.

Syllabus
The unit is divided into the following sections.

1. *Ethics and philosophy of conservation*
   Students will be made aware of the ethics and rationale applied to conservation of cultural materials. The process of selecting the most appropriate treatment and the cost effectiveness of the conservation technique will be evaluated.

2. *Documentation techniques*
   Techniques for documenting the nature of the deterioration and of the conservation measures taken to rectify the problems will be taught based on accepted conservation methods.

3. *Intervention and non-intervention conservation methods*
   Preventive and restorative conservation methods from both theoretical and practical perspectives will be described and discussed. The ethics and methods of retouching and repainting rock art will be presented.

4. *Conservation monitoring*
   The various techniques used to monitor deterioration and the results of conservation will be explained and applied in field classes.

Recommended Reading

GILLESPIEG, D. A. (ed.) 1983. *The rock art of Kakadu National Park.* ANPWS, Canberra.
PEARSON, C. (ed.) 1978. *Conservation of rock art.* Proceedings of the International Workshop on the Conservation of Rock Art, Perth, September 1977. ICCM, Sydney.
ROSENFELD, A. 1988. *Rock art conservation in Australia.* Special Australian Heritage Publication No. 2. 2nd Edition, Australian Government Publishing Service.
WAINWRIGHT, I. N. M. 1985. The state of research in rock art; Rock art conservation research in Canada. *Bollettino del Centro Camuno di Studi Preistorici* 22: 15-46.

FIELD CONSERVATION

Unit Value	:	3 credit points
Prerequisites	:	Graduate entry
Class Contact	:	This will be field trips and include lectures, tutorials and practical field work.

Aims and objectives
The aim of the unit is to put into practice the knowledge gained from first semester units such as Rock Art Recording, Conservation Procedures and Rock Weathering.

Syllabus
Students will be involved with the organisation of field trips, communicating with traditional owners and custodians and relevant local and state authorities.
Emphasis will be placed on surveying and recording sites, establishment of monitoring procedures, identification of rock types and deterioration agents and processes, and the applications of both non-intervention and intervention conservation techniques.
Information recorded and samples taken will be analysed in second semester units, such as Analytical Techniques and Rock Art Recording.

Timetable
The main field trip will be held in July during the semester break. Other field trips, to obtain information from traditional owners, carry out conservation works, devise a monitoring program and collect data as a basis for a visitor survey, will be arranged by the students during the second semester.

Recommended Reading

GILLESPIE, D. A. (ed.) 1983. *The rock art sites of Kakadu National Park.* ANPWS, Canberra.
GALE, F. and JACOBS, J. N. 1987. *Tourists and the National Estate.* Australian Heritage Commission, Canberra.
ROSENFELD, A. 1988. *Rock art conservation in Australia.* Special Australian Heritage Publication No. 2. 2nd Edition, Australian Government Publishing Service.

RESEARCH PLANNING

Unit Value	:	3 credit points
Prerequisites	:	At least two postgraduate units or permission of Principal Lecturer
Class Contact	:	2 hours of lectures per week and weekly seminars

Aims and objectives
Research and the use of the results from it are basic to science.

This capstone unit aims to develop a deeper questioning of scientific methods and philosophies to provide students with an ongoing approach to appraising scientific research. Further aims of the unit are to develop expertise in the planning of research projects and the ability to critically evaluate the literature and arguments on which research proposals are based. The following objectives have been set for students to achieve these aims.

1. To gain an understanding of the various philosophies of science and to evaluate them from the student's own standpoint.
2. To develop an understanding of what research and the scientific method are.
3. To develop literature review abilities including its synthesis and the development of arguments based on it, particularly with respect to planning research.
4. To gain expertise and critical abilities in problem identification and the formulation of testable hypotheses.
5. To be able to develop appropriate approaches and methods to test hypotheses and to critically evaluate them, especially in terms of the likely outcomes and research symmetry.

Format
The unit consists of three elements. The major task is the completion of a research proposal. The great majority of this work will be completed outside class time. Students will be expected to meet at least fortnightly with their supervisors. The second element is a series of two hour classes, which will be used for discussion of the philosophy of science, scientific method, research project selection, research skills and student presentations of progress reports. The final element is the Resource Science Seminar Series (Fridays, 1230-1330). Students of this unit will be expected to attend these seminars which provide a range of topics, problems and approaches to research.

RESOURCE MANAGEMENT

Unit Value	:	3 credit points
Prerequisite	:	Graduate entry
Class Contact	:	This will be a four-week unit held in conjunction with the Summer School of Part Management.

Aims and objectives
The aim of the unit is to introduce students to the general principles, practices and techniques of the management of resources, parks and sites, followed by more detailed consideration of the management and conservation of rock art as part of cultural resource management.

Syllabus
The unit is divided into the following sections:

1. *Literature review*
   Before the unit starts, students are required to prepare a paper, of not more than 10 000 words, on the significance of rock art in cultural resource management.

2. *Park management*
   The Summer School of Park Management covers topics such as models of management planning, infrastructure of resource management in Australia, strategies and techniques of visitor management, and communication skills.

3. *Field class*
   A seven day field class in an area containing a range of rock art will examine particular aspects of natural resource management relevant to cultural resource management. Students will prepare a management plan for the rock art in the field area.

4. *Rock art as a cultural resource*
   A series of lectures and tutorials will cover the ethics and standards of professional rock art management as a cultural resource, the problems of preventive conservation and visitor impacts, current Australian rock art management practices, attitudes of indigenous people to rock art conservation, and effective techniques of communication and negotiation with indigenous peoples and bureaucracies.

5. *Presentation on cultural resource management*
   Students will prepare a short presentation (15-20 minutes) on the fundamental issues and problems related to the management of a cultural resource with which they are familiar.

Recommended Reading

GALE, F. and JACOBS, J. H. 1987. *Tourists and the National Estate.* Special Australian Publication Series No. 6, Australian Heritage Commission, Canberra.
PIGRAM, J. J. 1983. *Outdoor recreation and resource management.* Groom Helm.
MERCER, D. C. (ed.) 1981. *Outdoor recreation - Australian perspectives.* Sorrett, Melbourne.
SULLIVAN, H. (ed.) 1984. *Visitors to Aboriginal sites: Access, control and management.* ANPWS, Canberra.

ROCK ART RECORDING

Unit Value	:	3 credit points
Prerequisites	:	Graduate entry
Class Contact	:	3 hours per week of lectures, tutorials and/or practicals, with some field work, over two semesters.

Aims and objectives
The aim of the unit is to provide students with the knowledge and skills for mapping and recording rock art sites.

Syllabus
The unit is divided into the following sections

1. *Surveying*
   Students will be taught the basic surveying techniques and learn to do site surveys using basic surveying equipment.

2. *Photography*
   Elementary and advanced photographic methods will be taught so that students are able to record art and conservation measures in a professional manner.

3. *Recording art and rock surface features*
   Lectures will be given on the methods used to record art and rock surface features. Practical sessions will be conducted to teach students the skills required for comprehensive recording of rock art, deterioration problems and conservation actions.

4. *Photogrammetry*
   Students will be instructed in the principles of photogrammetry and trained to use stereoscopic pairs of photographs for the precise recording of art.

5. *Solid modelling*
   The basics of solid modelling will be taught.

Recommended reading
CONNAH, G. (ed.) 1983. *Australian archaeology - A guide to field techniques.* Australian Institute of Aboriginal Studies, Canberra.
OGLEBY, C. and RIVETT, L. J. 1985. *Handbook of heritage photogrammetry.* Special Australian Heritage Publication No. 4., AGPS, Canberra.
FLOOD, J., JOHNSON, I. and SULLIVAN, S. (eds.) 1988. *Sites and bytes : A handbook on site recording and registration.* Australian Heritage Commission, Canberra.

ROCK WEATHERING

Unit Value	:	3 credit points
Prerequisites	:	Graduate entry
Class Contact	:	3 hours per week of lectures/tutorials and, or practicals. Field work sessions will also be held.

Aims and objectives
The aim of the unit is to provide students with an understanding of the primary controls, processes and products of rock weathering, as well as the petrology and mineralogy of the products.

Syllabus
The unit is divided into the following sections.

1. *Introduction to minerals and rocks*
   Students will be provided with an introduction to the basic rock forming minerals (quartz, feldspars, pyroxenes, amphiboles, olivine, garnet, layer silicates), and the different rock types (sandstone, quartzite, silcrete, limestone, basalt, dolerite, granite, shale).

2. *Components of the weathering cycle*
   The weathering cycle will be examined in terms of the parent minerals, the agents of weathering, the environments of weathering and the minerals produced by the weathering processes.

3. *Processes of rock weathering*
   Mineral reactions, micro-environments and weathering rates will be examined.

4. *The products of weathering*
   A study will be made of the organisation of bulk weathered materials in the regolith, controls on the transfer and dispersion of weathered materials by water, wind and biota. Case studies will examine the chemical and mineralogical changes and geomorphological features of weathered profiles in granite, basalt, limestone and sandstone terrains.

Recommended Reading
AMOROSO, G. G. and FASSINA, V. 1983. *Stone decay and Conservation.* Elsevier, Amsterdam.
LOUGHNAN, F. C. 1969. *Weathering of the silicate minerals.* Elsevier, New York.
OLLIER, C. D. 1984. *Weathering.* Longmans, London.
WINKLER, E. M. 1973. *Stone: Properties, durability in man's environment.* Springer Verlag, New York.

PART TWO

RECORDING AND STANDARDISATION IN ROCK ART STUDIES

Proceedings of Symposium E

First AURA Congress

29 August to 2 September 1988

Darwin, Australia

Edited by

B. K. SWARTZ, Jr

With contributions by Robert G. Bednarik, Roberto Bosio, Mauro Cinquetti, John Clegg, Paul Faulstich, Tommaso Guiot, William D. Hyder, Elanie A. Moore, Piero Ricchiardi, M. J. Rowland, Dario Seglie, B. K. Swartz, Jr and John K. Zancanella.

19

INTRODUCTION

B. K. Swartz, Jr.

An important aspect of international congresses is the opportunity afforded for free interchange of ideas by means of volunteered papers on what the presenting scholars think is significant. Symposia are devices for ordering the plethora of information being presented. Symposia set parameters, but do not necessarily establish themes of thought. This can be particularly true of symposia dealing with universal aspects, such as the subject of Recording.

Originally symposia on the topics of Recording and Standardisation were proposed. The Chairmen of the Congress and of the symposium for this volume were surprised at the weak response of offerings and the symposia were combined and ultimately entitled Recording. The total response of the integrated symposium was not strong. Nine manuscripts from symposia presentations were submitted, producing eight contributions (two papers being merged in the initial debate section).

I, as symposium editor, was confronted with two difficulties, (1) a limited number of papers were submitted, and (2) the content of the contributed papers was quite diverse - no structured theme emerged. The editor does hope that reviewers of this volume of the AURA proceedings will be cognizant of the problems involved.

There is topic overlapping, in general the papers can be characterised as follows: two papers deal with broad or universal topics (Clegg vs Swartz/Zancanella - debate and Bednarik), two papers deal with specific and/or high-tech recording methods (Seglie et al. and Swartz), one paper deals with specific, but unique, problems of general recording (Moore), one paper deals with replication and the public (Faulstich), and two papers deal with regional research activity from recorded data (Hyder and Rowland). The dominant observation that comes out of the symposium is the lack of acceptance or awareness (depending on how you wish to look at it) of preserving petroglyph data by systematic recording. At the beginning I thought this concern, along with recent technical advances, would be the major foci of the originally proposed symposium on Recording. Only the Swartz/ Zancanella contribution (within the debate) is concerned with this issue. Problems of protection by management, and conservation seem to be of greater interest, both at the Congress and elsewhere. It can be said that the problem of preservation, whether by recording or protection, seems to be of little concern to most rock art researchers. Research may not be a productive activity in the future unless some concern as to resource maintenance is considered.

The lead off is two papers juxtaposed to highlight a debate and set off a spark for the following presentations. The editor is not impartial in this debate as can be seen in the preceding paragraph. It would be improper for me, therefore, to make further comments here.

Bednarik's contribution puts into stark terms the cavalier attitude rock art scholars have demonstrated in establishing the terminology of the field. In the area of content description there are serious barriers to be overcome in proposing 'objective' (an adjective Bednarik also inquires into) terms on subjective data. It is curious that one term Bednarik does not deal with is *style*. On the basis of personal correspondence I have been informed of his complete concurrence with my conclusions on the subject in comments given on the Hyder paper following.

The bland use of the term 'rock art' for our field of study bothers me. Use of this term has been resisted by older professional archaeologists in North America (a category which includes your editor) who prefer following Steward's (1929) usage of 'petroglyph'. The term 'rock art' was first significantly used in the United States by Heizer in 1962 as the title of his landmark monograph *Prehistoric rock art of Nevada and eastern California* (Heizer and Baumhoff 1962). This is ironic in that intellectually Heizer is in the mold of Steward and was in the lead in formulating information system hypotheses with 'rock art' data. The appellation gained wide acceptance only after Grant's (1967) publication *Rock art of the North American Indian*. The trouble with the term is that it imposes a preconceived interpretation of the subject matter it labels - much of which is NOT art. To this day I am not comfortable with the term.

The resolution to Bednarik's dilemma about the widespread occurrence of petroglyph pecking and the absence of stone chisels may, perhaps, be found in an observation on replication experiments by Busby, Fleming, Hayes and Nissen. They state the 'direct pecking [i.e. with another rock] is quite capable of great precision' (1978: 99).

We must improve our terminology through critique and debate. It is hoped that more uniformity and appropriateness will emerge. Bednarik mentions Flood's (1987) glossary as a starting point for future work. I would like to suggest Nissen's (1981) extensive motif 'dictionary' as a good starting point for content terminology.

The Swartz and Seglie et al. papers bring contemporary technology into rock art recording and conservation. Both contributions have sections that deal with the use of photography and computers in petroglyph research. It is important that closer ties be established between technical experts and rock art researchers. There were additional papers on technical aids presented at the symposium, but they were not submitted for publication. The world-wide data network Swartz envisions should be a high priority for IFRAO.

Rowland's paper is a regional model on the procedures and problems for establishing a rock art programme. The tack taken is to organise the program about an environmental rock art study approach. This differs from Swartz and Zancanella, who attempt a recording programme to fulfil assessment needs; although Rowland certainly considers assessment, and also protection, problems.

Originally the Moore paper was a multi-authored regional study, but was adapted for inclusion in this volume by elaborating the recording methods employed by the investigation. The problem of effective photography in unremitting terrain has not been dealt with in rock art literature. Also, little has been written on facsimile recording in North America. The editor is in complete accord with Moore's plea for describing specific procedures used in recording rock art.

Faulstich's well-composed essay does not fit in any of the AURA symposia, including this one. The focus of the presentation is on education - not research. His distinction between facsimiles (exact copies as research aids) and replicas (meaningful reproductions that are not necessarily accurate in every detail as instructional aids) is useful and contrasts this contribution with Moore's facsimile study.

The topic of Hyder's paper differs from all others in this volume in that it does not deal with recording procedures in a general sense or with collecting and preserving data. Hyder's paper explains a method of rock art research.

Despite his phraseology I believe that Hyder is not scaling style in that units must be used in analysis. Style and type form the opposite poles of a continuum. Style is expressed, is inexplicable, and cannot be effectively described, but only perceived. It cannot be quantified. Type has all the opposing qualities. The use of the term 'style' is far too general in the rock art literature. This is not to say that spatio-temporal or functional analyses of petroglyphs cannot be undertaken, but not on the basis of ideal style. Rock art analysis can be conducted in the middle range of the continuum I have just formulated.

I wish to acknowledge the editorial assistance of Lisa A. Adams in the preparation of this volume.

REFERENCES

BUSBY, C., R. FLEMING, R. HAYES and K. NISSEN 1978. The manufacture of petroglyphs: additional replicative experiments from the western Great Basin. In 'Four rock art studies', C. W. Clewlow, Jr. (ed.) *Ballena Publications on North American Rock Art*, No. 1, pp. 89-108. Socorro.

FLOOD, J. M. 1987. Rock art of the Koolburra Plateau, north Queensland. *Rock Art Research* 4: 91-126.

GRANT, C. 1967. *Rock art of the North American Indian.* Thomas Y. Crowell (Apollo Editions), New York.

HEIZER, R. F. and M. A. BAUMHOFF 1962. *Prehistoric Rock Art of Nevada and Eastern California.* University of California Press, Berkeley.

NISSEN, K. M. 1981. Images from the past: An analysis of Western Great Basin petroglyph sites. Unpubl. Ph.D. thesis, University of California, Berkeley.

20
DEBATE

SHOULD PREHISTORIC PICTURES BE SITUATIONALLY DESCRIBED?
OR
SHOULD PETROGLYPHS BE SYSTEMATICALLY RECORDED?

The editor has taken the liberty to pair these two contributions (one senior-authored by the editor) to this volume, that present rather striking opposing views to rock art, in order to bring forth a fundamental debate. The two essays were composed independently of each other.

The terminology of prehistoric pictures vs. petroglyphs (favoured by the respective authors; see also Bednarik this volume) aside, two issues seem to emerge (1) problem-solving vs vanishing data, and (2) interpretation vs assessment. The first issue is an old one in American anthropology, going back at least to the time of Franz Boas' early ethnographic fieldwork. To conduct scientific research, a problem must be formulated (usually phrased as a question), then data relating to the problem is collected and then analysed so an hypothesis can be formulated that gives an explanation to the problem.

On the other hand, scholars concerned with vanishing data, like Boas, are primarily focused not on the solution of a problem, but upon immediate measures to preserve the data base, particularly a non-generating, limited and disappearing data base. Rock art specialists of this tradition maintain that the expense and effort of undertaking field work requires thorough recording and, if feasible, preservation measures. They view as extravagant the ignoring of data easily preserved and recorded that is before the field worker.

The issue of interpretation vs assessment is a post-Contract Archaeology dispute. Is the purpose of collecting data to enable the investigator to do research? Or is there an additional dimension, the acquiring of data to enable the land manager to assess resources and prepare planning documents?

CLEGGNOTES ON RECORDING PREHISTORIC PICTURES

John Clegg

Over the years I have concluded that there is no point in attempting to standardise recording methods or techniques, except for (1) professional ethics, (2) uniformity within one particular project, or (3) the development of a particular technique. If standardisation is attempted for the latter, continual updates are necessary to incorporate improvements.

Ethics are involved in recording. We can take a leaf from conservation ethics, and give primary consideration to doing nothing which is irreversible or could damage the artefact. Secondary in priority to our duty to the artefact is our duty to its owners, and thirdly to those responsible for its management. Only fourth in priority is our duty to ourselves, or our discipline.

Instead of standardising, I advocate the cook-book approach, which provides many recipes for recording methods. The idea is to allow the recorders to choose those methods which best suit their needs and capacities. As with all cook-books, a period of learning or experiment may be needed before a practitioner becomes competent at, say, making toast or reading a compass.

The need for many different methods stems from the three main sources of variation:

1. Every individual picture is different and presents its own problems.
2. Every recorder or recording situation has different resources of time, materials, equipment, money and skill.
3. Different tasks or archaeological 'problems' (but I do not wish to exclude non-archaeological interests) have different aims, which need suitable methods.

These three points indicate how to set about choosing a menu from the cookbook. Work out (1) what the picture is like, (2) what resources you have, and (3) what you want to do and how you will use the record, not necessarily in that order, then use the results of your working out to choose a suitable recipe.

Some methods do much the same things as others, and some are better than others. For instance, tracing on polythene is easier, quicker, and far more direct than making a reduced drawing using line-and-offset or a grid, but the tracing may need to be photographed for publication. I believe that almost all recipes have some advantages; we should only exclude methods (such as lighting large fires near engravings) which could damage the artefact.

It is good to choose a menu of recording recipes which begins with works from the simplest and most direct observation, 'There were some marks on the rock. I thought someone had scratched a groove', and only at the end introduces iconographic interpretation, 'There's a picture of a Martian!'. The very minor point of distinguishing clearly between what a pictures looks like (my !notation) and what it represents 'There's a !frog picture of the premier of New South Wales[1] in the local paper again' can save a great deal of wrangling and misapplication of what should be definable technical terms - like anthropomorph. Terms like 'martian' or 'anthropomorph' are predicated upon a system of classes, which may be natural or artificial, agreed or in dispute. A project to standardise such classes would be large as it deals with a Pandora's can of worms, and of questionable validity, let alone use.

Another hierarchy is desirable, one which deals with detail and accuracy. Accuracy can be measured by determining the extent to which repeated recordings of the same picture resemble each other (all within perhaps two centimetres in length, but of completely different colours). Detail can be measured by specifying the size of the smallest item recorded, in microns or millimetres or larger units. Thus one might want to know the forms of a small picture accurate to one centimetre, but its location within a couple of metres. Some specification of the accuracy and detail required removes a great deal of worry and unnecessary work; commonly, people give no thought to the accuracy they require, and are therefore continually anxious that their work or method is NOT ACCURATE ENOUGH (to maintain their status, or earn the number of marks or candies they may be apportioned by some judgmental system).

There are already several such cook-books (for example McCarthy 1972; Bain 1975; Anati 1977; Clewlow and Wheeler 1978; Clegg 1983) to which new recipes could be added, and superseded or practically useless ones removed. Perhaps a confederation of the various national RARs might 'anthologise' them into a useful international volume, which should include a section which tackles the problems of deciding which method to use.

REFERENCES

ANATI, E. 1977. Methods of recording and analysing rock engravings. *Camunian studies* vol. 7 (1st English edition). Bresca.

BAIN, J. G. 1974. *Techniques and procedures for rock art recording.* Center of Anthropological Study, Monograph No. 3. Las Cruces.

CLEGG, J. 1983. Recording prehistoric art. In Graham Connah (ed.) *Australian field archaeology: A guide to techniques,* pp. 87-108. Australian Institute of Aboriginal Studies, Canberra.

CLEWLOW, C. W. Jr. and M. E. WHEELING 1978. *An introductory recording manual for California and the Great Basin.* Institute of Archaeology, University of California, Los Angeles.

McCARTHY, F. D. 1972. Recording art on rock surfaces. In D. J. Mulvaney *Australian archaeology: A guide to field and laboratory techniques,* pp. 53-70. Australia Institute of Aboriginal Studies Canberra.

NOTE

1 The Right Honourable Nick Greiner is irreverently nicknamed Kermit - a reference to the frog in a children's' television program, *Sesame Street.*

THE MANAGEMENT OF RECORDING PROCEDURES IN THE PRESERVATION OF PETROGLYPH RESOURCES

B. K. Swartz, Jr. and John K. Zancanella

The authors of this paper contend that the most reasonable approach to the preservation and management of petroglyphs is intensive and systematised recording. In the long term, protection cannot be an effective approach to preservation since publicly known petroglyph exposures must be continuously monitored (a prodigious expense) because of their accessibility and visibility.

A large number of prehistoric and historic petroglyphs and pictographs occur in that portion of the western United States known as the Great Basin. The majority of this arid to semi-arid landscape is public land administered by several federal agencies, including the Bureau of Land Management and the Forest Service. As wards of the public domain, these agencies are responsible by law to identify, manage, and protect all cultural resources under their jurisdiction, including rock art. This is no easy task considering the amount of time and effort involved in documenting this type of archaeological manifestation and the fiscal realities most cultural resource managers must cope with from year to year. The larger and more impressive sites, and a growing number of smaller ones, are known to local inhabitants and archaeologists. For a variety of reasons most of these are in various stages of recording. This situation creates a need to organise and manipulate the available data in a comparable fashion for management, as well as analytical (scientific) purposes.

As a result of the senior author's fieldwork in Nevada, U.S.A., we offer as a management tool a set of recording levels to rank the kinds of data gathering performed at individual sites. Such rankings will allow agencies (and other researchers) to readily identify and set priorities and objectives for incompletely or inadequately recorded sites. The apparent indifference to recording is not only a problem in the Great Basin, and North America generally, but also internationally.

Petroglyph researchers have confused the operations of recording and explanation. An example of this confusion is stated by Schaafsma in Volume 8 (1985) of Schiffer's high-powered series 'Advances in Archaeological Theory and Method'. In defending problem-oriented research and attacking systematic recording she evokes the well-known philosopher Thomas S. Kuhn and quotes him to wit:

> In the absence of a paradigm or some candidate for

paradigm, all of the facts that could possibly pertain to the development of a given science are likely to seem equally relevant. As a result, early fact-gathering is a far more nearly random activity than the one that subsequent scientific development makes familiar (Schaafsma 1986: 241).

From this she concludes that the paradigm of study will determine what records (not data, BUT RECORDS), should be collected. Schaafsma is, perhaps, the foremost leader in what the senior author chooses to call, using Kuhn's phraseology, the Rock Art Paradigm. If petroglyphs are rock art, one is concerned with their 'style of rendering', but need not be concerned with utilitarian relationships - an important matter for interpreting petroglyphs as part of informational systems. Locational records are of little significance for the former, but become crucial for the latter. Anyone is entitled to pursue his or her paradigm, but not to warp the database by limiting recording on the basis of their theoretical outlook. In stark contrast to Schaafsma, a plea for systematic recording and inventorying has been made by some (e.g. Turner 1963).

Situational recording of petroglyphs can also distort petroglyph interpretation research in general, in both contextual and quantitative analysis. If design elements are recorded independently and not as natural groups, the relationship between the elements is lost. Additional contextual information disappears if the position of panels within a site is not recorded. If only certain design elements are chosen for recording over others, e.g. aesthetic quality, elaborateness, identifiability, clarity, etc. the sampling becomes skewed and quantitative comparisons from naturally surviving data becomes impossible.

The purpose of this paper is not to present a checklist of things to record, the ACASPP minimum recording standards (1980, 1981) does this admirably, but to establish procedure to organise and systematise recording activity so that the results can be effectively inventoried and managed. The concept of inventory as an activity in public land management is stated in Section 201a of U.S. Public Law 94-579 - Federal Land Policy and Management Act of 1976: 'The Secretary [of the Interior] shall prepare and maintain ... an *inventory* [our emphasis] of all public lands and their resources and other values ... giving priority to areas of critical environmental concern'.

Unfortunately, this act does not prescribe procedures for undertaking such an inventory. The critical operation for effective inventorying is recording (or in federal neologistic jargon - recordation). Ritter (1978: 2) lists additional U. S. federal legislative acts pertinent to rock art protection. But beyond specific recording, comparable units must be established so that assessment and evaluation can be done. The cook-book approach precludes proper inventorying procedures.

The following set of recording levels is presented as an aid to this end:

Level 1 - Noting the existence of petroglyphs at a site which is located on a map (in the United States presumably, but not necessarily USGS).

Level 2 - Photographing of petroglyphs. This is the usual level of avocational recording.

Level 3 - Sketching a map of the site which locates panels, and the sketching of individual panels. Recording at this level is that of past traditional professional practice.

Level 4 - Recording individual panel information, e.g., panel number, direction, dimensions, inclination, distance to other site panels, relation to surroundings, establishment of site and panel points.

Level 5 - Recording of specific design elements of a panel, e.g. size and relationship to other elements. Recording up to this level requires no highly technical equipment and conforms to the American Committee to Advance the Study of Petroglyphs and Pictographs minimum recording standards (though almost never done by either avocational or professional archaeologists).

Level 6 - Recording petroglyphs with highly technical equipment. This usually entails problem-oriented research such as chronometric calibration etc., rather than comprehensive recording. An exception is videotaping which expands Level 2 performance, but with lower resolution imaging. Recording at this level would also include mapping by use of an alidade or transit. This is seldom done since petroglyphs by themselves need not be excavated.

REFERENCES

ANONYMOUS 1976. Federal Land Policy and Management Act of 1976. Public Law 94-579 enacted by the 94th Congress of the United States of America.

RITTER, E. W. 1978. Management of rock art in the California desert. In E. Snyder, A. J. Bock and F. Bock (eds) *American Indian Rock Art*, Vol. 4, pp. 1-8. Papers presented at the IVth Annual American Rock Art Research Association Symposium, El Toro.

SCHAAFSMA, P. 1985. Form, content, and function: Theory and method in North American rock art studies. In M. B. Schiffer (ed.) *Advances in Archaeological Method and Theory*, pp. 237-77. Academic Press, Orlando.

SWARTZ, B. K., JR. 1986. Comment on 'Identifying high-risk visitors at Aboriginal art sites in Australia', by F. Gale and J. Jacobs. *Rock Art Research* 3: 13.

SWARTZ, B. K., JR. and others 1980-81. American Committee to Advance the Study of Petroglyphs and Pictographs (ACASPP) Minimum Recording Standards (various publications).

TURNER, W. G. 1983. The importance of systematic rock inventories, pp. 37-43. In F. G. Bock (ed.) *American Indian Rock Art*, Vol. 9. Papers presented at the IX Annual American Rock Art Research Association Symposium, El Toro.

21

STANDARDISATION IN ROCK ART TERMINOLOGY

Robert G. Bednarik

Abstract. There have been various efforts to achieve some degree of standardisation in the terminology used in rock art research, but generally with little success. Some reasons for this are explored, and it is shown that the current developments in rock art studies, the formation of a global research community, should facilitate a degree of standardisation.

The vexing problem of terminology is certainly not new in rock art studies; it has been raised many times and in many contexts. Proliferation of idiosyncratic terminologies continues, although the currently developing cosmopolitanism, of which this congress is a symptom, is beginning to have an effect. Some writers have expressed support for pluralism, fearing that the straightjacket of consistency might stifle initiative and intellectual progress (Maynard 1988). I shall therefore examine some of the issues involved.

The argument in favour of standardisation of terminology is squarely based on the need to communicate effectively and precisely. This need is especially relevant to a discipline with so few, widely scattered adherents who rely heavily on global contact. Rock art students, in contrast to conventional archaeologists who deal largely with regional ecological responses of populations, are concerned with cognitive and cultural archaeology, an extraordinarily complex subject that should always be seen in its global context.

Ideally, words should be immutable expressions of crystal-clear concepts, in a positivist model, but they are in fact merely vague tools of verbal communication. Their meanings are imprecise, dictionaries notwithstanding, and their interpretations by individual users differ as much as do the perceptions of the world held by the respective speakers. Therefore, even if there were such a thing as an 'objective observer' - which is not only preposterous but in fact self-contradictory - one would not be able to convey their observations with precision to other people. Examination of the questions relating to standardisation in rock art terminology reveals rather complex issues, and I shall try to illuminate some by using specific examples.

Words are definitional tools, and if misused can be the cause of major errors in empirical research. A case in point is the substitution of the term 'pictures' for 'art' when referring to rock art. Clegg introduced this practice for the sake of objectivity, reminding us quite correctly of the problems in defining 'art' scientifically. David and David (1988), who adopted this usage, have just provided a statistical analysis of a fascinating petroglyph body at Chillagoe, Queensland, from which they excluded all the abraded grooves they found at their sites. They suggested that these might be stone tool sharpening grooves. Most other researchers would include abraded grooves in a site description of rock art, and it seems to me that what prompted David and David to omit them were their terminological bias: abraded grooves hardly qualify as 'pictures'.

The preference for the term 'prehistoric pictures' raises other issues. For instance, is a plain, coloured rectangle a picture? It could certainly be art, as are many other things that are not pictures. Moreover, in many parts of the world there is no clear division between prehistoric and historic rock art; the two often occur together. Wright (1986: 133) even rejects the term 'prehistoric' in Australia as a misnomer, considering it Eurocentric and chauvinistic.

There is a second group that objects to the term 'rock art,' the people studying supposedly epigraphic rock art in the United States. Their claim that petroglyphs are picture writing, not art, is of interest in the context of local ethnography, but not in a terminological sense. Ideographs are derived from pictures; they are abstractions, and therefore they are formalised art. Anati (1981) even used the term 'ideogram' to name one of his three major classes of rock art, and the alleged function of certain North American petroglyphs has in my view no bearing on their status. Art is not defined in terms of its use.

Clegg's interpretation of the term 'picture' as the opposite of what it seems to mean to most other people is fascinating because I often experience a similar ambivalence, which suggests a certain relativity in terminology. A good example is provided by the innocent little word 'abstract', often used in rock art publications as the opposite of 'figurative'. Some writers refer to any geometric motif as 'abstract'. In art, to abstract means 'to reduce', for example to reduce a three-dimensional object of the physical world to a two-dimensional depiction. Hence figurative depiction is an abstraction. On the other hand, geometric motifs in rock art may represent archetypes, depicting perhaps phosphene types (Bednarik 1987), in which case they would be truly figurative or iconic. Consequently, what is described as figurative or representational art is abstract, whereas the motifs often labelled as abstracts may in fact be figurative.

Such ambiguities and contradictions should be of concern; they suggest fundamental inconsistencies in our terminology, and a lack of objectivity. Now there is a fascinating pair of words: 'objective' and 'subjective'. Positivism is sometimes seen as the paragon of objectivity, dealing as it does exclusively with 'positive facts and phenomena'. Yet it is using a completely subjective frame of reference in this pursuit. Objective truth in the real world is not accessible to us; we only have the perceptions of our mind. On the other hand, the word subjective refers to what exists in the mind, which is in fact the only truly objective information we are capable of gaining. It follows that the terms objective and subjective can be construed to mean the opposite of what they are supposed to mean.

The term 'rock art' is used in all major languages and I can see no reason to reject it, particularly as we lack a satisfactory empiricist definition of 'art'. At the Valcamonica Symposium of 1987 a debate on the subject remained inconclusive and participants and others were invited to send their definitions of art to Capo di Ponte. Clegg (1987: 34) defines art by its 'current meaning in our society', and typical dictionary definitions range from 'products of man's creativity' to 'conscious production or arrangement of sounds, colours, forms, movements or other elements in a manner that affects the sense of aesthetics.' To my mind such explanations are unsatisfactory; they might tell us what forms art takes, but not what it actually is or what it does, except to refer us to aesthetics. But surely aesthetics are derived from art production, not vice versa.

Art, I believe, is the medium conveying the artist's awareness of perceived reality to the sensory perception of the beholder. It simply externalises concepts of reality. Accordingly, even the most rudimentary intentional markings are an expression or result of consciousness. The substitution of the word 'picture' is therefore pointless, as art must have been around much longer than pictures, even 'non-iconic pictures'. Besides, it would create some linguistic problems to maintain consistent definitions for words such as picturesque or pictorial.

Another type of terminological pitfall can be demonstrated with the all-too-frequently-used adjective 'anthropomorphic'. Etymologically derived from 'anthropomorphite', which describes a person ascribing human form to God, the word 'anthropomorphic' has acquired an extended meaning to include inanimate objects, animals or natural phenomena to which human motivations, characteristics or behaviour are attributed. But 'having human form or appearance' is expressed by another term, the word 'anthropomorphous', and this is the correct adjective for anthropomorph. Typical anthropomorphic statements would be to ascribe human motivations to a pet or to the wind, or to credit a deity with human form. Only the latter use could be relevant in rock art, and only in rare cases; for instance, some figures of deities in Asian rock art are truly anthropomorphic. But I doubt that we could confidently pronounce any rock pictures in Australia or the United States as depicting deities, while there are immense numbers of anthropomorphous figures in these countries.

The careless use of terminology in this example may betray a casual research attitude. It is important to appreciate that the difference between using the terms 'anthropomorphic' and 'anthropomorphous' corresponds to the difference between claiming to know what a picture depicts, and simply stating what it resembles in the eyes of the uninitiated. The first term is highly specific, the second is blatantly subjective, and can be used to denote any picture the observer considers to resemble a human form.

One of the most common terminological problems in rock art concerns the naming of motif types, where the unbridled enthusiasm of generations has left us with a legacy of jargon much of which deserves to be discarded. The cultural and scientific bias of researchers is well illustrated by the following example. The three-pronged convergent lines motif, which appears in the logo of AURA, occurs in many parts of the world. When it is found in Australia it is inevitably called an emu track. If it is found 'engraved' on an Indian ostrich eggshell it might be described as the fin of a fish. In Europe it becomes either a vulva or an arrow, depending on which of the two dominant notions the ethnocentric observer subscribes to, the sexual or the hunting obsession. But if the same motif occurs in South America it is described as a nandu track, and in North America, for example in Wisconsin, it passes as a turkey track. Yet in all these cases the motif may simply represent a non-iconic archetype. Rosenfeld (1981) proposed the term 'trident' as an objective alternative to describe the motif, but acceptance of this brilliant improvement has not been enthusiastic so far.

Finally, even complete agreement on the definition of a term does not necessarily guarantee that it will be correctly applied by all users. In fact, some agreed definitions are not adhered to by most writers. This is then surely not a problem of terminology, but one attributable to the inability of researchers to articulate theoretical concepts and their practical applications. Again I use an example to illustrate. The term 'pecking' is now generally agreed, not only in Australia but also abroad, to refer to a method of producing petroglyphs by indirect percussion. The term is used extensively in Australia, and we often hear of 'fully pecked' or 'pecked and abraded' motifs. Each petroglyph so produced would result in a number of discarded stone chisels or punches with distinct bipolar damage, and in view of the huge numbers of such petroglyphs we would expect to find worn stone chisels in their thousands. I have in fact looked for them but have failed to find them.

The stone implements clearly used in the production of petroglyphs which I have found at sites in many different countries are surprisingly uniform, but they were certainly not used for indirect percussion. I am not suggesting that the technique of pecking was never used, merely that its significance is greatly overrated. My guess is that between 25 000 and 100 000 nicks are required per square metre of pecked petroglyph, and some figures occupy more than that. This would seem to require hundreds, possibly thousands of stone chisels, and upon completion the artist's fingers would presumably require amputation. Should we not ask how we ourselves would go about the task of producing a petroglyph, instead of making unsupported, academic pronouncements? What connects us to prehistoric art more than anything else is our humanness, which often seems a more reliable interpretive aid than mindless application of oversimplified empiricist method. Perhaps in future authors using the term 'pecking' should be required to demonstrate the inferred technique by replication.

At the beginning I posed the question of how desirable it is to standardise terminology. By now it has probably become apparent that such endeavours would be of limited effectiveness unless they are accompanied by a refinement in research philosophy. Globally, rock art studies are entering a phase of cleansing and renaissance which will result in the rejection of countless prejudices, misidentifications and unsubstantiated assumptions that have been allowed to accumulate. Writers should be cautious in adopting previously used terms. The mere fact that they have been published in certain applications is not necessarily an indication that they were used correctly, and the exercise of healthy scepticism is most commendable. A good example is provided by the enthusiasm of so many writers in telling us what rock art depicts. Clegg (e.g. 1985) has been trying to tell us for many years that the iconographic content of rock art is frequently inaccessible to us, but his admonition seems to have had little effect.

Standardisation of terminology alone will not necessarily result in our escape from the terminological quagmire; to be effective it must be combined with a considerably more critical review of all aspects of rock art. Propositions that are not testable, in practice or at least theoretically, are not scientific, and if it is our intention to study prehistoric art scientifically, a massive refinement of terminology would be inevitable. If we decide that such a process might be too painful for us, then perhaps there is not much to be gained from standardisation. By itself, it will do little to improve our work. But if we are willing to take the hard road to scientific relevance, then we will definitely need a tightly designed, effective terminology. It is my belief that if a standard terminology were endorsed by an influential group it would quickly find acceptance internationally, because the global research community, which has only taken shape in the last few years, is ready to accept such a lead. For my part, the most impressive rock art glossary so far provided is that by Flood (1987: 120). Among her 28 definitions I only find one I disagree with, that of the word 'engraving'. I agree with Maynard (1977) that if this word is used as a major term for all petroglyphs we are left with no name for true engravings; i.e. those that were produced with a burin or graver. I think that Flood's glossary could be a starting point for uniform terminology, and I invite debate of this general subject.

REFERENCES

ANATI, E. 1981. The origins of art. *Museum* 33: 200-10.

BEDNARIK, R. G. 1987. Engramme und Phosphene. *Zeitschrift für Ethnologie* 112: 223-35.

CLEGG, J. 1985. Comment on D. Groenfeldt: The interpretation of prehistoric art. *Rock Art Research* 2: 35-45.

CLEGG, J. 1987. Human picturing behaviour and the study of prehistoric pictures. *Rock Art Research* 4: 29-35.

DAVID, B. and M. DAVID 1988. Rock pictures of the Chillagoe-Mungana limestone belt, north Queensland. *Rock Art Research* 5: 147-56.

FLOOD, J. M. 1987. Rock art of the Koolburra Plateau, north Queensland. *Rock Art Research* 4: 91-126.

MAYNARD, L. 1977. Classification and terminology of Australian rock art. In P. J. Ucko (ed.), *Form in indigenous art: Schematisation in the art of Aboriginal Australia and prehistoric Europe*, pp. 385-402. A.I.A.S., Canberra.

MAYNARD, L. 1988. Comment on F. D. McCarthy: Rock art sequences: a matter of clarification. *Rock Art Research* 5: 30.

ROSENFELD, A. 1981. Rock engravings in the Laura area. In A. Rosenfeld, D. Horton and J. Winter, *Early Man in north Queensland*, pp. 50-89. Terra Australis Series No. 6, A.N.U. Canberra.

WRIGHT, B. 1986. Comment on N. R. Franklin: Stochastic vs emblematic: an archaeologically useful method for the analysis of style in Australian rock art. *Rock Art Research* 3: 131-3.

22

ARCHIVAL STORAGE, DIGITAL ENHANCEMENT AND GLOBAL TRANSMISSION OF ROCK ART PHOTOGRAPHS: A PRACTICABLE DESIGN FROM EXISTING TECHNOLOGY

B. K. Swartz, Jr.

Abstract. The crucial approach to rock art recording is high-resolution photography and non-degradable archival storage. Digitised image data can be compactly stored on optical or laser disks. The considerable benefits of such systems are discussed.

The combination of four technological breakthroughs in recent years has provided the possibility of establishing an extremely effective archival system for rock art records, namely:

1. Compact image storing on optical disks
2. Digital camera scanning
3. Digital data storing on compact disks (CD-ROM)
4. Integrated image transmission

By use of these technical opportunities, non-degradable, high-resolution colour photographic images of all known rock art on the planet can be kept in a small room. Each individual example can be quickly retrieved and enhanced, and all this information can be transmitted throughout the world. With the exception of technical development four, the only factors that preclude such an archive, other than a modest institutional commitment of expense, are ignorance and special interests.

Establishment and maintenance of the highest resolution of images is the recording goal of data generation for rock art archives. Traditional 35 mm photography is the best field recording procedure. I recommend Kodachrome as the ideal film type (at least in the United States) for photographically recording petroglyphs. It possesses the three needed characteristics for such use (1) good resolution, (2) image stability, and (3) colour fidelity. Tungsten motion picture film has high resolution, but poor stability. Kodachrome film will remain stable for a century if stored in a dark environment in temperatures near 25°C and relative humidity near 58%. Kodachrome differs from all other film types in that it has a built-in externally coupled exposing system. The film is composed of five black-and-white coated emulsion layers. When the film is processed after exposure three light sensitive layers initially develop as in black-and-white film. The resulting silver-based exposure is replaced with colour dyes from the layers that correspond to the three light sensitive layers. These displacing dyes are not amorphous in nature, but are crystalline-formed and chemically resistant (Eastman Kodak 1987). A film that exceeds Kodachrome in colour fidelity is Ektochrome 100 (Professional), but its stability is only 50 years. Fujichrome film is colour enhanced, reducing fidelity. A colour checker should be used, especially for pictographs. Rapid petroglyph recording can be accomplished by videotaping, but frame image resolution is poor. In fact, any photograph or drawing, transparency or print, can be recorded on an optical disk.

The analogous two-dimensional image data of pictures is converted into numerical digital data and inscribed on an optical disk by a scanner with an electronic camera fitted with a charge couple device and a targa board. Scanning divides picture imagery into discrete element units, or pixels, of varying electrical energies caused by differences in brightness. This numerical information is then transmitted from the optical disk into a heavy-duty graphics work station terminal. The digitised data can then be compressed for compact storage and uncompressed when needed for processing in the computer. The image is then displayed on the terminal monitor screen in graphic form for image enhancement with a computer-aided pixel painting software package such as ARTRONICS.

The use of optical disks for storage of rock art data has been advocated by Dickman (1984). These disks are manufactured so that a graphic representation (or picture) is encoded by imprinting arranged microscopic pits along a concentric ring. The encoding can be 'read' by a finely projected laser beam. Up to 26 000 picture frames can be inscribed, stored and monitored on a single 8-inch disk by using a personal computer. Image data can then be stored for over 50 years. Optical disks can be erased, hence they also can be used as work disks in addition to archival storage. Quality copywork can then be reproduced on the computer printer using film writer paper. Larger 12-inch disks with 54 000 frame capacities can be produced commercially. Each frame can be instantly recalled if its location on the disk is recorded and known. Pictures can include maps and catalogue listings. Disk data can also be 'flagged' according to some shift or break in the information pattern unrelated to the order of presentation, e.g. continental location or some factor of panel content. This feature allows the user to skip from place to place on the disk to order data.

It is possible to transmit digitised data from optical disks to outlet stations established world-wide by using the Wang Integrated Image System (WIIS). Presently only black and white images can be entered into the system, but colour transmission is feasible.

Rip (1983) describes several image enhancement operations that can now be accomplished with computer stored digitised image data:

1. Constant Value Shift: a constant brightness factor can be added to each pixel. This corrects illumination for

under- or over-exposed photographs.

2. Histogram Modification: image mapping can be done by frequency occurrence histogram pixel plotting. General picture contrast enhancement can be done by mapping brightness values.
3. Lineal Stretch Transfer Function; scaling can be increased for pixel variation. Specific image contrast enhancement can be done by increasing variation of brightness values.

Operations 2 and 3 can be combined to further increase contrast enhancement.

4. Principle Components Analysis: comparison of images in each basic colour reduces brightness complexity and sharpens image edges.

Rip (1989) describes the possibilities of digital colour manipulation for hue, saturation, and brightness. If colour chart data are included on photographs, it is possible to correct digital image coding with known original chart colours by using a suitable computer palette program.

After maximal image enhancement is accomplished, a copy can be then reproduced on a colour laser copier for quality verification and professional use such as published illustrations.

The establishment of an archive as herein proposed could provide the following useful services to data supplying contributor clients desiring relating geographical, topical or comparative data:

1. Computer printout of archive accessions lists.
2. Computer printout of pertinent catalogue lists of specific pictures.
3. Digitised image data put on contributor-furnished disks or tapes. Read Only Memory compact disks (CD-ROM) would be the most suitable. Their high capacity allows up to 250 digitised images per disk.
4. Computer printouts of pictures.
5. Imaged-enhanced laser prints (availability would probably have to be limited, expensive and time-consuming to archive personnel).
6. Comparative graphic image data (possible future capability).

Acknowledgments

I wish to acknowledge explanation of technical information supplied me by David A. Schoen, Clifford L. Ogleby, Karen Eilene Saenz and David J. Caucci.

REFERENCES

DICKMAN, J. L. 1986. An image digitising and storage system for use in rock art research. *Rock Art Research* 1: 25-32. Melbourne.

EASTMAN KODAK 1987. Yes it's Kodachrome Professional Film (PC-2). Rochester.

RIP, M. R. 1983. Digital recording and image processing of rock art by computer. *South African Archaeological Bulletin* 38: 77-79.

RIP, M. R. 1989. The use of colour space transformations for the enhancement of rock art images by computer. *Rock Art Research* 6: 12-16.

23

PETROGLYPHS: FROM TRIDIMENSIONAL RECORDING IN THE FIELD TO COMPUTERISED FILING

Dario Seglie, Piero Ricchiardi, Mauro Cinquetti, Tommaso Guiot and Roberto Bosio

Abstract. The method of thermoplastic mono-matrix casting is briefly described, which replaces earlier casting methods. Also, the storage and manipulation of large quantities of data and images on video disks is discussed.

Research and survey methods

The first records of rock, dolmen and menhir engravings date back some centuries, even though there is good reason to believe that some have always been known to local populations. In the early years of this century the systematic survey of rock engravings was seriously undertaken in the western Alps. The English researcher Clarence Bicknell, working at Mount Bego in the Maritime Alps, near Tenda, made numerous drawings and casts, mostly of wet paper and glue. About 1930, when the first systematic surveys in Valcamonica and in the Cottian Alps were initiated, some researchers obtained plaster casts directly from engraved rocks. The taking of photographs, now irreplaceable, gave a whole vision of the surroundings even then, and drawings proved to be useful.

Since World War II enormous progress has been made. A system of survey that embraces all the rock engravings of a defined archaeological site has been employed for years by E. Anati at Valcamonica. The rock is thoroughly washed with water and traces of moss and lichen are eliminated by means of soft vegetal fibre brushes that do not abrade the rock. When the surface has dried, a coat of contrasting paint is evenly applied so that the distinct colour contrast between the original surface and the hollow parts highlights natural striations and engravings. The next step is recording by minutely copying the figures with a special ink pen on transparent polythene. By use of this method, the whole area is recorded.

Phosphorescence photography of rock engravings

In 1970, H. Kuhn employed fluorescence photography to record engravings of Greek stelae. However, this procedure may be contraindicated because fluorescent substances are used instead of phosphorescent ones to photograph open air rock engravings. The use of a battery Wood lamp, whose low intensity is required, makes long exposure necessary. Moreover, its use causes a strong blue chromatic shift on the photogram with colour temperatures of 30 000α K. In fluorescent shots of lichen-colonised engraved rocks, the lichens themselves show a particular fluorescence that disturbs the image.

In 1986, the Centro Studi e Museo d'Arte Preistorica of Pinerolo developed a photographic reproduction method based both on phosphorescence and on a *two-coloured* technique developed through ten years of experience at the Centro (Figure 1). Its basic principle is to employ the phosphorescence emission of some substances in contrast to fluorescence. Phosphorus persists even after the excitation of substances by a source of light has ceased. At the end of the operation, both the phosphorescent liquid and the black covering coat (used in the *two-coloured* technique) can be sponged out.

Tridimensional recording of rock engravings

The earliest method used is plaster casts directly made on the surface of the monument we want to reproduce. However, the water and plaster mixture solidifies on the rock anchoring to asperities, causing undercutting. This method requires the transport of a large amount of plaster and water in order to prepare the mixture. Another method, formerly used by the Prehistoric Art Study Centre in Pinerolo, employs a sculpture plasticine board. A plasticine foil is pressed on the engraved rock surface and then removed by means of a special perforated slab acting as a sucker. Even in this case, some negative aspects greatly limit its employment. Therefore, we began broad experimentation to establish a more reliable procedure that avoided monument damage, the negative cast with thermoplastic mono-matrix (SRB System). The first step is to wash and clean the rock; then it must be moistened. The thermoplastic resin, melted by suitable heat (the melting point of the material is about 80°C) is applied in thin layers. As the resin comes into contact with the water on the rock surface it reticulates and solidifies almost immediately. The essential requirement of the thermoplastic resin employed is its completely neutral chemical reaction which neither alters nor damages the rock. The molds are packed and transported to the laboratory where a positive cast is made which precisely reproduces the original rock surface. Such casts become scientific documents and serve as aids in future research.

Computers in rock art studies

In the western Alps, as in other areas, there are problems due to lack of records, or the incompleteness of records. Hence, there is a need to collect the largest possible amount of data in a suitable centre and to adopt computer-aided techniques for data recording, organisation and processing.

Some basic requirements of a recording system for rock art are:

1. Capability of storing and managing large quantities of descriptive and codified data.
2. Capability of recording images and keeping them asso-

ciated with the data.

3. Adequate speed in finding images and data.
4. Assembling of data according to their characteristics or groupings.
5. A calculation of phenomena frequency and of statistical parameters.
6. Easy updating, including image recording.

The best solution is to choose a video disk system. As a suitable hardware we suggest the existing data image recording system based on TQ2026-F, the IDBS system (Figure 2). Two important IDBS capabilities are: (1) recordings of 96 000 images with 150 characters to describe the image, and (2) an image finding-time of less than two seconds.

BIBLIOGRAPHY (not cited)

ALMGREN, O. 1934. *Hällristningar och Kultbruk*, Stockholm.

ANATI, E. 1966. Metodi di archivio e di analisti dell'arte rupestre. *Bollettino del Centro Camuno di Studi Preistorici* 2: 133-55.

ANON. 1975. La photographie en archéologie - Les techniques spéciales-ultraviolet et fluorescence (UV). *Les dossiers de l'archéologie* 13: 110-11.

BESSONE, G., R. FONTANINI, P. RICCHIARDI and D. SEGLIE. 1968. Ritrovamenti di incisioni rupestri nel Pinerolese. *Actes du Congrès des Sociétés Savantes de Savoie, Saint-Jean-de-Maurienne*, pp. 46-56.

BORGNA, G. C. 1980. *L'arte rupestre preistorica nell'Europa Occidentale*. Pinerolo.

BREUIL, H. 1952. *Quatre cents siècles d'arte pariétal*. Montignac.

CINQUETTI, M. 1985-86. Survey sulla Rocca di Cavour. *Survey - Bollettino del Centro Studi e Museo d'Arte Preistorica di Pinerolo* 1-2: 7-16.

COïSSON, O. 1973. Caratteristiche delle incisioni rupestri della Val Pellice. *Atti del 1º Convegno di Studi preistorici*, pp. 59-74. Pinerolo.

COïSSON, O. 1980. Incisioni rupestri del versante italiano delle Alpi Marittime e Cozie. *Ampurias* 43: 79-95.

FOZZATI, L., and R. NISBET. 1985. Rilevamento archeologica. 1983-84 (tavv. XXI-XXII). *Quaderni della Soprintendenza Archeologica del Piemonte* 4: 33-34. Cavour, Rocca.

FOZZATI L., R. NISBET, and D. SEGLIE 1984. Rilevamento archeologico. *Quaderni della Soprintendenza Archeologica del Piemonte* 3: 272. Cavour, Rocca.

FREDSIÖ, A. 1956. *Göteborgstraktens hällristingar*. Göteborg.

FREDSIÖ, A., and A. WESTHOLM. 1964. *Images de l'Age du Bronze*, Göteborg.

GAMBARI, F. M., and M. VENTURINO GAMBARI. 1983. Rapporti tra le culture preistoriche piemontesi e le aree transalpine: nuovi dati e proposte interpretative. *Bulletin de la Société de Recherches et d'Etudes préhistoriques alpines d'Aoste - Actes du IIIe colloque sur les Alpes dans l'Antiquité, Aoste 10-12 sept. 1982* 15: 99-124.

GERSAR (ed.) 1980. Apercu sur l'art rupestre de Haute-Maurienne. *Les Cahiers du GERSAR* 2.

GERSAR (ed.) 1983. *Art Rupestre*.

GRAZIOSI, P. 1956. *L'arte dell'antica età della pietra*. Firenze.

JALLA, J. 1933. Alcuni dati sulle antichita rintracciate nel Pinerolese. *Bollettino della Società Piemontese di Archeologia e Belle Arti* 18(1-2): 30-34.

KUHN, H. 1970. *Photographie par fluorescence de stèles funéraires grecques*. München.

LAMING-EMPERAIRE, A. 1962. *La signification de L'arte rupestre paléolithique*. Paris.

LEROI-GOURHAN, A. 1971. L'Homme et la matiére. *Evolution et techniques*, I (3rd ed.).

LEROI-GOURHAN, A. 1973. Milieu et techniques. *Evolution et techniques*, II (3rd ed.).

LEROI-GOURHAN, A. 1978. L'expression du temps et l'animation des figures au Paléolithique. *Systèmes de signes. Hommage à Germaine Dieterlen*, pp. 359-67. Paris.

NELH, G. 1983. Pierres à cupules et roches gravées en Savoie. *L'histoire en Savoie* 17 (71): 1-24.

NISBET, R., and R. BOSIO. 1986. Archeologia e preistoria alpina. *Introduzione all'antropologia e all' archeologia territoriale: le Alpi Occidentali*. Regione Piemonte, Torino 49-81.

NISBET, R., and D. SEGLIE. 1983. *Balm'Chanto - Archeologia della Val Chisone*. Pinerolo.

NISBET, R., and D. SEGLIE. 1983. Cavour, Rocca. Rilevamento archeologico (tav. LVI). *Quaderni della Soprintendenza Archeologica del Piemonte* 2: 177-78.

PIVA, A. 1931. Figure e incisioni rupestri. Richerche preliminari eseguite sulle montagne Pinerolesi. *Bollettino della Società Piemontese di Archeologia e Belle Arti* 15 (3-4): 79-83.

PONS, S. 1929. *Un antico disegno nella valle della Germanasca: il disegno di ponte Raut, Relazione presentata al Convegno Nazionale degli Speleologi d'Italia*, Istituto Italiano di Paleontologia Umana. Firenze.

PRIULI, A. 1983. *Incisioni rupestri nelle Alpi*. Ivrea.

RICCHIARDI, P. 1982. Il masso altare di Cro da Lairi. *Novel Temp* 19: 21-25.

SANTACROCE, A. 1968-69. Brevi notizie sulle incisioni rupestri ed alcuni suggerimenti per la loro ricerca. *Bulletin d'Etudes préhistoriques alpines*, pp. 122-67.

SEGLIE, D., and P. RICCHIARDI. 1975. Incisioni rupestri preistoriche nelle Alpi occidentali italiane. Metodi di rilevamento e di analisi. *Atti del congresso: Il Moncensisio e la sua Regione, Lanslebourg-Susa 5-7* 401-34.

SEGLIE, D., and P. RICCHIARDI. L'arte rupestre preistorica nelle Alpi Cozie (Sistemi di e rilevamento di studio). *Atti del congresso sul bimillenario della città di Aosta, 5-20 ottobre 1975*, Istituto Internazionale di Studi Liguri, Bordighera. 119-33.

SEGLIE, D., and P. RICCHIARDI. 1980. New archaeological methods on the three-dimensional reproduction of the prehistoric rock engravings (petroglyphs). *Actas del Congresso Nacional de Argueologia*. Lisbon.

SEGLIE, D., P. RICCHIARDI, M. CINQUETTI, and J. LIEGEOIS. 1975-76. Le incisioni rupestri della Punta di Marato (Sartene) - Inventario dell'arte rupestre della Corsica. *Giornale Storico della Lunigiana* 1-4, Instituto Internazionale di Studi Liguri, pp. 117-30.

SEGLIE, D., P. RICCHIARDI and O. COïSSON. 1978. Decouvertes recentes de gravures rupestres sur le versante italian des Alpes Cottiennes. *Actes du XXVII Congrès des Sociétés Savantes de Savoie, Thonon Les Bains*, pp. 11-29.

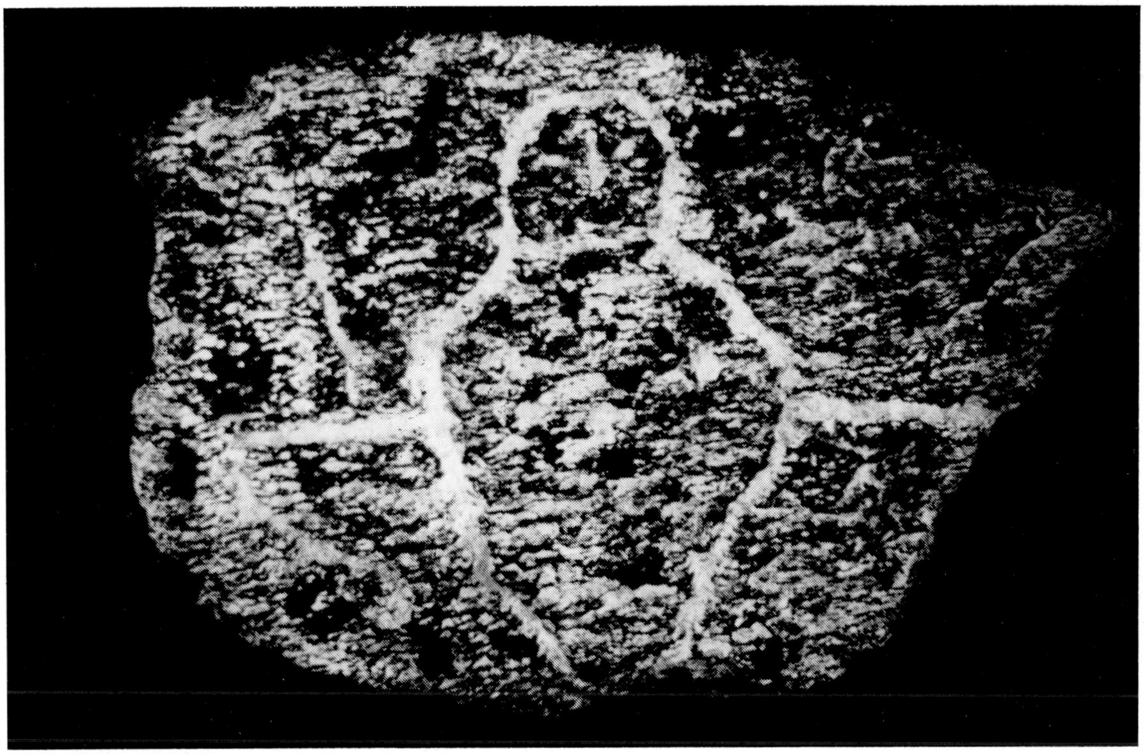

Figure 1. Phosphorescence photography of petroglyph. Eneolithic stele, Miandassa, Cottian Alps, Italy.

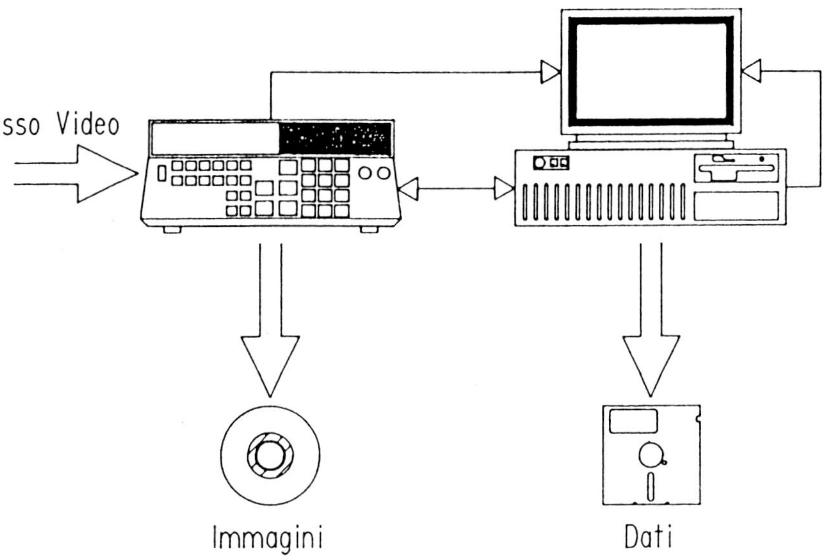

Figure 2. Data image recording system: the IDBS system for rock art.

24

GETTING ORGANISED: RECORDING OF ABORIGINAL ART SITES IN QUEENSLAND, AUSTRALIA

M. J. Rowland

Abstract. Prior to 1981 the Archaeology Branch (Queensland) site files were developed and managed on an ad hoc basis. Since then a new series of site cards have been developed and site information computerised. This has enabled some basic assessment of site types, site distributions and effectiveness of site cards to be made. Problems identified include an over-representation of art sites relative to other site types, a geographical bias in sites recorded and a lack of information on site preservation.

Introduction

Government involvement in archaeology in Queensland commenced officially in 1967 with the proclamation of the Aboriginal Relics Preservation Act and establishment of an Archaeology Branch within the Department of Aboriginal and Islander Affairs (now Heritage Section of the Department of Environment and Heritage). However, a Site Register was not commenced until 1971 and a permanent Site Registrar was not employed until 1989 following replacement of the 'Relics Act' by the Cultural Record (Landscapes Queensland and Queensland Estate) Act, 1987.

A review of the existing Site Register in 1981 revealed that it was poorly organised and managed and inadequate as a data base for either research or assessments of significance, although based on the Register and personal experience, Morwood (1984) was able in 1980 to undertake an overview of 969 recorded art sites.

The 1981 review recommended a restructuring of the Site Register focusing on a reorganisation of the site cards and filing system and secondly on the need to develop a system for assessing site significance. Implementation of these recommendations has resulted in the development of a comprehensive set of site cards, an organised archival system, computerisation of all data and adoption of a biogeographical zonal system as a basis for assessment of significance (Rowland 1986, 1989). It is now possible to manipulate the data stored in the Site Register for research and management purposes and the following discussion attempts to do that in the case of art sites.

The site cards

The SITE INDEX form and ART SITE form are shown in Figure 1. These and other specialised recording forms are discussed in more detail elsewhere (Rowland 1986, 1989; Johnston and Rowland 1987). It is important to note that only two site TYPES are defined; i.e. either an open site or cave/rockshelter site. Paintings and engravings are not defined as site types but as SITE ATTRIBUTES, i.e. a particular rockshelter may have a number of attributes (e.g. paintings, engravings, a burial, stone or shell deposits etc.). It should also be noted that the SITE INDEX sheet includes a substantial section requiring information on site condition and recommendations regarding preservation and management. The option to provide this sort of information was lacking from previous site cards and unfortunately was rarely supplied.

Descriptive aspects of SITE ATTRIBUTES are placed on a series of 8 separate colour-coded cards designed to obtain basic information relevant to research and management objectives and can be modified as field experience and changing values necessitate.

The ART ATTRIBUTE card seeks some basic information on site dimensions, area of art, state of preservation and a basic census of motifs. With some notable exceptions, sites recorded prior to 1983 did not contain this information.

Art sites recorded in Queensland

Under Section 23 of the Aboriginal Relics Preservation Act all Aboriginal relics located since 1967 were supposed to be reported to the Heritage Section. In reality this has not happened and the following discussions which are based only on the Heritage Section Site Register may therefore be at variance with the distribution of art sites known to some researchers. It is not possible to estimate how many sites have actually been recorded but not reported to the Section.

Number of sites

There are currently 4427 sites with 5591 attributes on the Site Register (Table 1). Art (paintings and engravings) is clearly the commonest attribute recorded (total 26.4%) followed by artefact scatter (22.3%) and midden (14.2%). However, neither the total number of art recordings nor the relative percentage of art sites can be considered to be representative. For example, in respect to numbers of sites, Morwood (1984: 130) has estimated that there may be well over 50 000 art sites in Queensland. Nevertheless it appears that obtrusive archaeological features such as art sites may have been recorded more frequently than unobtrusive features such as burials. These apparent biases in types of sites recorded can be further demonstrated by an analysis of site types recorded per category of recorder.

Recorder types

The Heritage Section Site Register developed throughout the 1970s and early 1980s via field recordings by Heritage Section staff, researchers, Honorary Wardens and members of the public. Table 2 shows the site types recorded by each of these categories of site recorder.

This table demonstrates that rangers, Wardens and the

public have tended to record art sites at the expense of other site types while researchers and consultants have tended to record middens and artefact scatters to a greater extent. Neither group can be seen to have recorded a representative cross-section of sites. Art sites have been over-represented because of their obtrusiveness and midden and artefact scatters because of their tendency to be more numerous in areas where Environmental Impact Surveys have been undertaken.

Distribution of art sites

Figure 2 shows the distribution of art sites (paintings and engravings) by 1:250 000 map sheet and demonstrates the relative concentration of art in particular areas. However, it is difficult to know to what extent the distribution is real or a reflection of recorder bias. As a rough guide it is suggested that paintings and engravings should not represent more than 20% (in reality probably far less) of the recorded sites for a map sheet (Fig. 3). Using this figure, 38 map sheets are shown to have an over-representation of art sites while only 22 have the 'right' number. Figure 3 also indicates those map sheets for which 10 sites or more (32 map sheets) have been recorded but none of which are art sites, thus giving some idea of 'non-art' areas. Thirty-five map sheets have insufficient sites recorded to make any assessment.

Analysis at the map sheet level gives some idea of art site distribution and biases in distribution. However, analysis at this level does not assist in assessing significance. Therefore, an attempt was made to develop a system for assessing site significance.

Assessing significance

Archaeological significance can be assessed at a variety of levels from local to regional, to national and international. Until recently sites were assessed in Queensland on an ad hoc basis in terms of their significance state-wide. Since Queensland is so large (1.7 million sq. km.) and its archaeological universe so poorly known this has proved to be an inadequate basis for assessment and an attempt has therefore been made to develop a more appropriate basis for assessment based on biogeographic zones (Johnston and Rowland 1987).

Significance assessments in Australia have come to rely on the concept of 'representativeness' (Bowdler 1981, 1983, 1984; Flood 1984; Witter 1984). Definitions of the spatial limits in which such assessments should be made have been limited (Flood 1984; Witter 1984; Blain and Davis 1984).

Representativeness rose to prominence in Australian archaeological literature in 1981 when Bowdler identified it and 'timely and specific research questions' as the two criteria most important in assessing *scientific* significance of archaeological sites. Bowdler used a statistical definition of representativeness which she then translated into a workable archaeological definition as follows:

> Representativeness is the degree to which sites in the investigated (perhaps threatened) area are representative of sites known elsewhere, and where they might be better protected (Bowdler 1981: 128-29).

Representativeness in this sense, however, is a crisis management and assessment tool. It also contains an inherent problem in that the parameters of the sample universe are poorly known. For example, at the local level we may have some success in obtaining a representative sample but as the unit of analysis gets larger the reliability of the sample must inevitably be reduced. This problem was resolved by Bowdler (1983: 40) in introducing 'context' into the definition of representativeness. However, while her examples provided a high level of resolution at the local site level, few attempts have been made to assess context at the regional or national level (Flood 1984; Witter 1984).

Hughes and Sullivan (1982, 1984) have shown that the types, frequency and preservation potential of Aboriginal archaeological sites in Australia is strongly correlated with local environment and specifically with local bedrock geology, landforms, soils, vegetation and climate. The approach probably owes much to American developments (Butzer 1971, 1982) and more specifically to sampling techniques based on ecological and geomorphic zones (Judge, Elbert, Hitchcock 1975) and ultimately to Binford (1964: 433) who defined the region as the unit most applicable to analysing populations of sites. Biocultural areas may also correlate to some degree with such regions though deterministic connotations need to be avoided.

To ensure representation and conservation of significant sites in Australia, Flood (1984: 58) has proposed that the Australian Heritage Commission obtain representation of sites in every different region of Australia using one of the 300 biophysical regions defined by Laut et al. (1975). The Laut et al. system, however, divides Queensland into 104 regions, some of which recur up to 7 times in different locations, giving a total of 181 separate classified areas. This is further complicated by the fact that the areas may be up to 600 km apart. Choosing a system appropriate for Queensland thus involved a choice between the complexity of Flood's (1984) scheme and a broader scheme that at the same time was sensitive enough to discriminate between areas with fundamentally different environments and archaeology.

Queensland biogeographical regions

Twelve major biogeographical regions have been defined for Queensland ranging from fairly uniform ones like the Desert Uplands, to regions like the Gulf Plains, with a number of similar yet distinctive sub-regions (Stanton and Morgan 1977: 3) (Fig. 4). A small area of the state on the New South Wales/Queensland border was not included in the report by Stanton and Morgan but is defined by Wills (1976) as a distinct region and is included here in line with the Queensland National Parks and Wildlife Service divisions (Sattler 1986).

The regions for the moment are broad enough to overcome problems of scale and yet include sufficiently different environments to be meaningful. However, it is obvious that in the future, regional sub-divisions will need to be made.

Table 3 indicates the number of art sites recorded per region as a percentage of total art sites recorded in Queensland and hence provides an order of regional significance. However, not all regions are the same size, and art site density may therefore be a more useful criteria. Table 4 shows the number of art sites per sq. km per region and this indicates a different order of significance.

It can be noted that there is a clear distinction between five regions with densities of one art site per 500-600 sq. km, three regions with a density of one art site per 2500-4500 sq. km and five regions with one site per 10 000-30 000 sq. km. When these regions are mapped (Fig. 5), a rather obvious point can be made - i.e. art sites are commonest along the Great Dividing Range, where suitable rockshelters are available, and in the North West Highlands where such locations would also be available. However, again it should be noted that art sites have been over-recorded in some regions while others have not yet been fully explored.

Table 5 shows the number of art sites recorded per region as a percentage of the total sites recorded for the region. Again, using the purely arbitrary figure of 20%, art sites may be said to be over-represented in 8 of the 13 regions (see Fig. 6), though this of course does not imply that they are not significant art regions.

Analysis of distribution and content of art sites

It has been demonstrated above that the numbers and distribution of art sites on the site file is a reflection of a number of biases and therefore they cannot be seen to be a true representation of real distributions. Nevertheless, sufficient sites have been recorded over a wide enough geographical area that some preliminary analyses are possible.

Distribution of painting sites vs engraving sites

Figure 7 shows the distribution of painting sites and engraving sites relative to one another. Of the 69 map sheets with recorded art sites, paintings and engravings are recorded together on 27 (39%) occasions. Art sites only are recorded on 32 occasions (46%) and engravings only, on 10 (14%) occasions. Due to the biases in the data base these differences cannot be pursued too far. However, Figure 8 summarises these trends in the relative distribution of painting and engraving sites which indicates a trend towards a predominance of paintings in coastal areas, a relative balance of paintings and engravings over a substantial part of inland Queensland, and a predominance of engravings in south-west Queensland.

Content of art sites

The new Heritage Section site recording forms contain a number of variables relating to site dimensions, conservation, motifs, techniques etc. However, these details have not yet been computerised. Furthermore, the majority of art sites were recorded prior to the introduction of these site cards so information is generally not available for these sites. In 1980, Morwood (1984) was able, from the site records and his own extensive field experience, to identify 4 distinct art areas and these are summarised in Figure 9. Considerably more work has been completed since Morwood's review (David 1988; Franklin 1988; Flood 1987; Flood and Horsfall 1986; Morwood 1985 and Morwood and Gibson 1984) and more is in progress (e.g. Elizabeth Hatte, Mike Morwood, Bruno David, Natalie Franklin and Jo Flood) so that further distinctions will be possible in the future. David (1988), for example, has defined a Chillagoe-Mungana art style which is distinct from Koolburra, Laura and the Mitchell-Palmer area.

Motifs

An attempt to abstract from the computer information on motifs is difficult due to the factors outlined above. However, sufficient information was obtained to provide a basic analysis in Figures 10 and 11. Insufficient information is available concerning engravings to suggest more than a tentative four part division (Fig. 10). More of a pattern appears to be apparent in the distribution of painting sites (Fig. 11); however, even this must be seen to be a tentative division to be tested by work in progress.

Conclusions

The Heritage Section site filing system is now organised to the extent that some basic analysis of its content can be undertaken. However, it is apparent from such an analysis that the sites on record cannot be viewed as a representative selection of either known site types or their geographical distributions. It is therefore difficult to assess site significance since one of the major criteria of significance is a site's representativeness (cf. Bowdler 1981: 129). It is apparent, for example, that art sites are over-represented at the expense of other sites both on a state-wide and regional level.

The analysis presented in this paper suggests a number of basic approaches are necessary to further advance the recording and assessment of art sites in Queensland. Firstly, it is apparent that a substantial number of recorded art sites have resulted from unsystematic recording by rangers, Wardens, Park Rangers and the public. These recordings have resulted in a bias in the distribution and in the relationship of these sites to other site types. This type of casual random recording will undoubtedly continue and this is to be encouraged in order to build up the data base of sites. Since Queensland is so large and contains a small dispersed population (a total population of 2.4 million at a density of 1.4 persons per sq. km), the above categories of people provide a network of 'eyes and ears' across such a large landmass that could not be achieved by professional archaeologists alone. Indeed, the magnitude of a systematic recording of any site type over the entire area of Queensland is beyond the scope of funding or personnel available now or in the foreseeable future. Nevertheless, some attempt needs to be made to direct casual recorders to record on a more systematic basis at the regional level.

Secondly, there is a need for more systematic work on non-art site types in areas of high art site density. This problem has only recently been addressed (Lilley 1986; Morwood and Godwin 1987) but needs to be expanded to predictive models of both art and non-art sites in particular areas. Researchers undertaking rock art studies should be made responsible for recording other site types in their area of investigation.

Thirdly, there is a need to obtain information on art sites already recorded but for which no site card is available. The Heritage Section site file is now the most accessible and organised in the State and should provide a useful tool for all researchers. However, for a variety of reasons, people have not provided the Section with site cards. In accordance with the legal requirements of the Cultural Record Act 1987 all sites should be reported to the Section. The computer data base has levels of access and permits are required to undertake research so that a high level of security and confidentiality can be provided.

Fourthly, sites recorded prior to 1983 contain serious deficiencies in two aspects of data recording. In these recordings a basic census of motif types and range of ochres used is generally absent and there is little discussion of site preservation. Consequently, information on these sites cannot generally be used for either research or site management programs. The new site cards now include space for recording of both these types of information. However, there is still a need to encourage the public to record this data and there is considerable work still to be done in revisiting sites for which we have little other information than 'art site'. Even archaeologists at times seem to lapse in their description of art sites. For example, many site records include in the form of descriptions a 'wide range of motifs', 'a few stencils' and no other information is provided.

Fifthly, there is a need to continue to test the relevance of the regional biogeographic system as a planning matrix for both research and management purposes. Two regions (the Mitchell Grass Downs and Desert Uplands) have been selected for study on the basis that little is known of their archaeological resources (see Table 6). Preliminary overview of these regions has been commenced with an emphasis on establishing research and management potentials and overcoming numerical and distributional biases in site data.

Finally, conservation/preservation of art sites in Queensland has so far been restricted to fencing, sign-posting, construction of walkways and low profile barriers. This has tended to be done on an ad hoc basis, i.e. sites that are impressive, or easy to get at, have been managed rather than a selection of those based on representativeness/significance. Due to lack of resources this will probably continue to be the basic approach followed. However, it is hoped that the more systematic surveys outlined above will allow for more valid assessments to be made.

Summary

Art sites in Queensland can be viewed in the same way as all other archaeological site types. Much of the recording of sites has been unsystematic and of an inadequate standard. Recently the Heritage Section has reorganised its files, redesigned its site cards and begun to manipulate the data available in such a way that it is now possible to recognise what we know or do not know about site types and site distributions. The potential is now available to the Section to direct researchers and others in more systematic recording, and when funds become available, to undertake more systematic research on its own initiative. While recent rock art research projects undertaken in Queensland have been of a high standard, there is a long way to go in establishing a Queensland-wide or even regional perspective on art sites. Caution is necessary in making comparisons between systematic local studies and the general Queensland distribution of art sites. Information on sites that have been recorded should be forwarded to the Branch as a matter of urgency so that an adequate data base for research and management purposes can be established.

Acknowledgments

Harvey Johnston made a substantial contribution in getting the computer system up and running and his help is gratefully acknowledged.

REFERENCES

BINFORD, L. R. 1964. A consideration of archaeological research design. *American Antiquity* 29: 425-41.

BLAIN, H. D. and J. R. DAVIS 1984. An archaeological site potential map of Australia. *Technical Memorandum 84/28*, CSIRO, Institute of Biological Resources, Division of Water and Land Resources, Canberra.

BOWDLER, S. 1981. Unconsidered trifles? Cultural resource management, environmental impact statements and archaeological research in New South Wales. *Australian Archaeology* 12: 123-33.

BOWDLER, S. 1983. *Aboriginal Sites on the Crown Timber Lands of New South Wales*. Forestry Commission of New South Wales, Sydney.

BOWDLER, S. 1984. Archaeological significance as a mutable quality. In S. Sullivan and S. Bowdler (eds) *Site surveys and significance assessment in Australian archaeology*, pp. 1-9. Department of Prehistory, Research School of Pacific Studies, Australian National University, Canberra.

BUTZER, K. W. 1971. *Environment and archaeology*. Aldine-Atherton. Chicago.

BUTZER, K. W. 1982. *Archaeology as human ecology*. Cambridge University Press, Cambridge.

DAVID, B. 1988. The rock pictures of the Chillagoe-Mungana and Mitchell-Palmer limestone belts of north Queensland: reconstructing cultural relations in Queensland's prehistoric past. Final report to Australian Heritage Commission, Canberra.

FLOOD, J. 1984. More or less significant. A national perspective on assessing the significance of archaeological sites. In S. Sullivan and S. Bowdler (eds) *Site surveys and significance assessment in Australian archaeology*, pp. 55-60, Department of Prehistory, Research School of Pacific Studies, Australian National University, Canberra.

FLOOD, J. 1987. Rock Art of the Koolburra Plateau, north Queensland. *Rock Art Research* 4: 91-126.

FLOOD, J. and HORSFALL, N. 1986. Excavation of Green Ant and Echidna Shelters, Cape York Peninsula. *Queensland Archaeological Research* 3: 4-64.

FRANKLIN, N. 1988. *Rock Art of the Laura region north Queensland*. Report to the Archaeology Branch, Department of Community Services.

HUGHES, P. J. and SULLIVAN, M. E. 1982. Geoarchaeology in Australia. A review. In W. Ambrose and P. Duerden (eds) *Archaeometry: An Australian perspective*, pp 100-11. Department of Prehistory, Research School of Pacific Studies, Australian National University, Canberra.

HUGHES, P. J. and M. E. SULLIVAN. 1984. Environmental approaches to assessing archaeological significance. In S. Sullivan and S. Bowdler (eds) *Site surveys and significance assessment in Australian archaeology*, pp. 34-47. Department of Prehistory, Research School of Pacific Studies, Australian National University, Canberra.

JOHNSTON, H. and M. J. ROWLAND. 1987. Representativeness, conservation and the Archaeology Branch (Queensland) Site Register. Report to the Archaeology Branch, Department of Community Services, Brisbane, Queensland.

JUDGE, W. J., J. I. EBERT and R. K. HITCHCOCK 1975. Sampling in regional archaeological surveys. In J. W. Mueller (ed.) *Sampling in archaeology*, pp. 82-123. University of Arizona Press, Tucson.

LAUT, P., MARGULES, C. and NIX, H. A. 1975 *Australian biophysical regions. A preliminary regionalisation*. Department of Urban and Regional Planning, Canberra.

LILLEY, I. 1986. An open site survey of Koolburra Plateau, Cape York Peninsula. *Queensland Archaeological Research* 3: 65-71.

MORWOOD, M. J. 1984. Aboriginal art sites in Queensland: A summation. *Rock Art Research* 1: 130-32.

MORWOOD, M. J. 1985. Facts and figures. Notes on rock art in the Mt Isa area, northwestern Queensland. *Rock Art Research* 2: 140-45.

MORWOOD, M. J. and M. GIBSON 1984. Incised stones from Glenormiston Station, south west Queensland. *Memoirs of the Queensland Museum* 21: 561-65.

ROWLAND, M. J. 1986. Archaeological research, site recording and management, Archaeology Branch, Department of Community Services, Brisbane. *Australian Archaeology* 23: 104-14.

ROWLAND, M. J. 1989. The current state of the site files, Department of Community Services, Queensland. In J. Flood, I. Johnson and S. Sullivan (eds) *Sites and bytes. Recording Aboriginal places in Australia*, pp. 262-94. Special Australian Heritage Commission Publication Series No. 8. Australian Government Publishing Service, Canberra.

SATTLER, P. S. 1986. Nature conservation in Queensland: Planning the matrix. *Proceedings of the Royal Society of Queensland* 97: 1-21.

STANTON, J. P. and MORGAN, M. G. 1977. *Report No. 1: The rapid selection and appraisal of key and endangered sites: the Queensland case study*. A report to the Department of Environment, Housing and Community Development. The University of New England, School of Nature Resources, Armidale.

WILLS, A. K. 1976. The granite and trap rock area of south east Queensland. A land inventory. Division of Land Utilization. *Technical Bulletin No. 13*. Queensland Department of Primary Industries.

WITTER, D. 1984. Providing a context for management archaeology.

In S. Sullivan and S. Bowdler (eds) *Site surveys and significance assessment in Australian archaeology*, pp. 48-54. Department of Prehistory, Research School of Pacific Studies, Australian National University, Canberra.

Paintings	1 240	22.2%
Engravings	236	4.2%
Burials	200	3.6%
Stone Circles	25	0.4%
Stone Arrangements	448	8.0%
Earthern Circles	14	0.3%
Earthern Arrangements	61	1.1%
Pathways	4	0.1%
Carved Trees	16	0.3%
Scarred Trees	290	5.2%
Fish Traps	45	0.8%
Wells	79	1.4%
Weirs	2	0.0%
Axe-grinding grooves	128	2.3%
Dwellings	19	0.3%
Heath/ovens	64	1.1%
Quarry	140	2.5%
Shell Middens	793	14.2%
Artefact Scatters	1 244	22.3%
Other	543	9.7%
No of attributes	5 591	100.0%
No of sites	4 427	

Table 1. Total sites and attributes on the Archaeology Branch Site Register.

	Rangers	Wardens	Researchers	Consultants	Public
Paintings	43.1	34.9	15.3	2.3	31.7
Engravings	9.2	3.5	4.4	0.4	4.0
Burials	6.1	5.4	1.8	1.4	6.1
Stone Arrangements	5.3	3.0	0.2	1.5	6.0
Earthern Arrangements	2.0	3.6	0.6	0.0	2.8
Pathways	0.0	0.0	0.1	0.0	0.0
Carved Trees	0.4	0.2	0.1	0.0	0.2
Scarred Trees	6.4	6.2	2.7	5.9	9.6
Fish Traps	1.8	0.7	1.1	0.2	0.5
Wells	1.1	3.2	1.6	0.0	1.7
Weirs	0.0	0.2	0.1	0.0	0.0
Axe-grinding grooves	3.5	2.2	2.0	0.7	3.7
Dwellings	0.7	0.3	0.3	0.0	0.1
Hearths	0.4	1.5	0.5	1.1	0.4
Quarrys	1.9	1.7	2.6	2.9	2.5
Shell Middens	3.9	3.5	29.3	17.0	7.7
Artefact Scatters	4.7	13.2	33.4	58.5	7.2
Other	9.5	16.7	3.9	8.1	15.8
	100.0%	100.0%	100.0%	100.0%	100.0%

Table 2. Percentages of site types recorded by different categories of recorders.

(11)	Brigalow Belt	528	42.2%
(9)	Einasleigh Uplands	233	18.6%
(3)	Cape York Peninsula	199	15.9%
(1)	North West Highlands	160	12.8%
(7)	Wet Tropical Rainforest	47	3.8%
(5)	Channel Country	18	1.4%
(2)	Gulf Plains	17	1.4%
(10)	Desert Uplands	14	1.1%
(12)	South East Queensland	11	0.9%
(4)	Mitchell Grass Downs	10	0.8%
(6)	Mulga Lands	7	0.6%
(8)	Central Coast Rainforest	4	0.3%
(13)	Granite Belt	2	0.2%

Table 3. Art sites per region and percentage of total art sites.

(7)	Wet Tropical Rainforest	1/ 477 sq km
(1)	North West Highlands	1/ 507 sq km
(3)	Cape York Peninsula	1/ 546 sq km
(9)	Einasleigh Uplands	1/ 563 sq km
(11)	Brigalow Belt	1/ 628 sq km
(13)	Granite Belt	1/ 2 591 sq km
(8)	Central Coast Rainforest	1/ 3 885 sq km
(10)	Desert Uplands	1/ 4 441 sq km
(2)	Gulf Plains	1/ 10 160 sq km
(12)	South East Queensland	1/ 10 363 sq km
(5)	Channel Country	1/ 12 090 sq km
(4)	Mitchell Grass Downs	1/ 25 735 sq km
(6)	Mulga Lands	1/ 29 609 sq km

Table 4. Art sites per square km for each region.

Einasleigh Uplands	67%
Granite Belt	66%
Cape York Peninsula	57%
North West Highlands	51%
Brigalow Belt	44%
Desert Uplands	37%
Wet Tropical Rainforest	36%
Gulf Plains	26%
Central Coast Rainforest	17%
Mitchell Grass Downs	14%
Channel Country	4%
Mulga Lands	2%
South East Queensland	1%

Table 5. Number of art sites recorded per region as a percentage of total sites recorded for the region.

	LAND UNIT	AREA (SQ KM)	NUMBER OF SITES No	%	SITES/ SQ KM
1	North West Highlands	81 178	215	5.5	1/ 377
2	Gulf Plains	172 720	59	1.5	1/2 927
3	Cape York Peninsula	108 814	222	5.7	1/ 490
4	Mitchell Grass Downs	257 353	35	0.9	1/7 352
5	Channel Country	217 627	347	8.9	1/ 627
6	Mulga Lands	207 264	272	7.0	1/ 762
7	Wet Tropical Rainforest	22 454	96	2.4	1/ 233
8	Central Coast Rainforest	15 540	23	0.6	1/ 675
9	Einasleigh Uplands	131 267	315	8.1	1/ 416
10	Desert Uplands	62 179	38	1.0	1/1 636
11	The Brigalow Belt	331 622	1 140	29.1	1/ 290
12	South East Queensland	113 995	977	24.9	1/ 121
13	Granite Belt	5 182	3	0.1	1/1 727
	Land Unit not determined	-	169	4.3	-
	TOTAL	1 727 195	3 911	100.0	(1/441 av)

Table 6. Number of archaeological sites per biogeographical region.

SITE INDEX FORM STATE FILE NUMBER

ARCHAEOLOGY BRANCH GPO BOX 2210 BRISBANE 4001

SITE TYPE ☐ OPEN SITE ☐ CAVE /ROCKSHELTER SITE

SITE ATTRIBUTE
☐ Painting(s)
☐ Engraving(s)
☐ Burial(s)
☐ Stone Circle(s)
☐ Stone Arrangement(s)
☐ Earthern Circle(s)
☐ Earthern Arrangement(s)
☐ Pathway(s)
☐ Carved Tree(s)
☐ Scarred Tree(s)
☐ Fish Trap(s)
☐ Well(s)
☐ Weir(s)
☐ Axe Grinding Groove(s)
☐ Dwelling(s)
☐ Hearth/Oven(s)
☐ Quarry(s)
☐ Shell Midden(s)
☐ Artefact Scatter
☐ Other ....

MATERIALS
☐ Stone
☐ Shell
☐ Bone
☐ Wood
☐ Vegetable
☐ Charcoal
☐ European material
Other ....

STRUCTURE
☐ Surface Scatter
☐ Buried Single Component
☐ Multi-Component or Stratified Site
Other ....

INDICATE WHICH OF THE FOLLOWING SITE CARDS ARE ATTACHED:
☐ Art Site (orange)
☐ Carved/Scarred Trees (brown)
☐ Burial (grey)
☐ Stone and Earthern Arrangements (yellow)
☐ Fish Trap
☐ Weir
☐ Axe Grinding Grooves
☐ Native Well
☐ Dwelling
☐ Hearth/Oven
☐ European material
(buff)
☐ Stone Materials (green)
☐ Plant and Animal Material (blue)
☐ Consultation Form (Red)

BRIEF DESCRIPTION OF SITE: ....

RECORDER
Name .... ☐ Ranger Recorders Reference ....
Address .... ☐ Warden Site Name ....
☐ Researcher Date Recorded ....
☐ Consultant
Telephone .... Other .... Permit Number ....

LOCATION Map Sheet Name .... Edition Number ....
Scale .... Grid Reference ....
Parish .... Portion Number .... Property Name ....
Airphoto run number .... Latitude .... Longitude ....
What other Sites are known in the Area ....

LAND TENURE ☐ Freehold ☐ Leasehold ☐ Reserve ☐ Crown Land
☐ National Park ☐ Forestry ☐ Mining lease ☐ Aboriginal
Other ....

Owner/Leaseholder
Name ....
Address ....
Telephone ....

Manager
Name ....
Address ....
Telephone ....

LAND USE ☐ Forested ☐ Cleared ☐ Ploughed ☐ Cultivated ☐ Grazed
☐ Built on ☐ Used for recreation ☐ Undeveloped
Other ....

A

ENVIRONMENT
Land Unit System (OFFICE USE) ....

SITE ENVIRONMENT
Topography
☐ River Flat ☐ Creek Bank ☐ Estuary ☐ Cliff Line ☐ Dune
☐ Lake Edge ☐ Swamp ☐ Plain ☐ Slope ☐ Hill Top
Other ....

Parent Rock
☐ Quartz ☐ Quartzite ☐ Silcrete ☐ Granite ☐ Porphory
☐ Sandstone ☐ Slate ☐ Shale ☐ Limestone
Other ....

Soil
☐ Clay ☐ Silt ☐ Sand ☐ Gravel ☐ Rock ☐ Humus
Other ....

Vegetation
☐ Grassland ☐ Heathland ☐ Shrubland ☐ Open Woodland ☐ Forest
Other ....
Dominant Vegetation types ....
Faunal Species noted ....

WATER SOURCE(S)
Nearest Drinking Water
☐ River ☐ Creek ☐ Lake ☐ Well ☐ Spring/Seep ☐ Claypan ☐ Waterhole
Other ....
Distance of Water from Site ....
Is the water ☐ Permanent ☐ Temporary ☐ Unknown

DESCRIPTION OF ENVIRONMENT
General Area: ....
Site Environment ....

B

Figure 1. 'Site Index Form' and 'Art Site' form of the Archaeology Branch, Queensland.

CONDITION OF SITE

Causes of Damage:

☐ Vandalism ☐ Camping ☐ Foot Traffic ☐ Vehicle Traffic

☐ Pastoral ☐ Animal ☐ Natural Erosion ☐ Mining

☐ Urban/Rural Development ☐ Insect ☐ Plants

Describe ....

SITE VISIBILITY

Estimate percentage of site covered by vegetation ....

Describe ....

DOCUMENTATION

Have items been collected from the site ☐ Yes ☐ No ☐ Unknown

Name of Collector ....

Location of Collection .... Qld. Museum No. ....

Location

☐ Photos .... ☐ Notes ....

☐ Tapes .... ☐ Manuscript ....

☐ Drawings .... ☐ Charcoal ....

☐ Aerial Photo .... ☐ Other ....

☐ Tracings ....

Published References ....

Recommendations Preservation/Management (condition of site, likely visitation, natural erosion, attitude of owner, etc)

....

C

MAP (access to site) Show distances, Directions and Features:

Description of Access to Site: ....

D

Figure 1. 'Site Index Form' and 'Art Site' form of the Archaeology Branch, Queensland.

ATTACH TO SITE INDEX FORM

ART SITE

STATE FILE NUMBER

☐ Stencils ☐ Prints ☐ Drawings ☐ Other ....
☐ Abraded ☐ Pecked & Abraded ☐ Pecked Only ☐ Other ....

Recorder's Name .... Recorder's Reference ....

DIMENSIONS OF SITE

Length (max) ....
Width (max) ....
Dripline (max) ....
Orientation ....

ART SURFACE

☐ Vertical
☐ Horizontal
☐ Sloping
☐ Broken
☐ Other ....

COMPOSITION OF FLOOR

☐ Bedrock .... %
☐ Rubble .... %
☐ Sand/Soil .... %
☐ Charcoal/Ash .... %
Depth of Deposit - ....
Stone Artefacts Present - ☐ Yes ☐ No

SITE CONSERVATION

Undamaged ☐
Damaged by: ☐ Pigs
☐ Cattle
☐ Macropods
☐ Brumbies
☐ Insects
☐ Birds
☐ Vandals
☐ Other ....

TYPE OF DAMAGE

☐ Rubbing
☐ Rooting
☐ Leaching
☐ Dust
☐ Erosion
☐ Nests
☐ Droppings

SURFACE DETERIORATION

☐ Stable
☐ Flaking
☐ Granular Disintegration
☐ Honeycomb Weathering
☐ Block Collapse
☐ Mineral Staining
☐ Fungus
☐ Other ....

EXPOSURE

☐ Direct Exposure to Sunlight
☐ Direct Exposure to Rain
☐ Water Runoff/Seepage Across Surface

Estimate Percentage of Art Surface Damaged .... %

SKETCH

ADDITIONAL INFORMATION ....

ART CENSUS

For each artistic motif record the design (diagrammatically); the technique used; the colour; and (for hand and foot stencils) the side represented (i.e. left 'L', right 'R', or indistinct 'I'). Where time does not permit a full count select a representative range of motifs and indicate the approximate number. Abbreviations are acceptable if a key is provided (e.g. St = stencilled).

THIS COUNT IS: ☐ A Total Census
☐ A Reasonably Complete Census
☐ A Selection of Representative Motifs

MOTIF	TECHNIQUE	COLOUR	SIDE	COMMENTS	PHOTOGRAPHED	
					YES	NO

Figure 1. 'Site Index Form' and 'Art Site' form of the Archaeology Branch, Queensland.

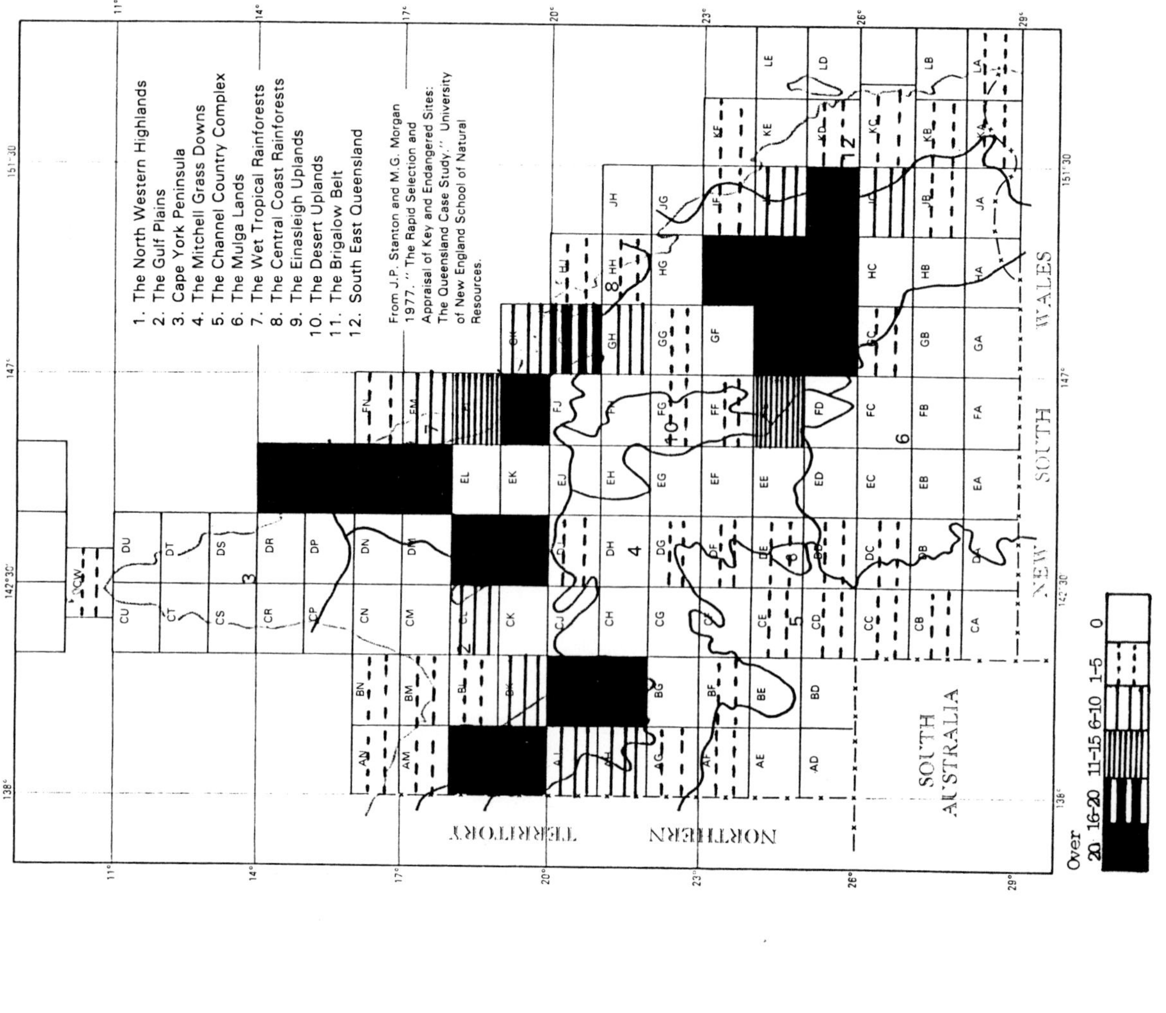

Figure 2. The distribution of registered rock art sites in Queensland.

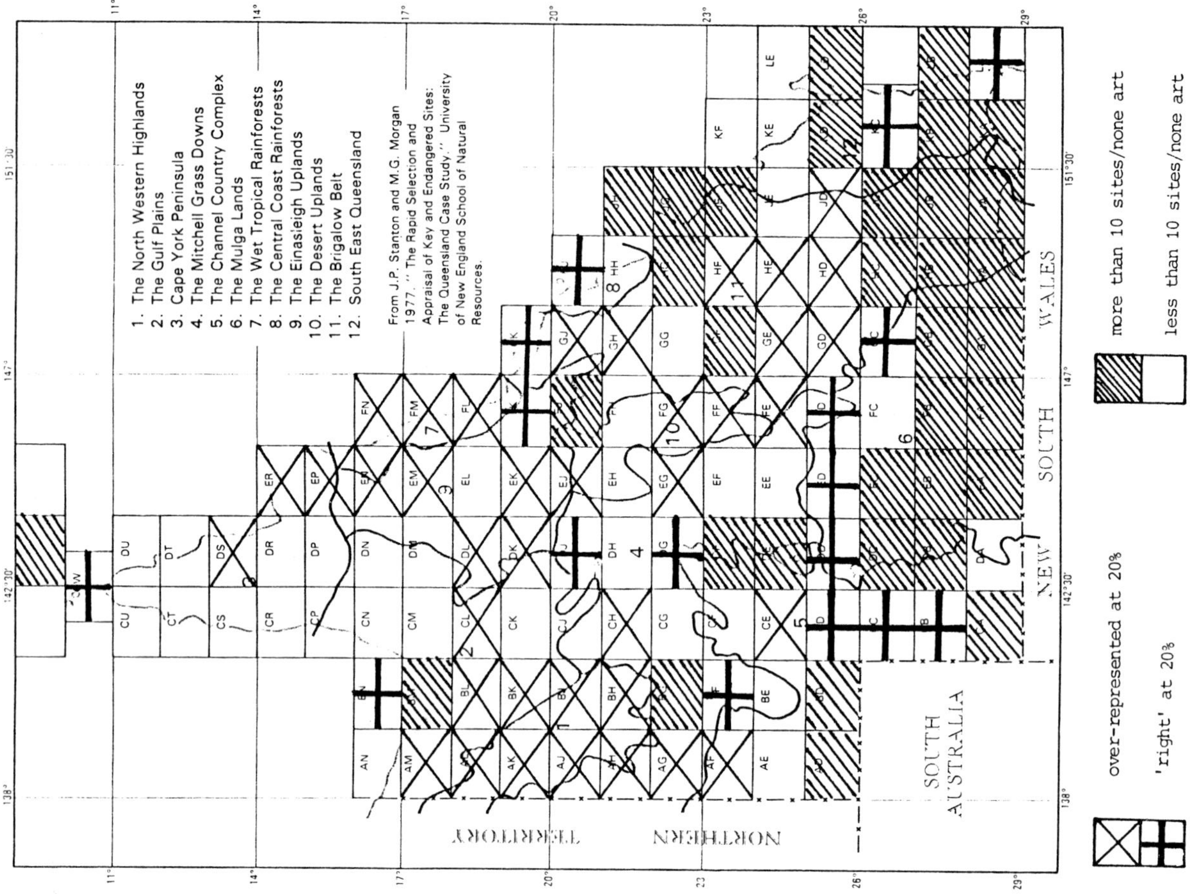

Figure 3. Representation of rock art sites in Queensland.

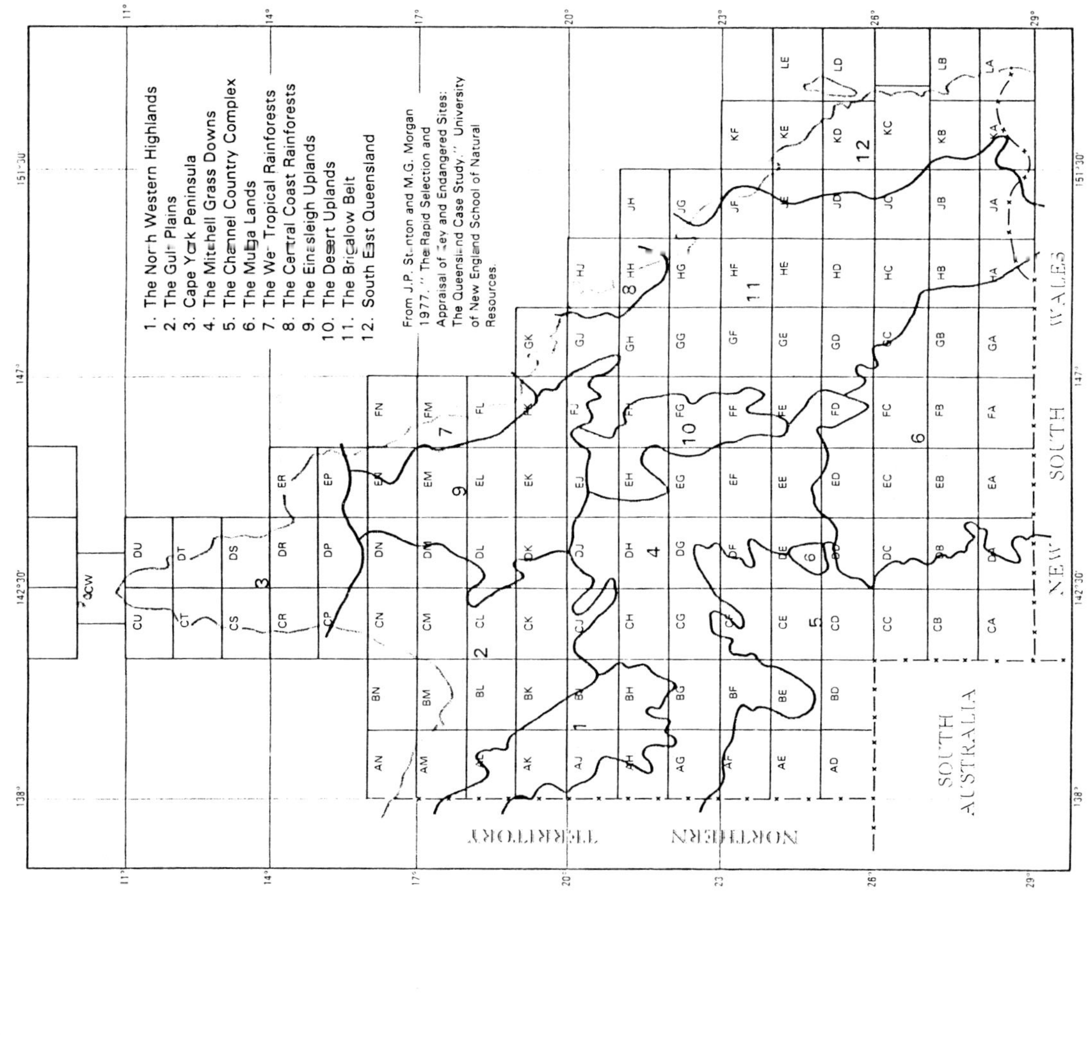

Figure 4. The 12 major biogeographical regions of Queensland.

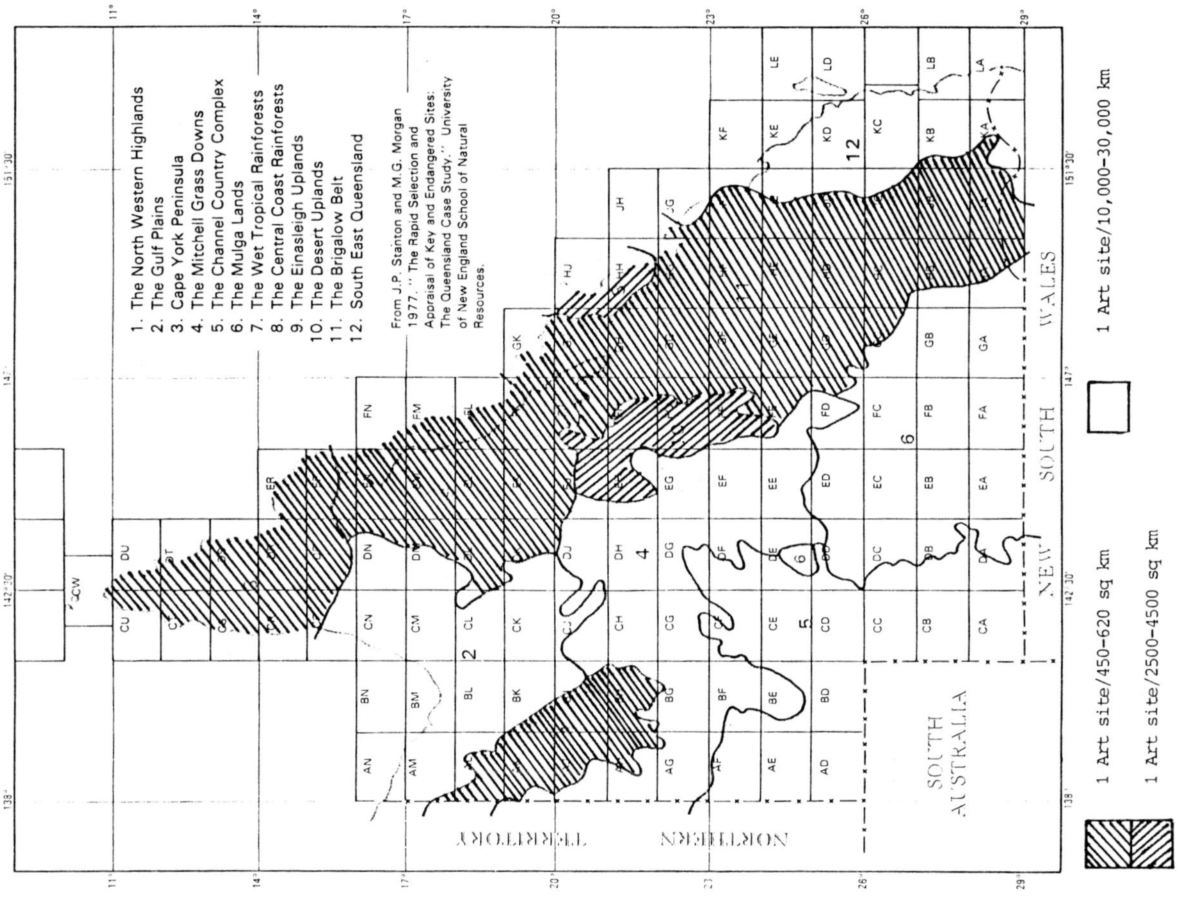

Figure 5. Recorded densities of rock art sites.

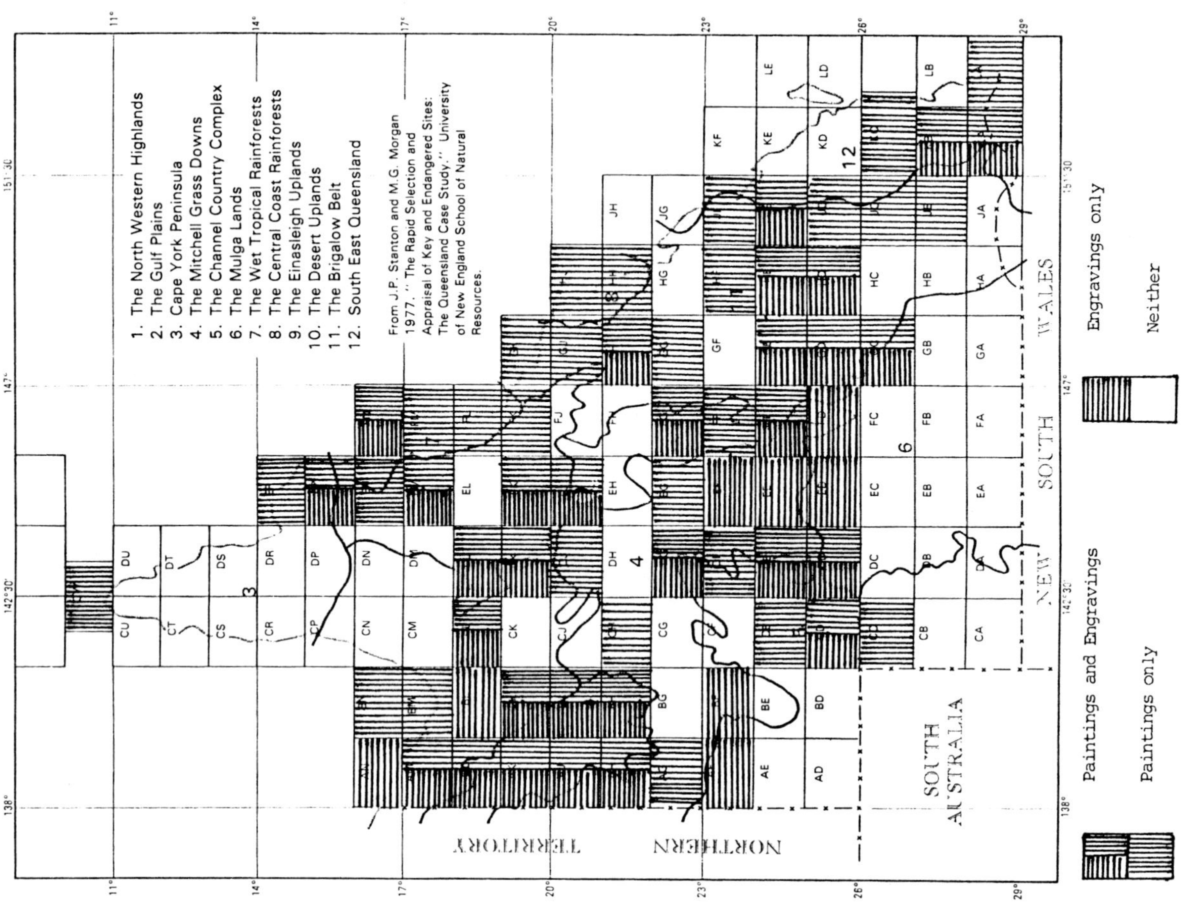

Figure 7. *Distribution of painting sites and engraving sites relative to one another.*

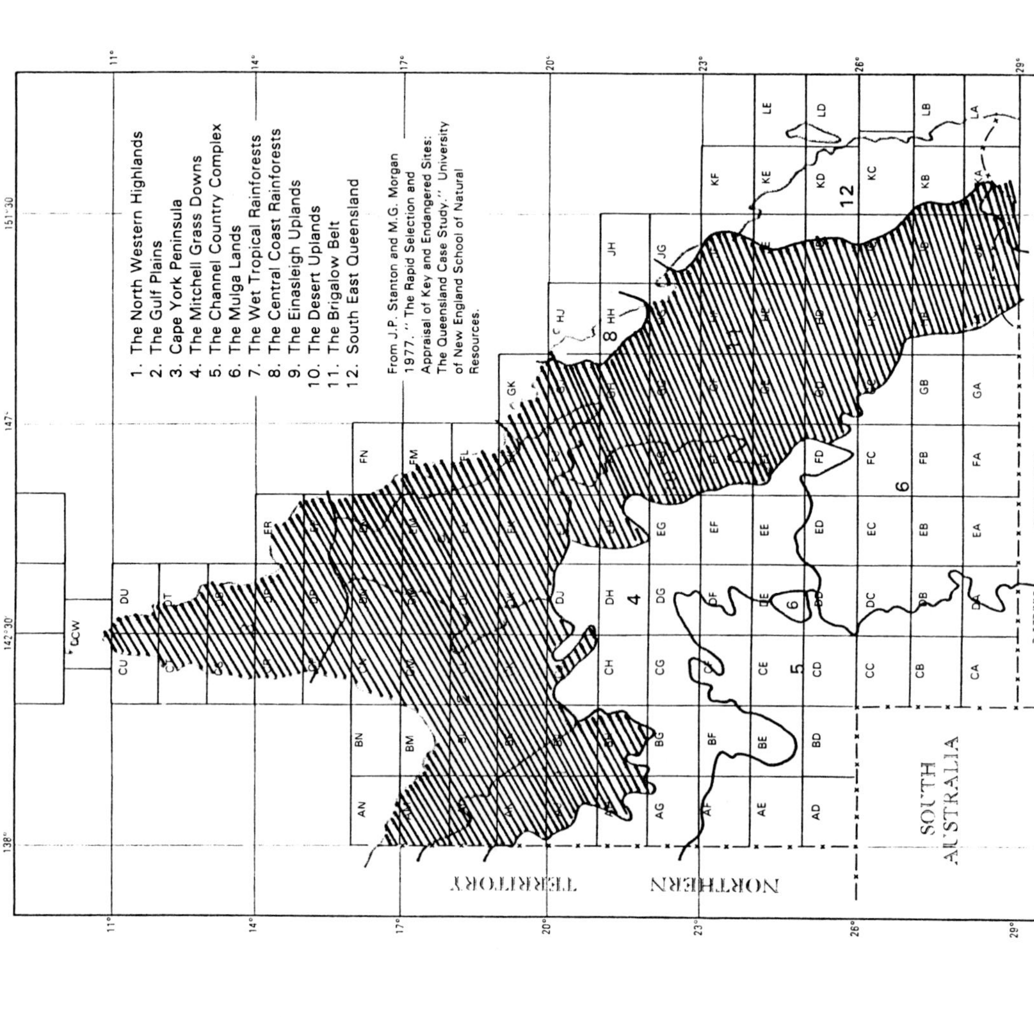

Figure 6. *Apparent over-representation of rock art sites.*

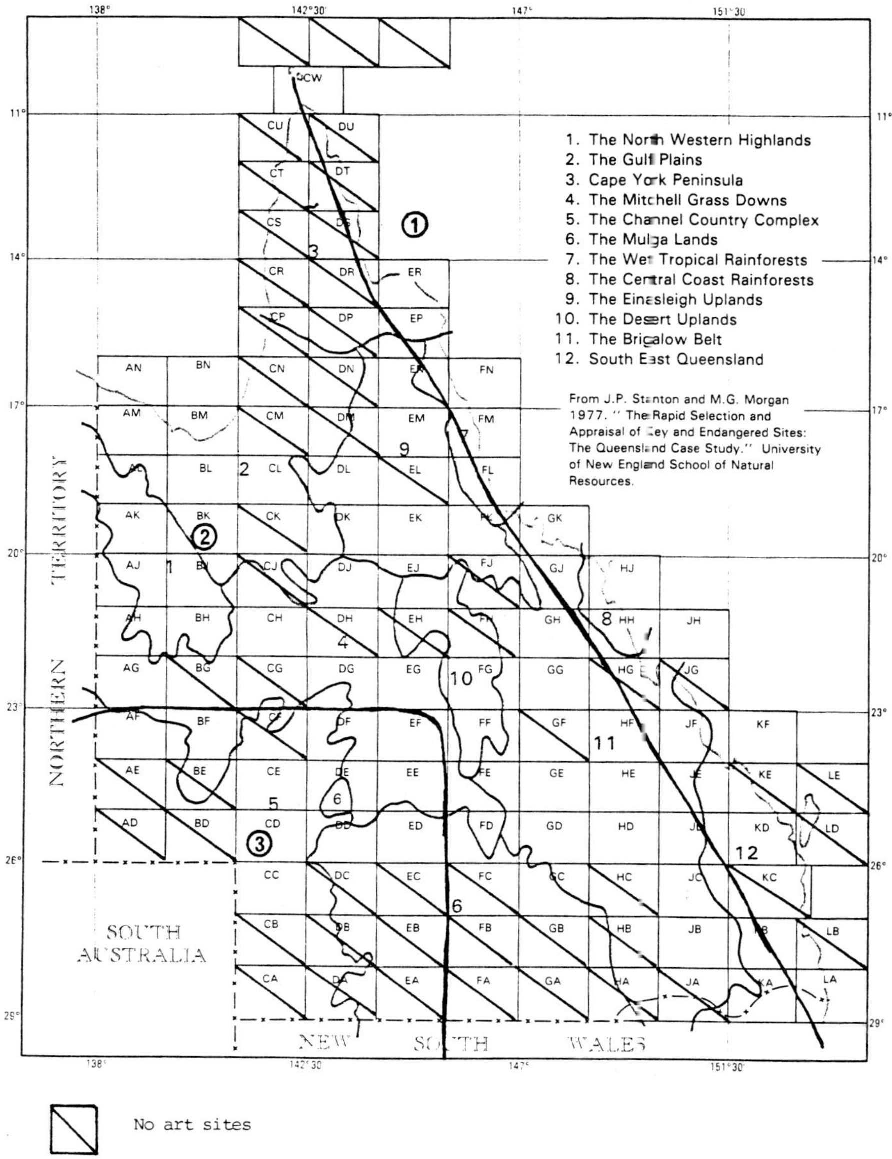

Figure 8. *Relative distribution of painting and engraving sites.*

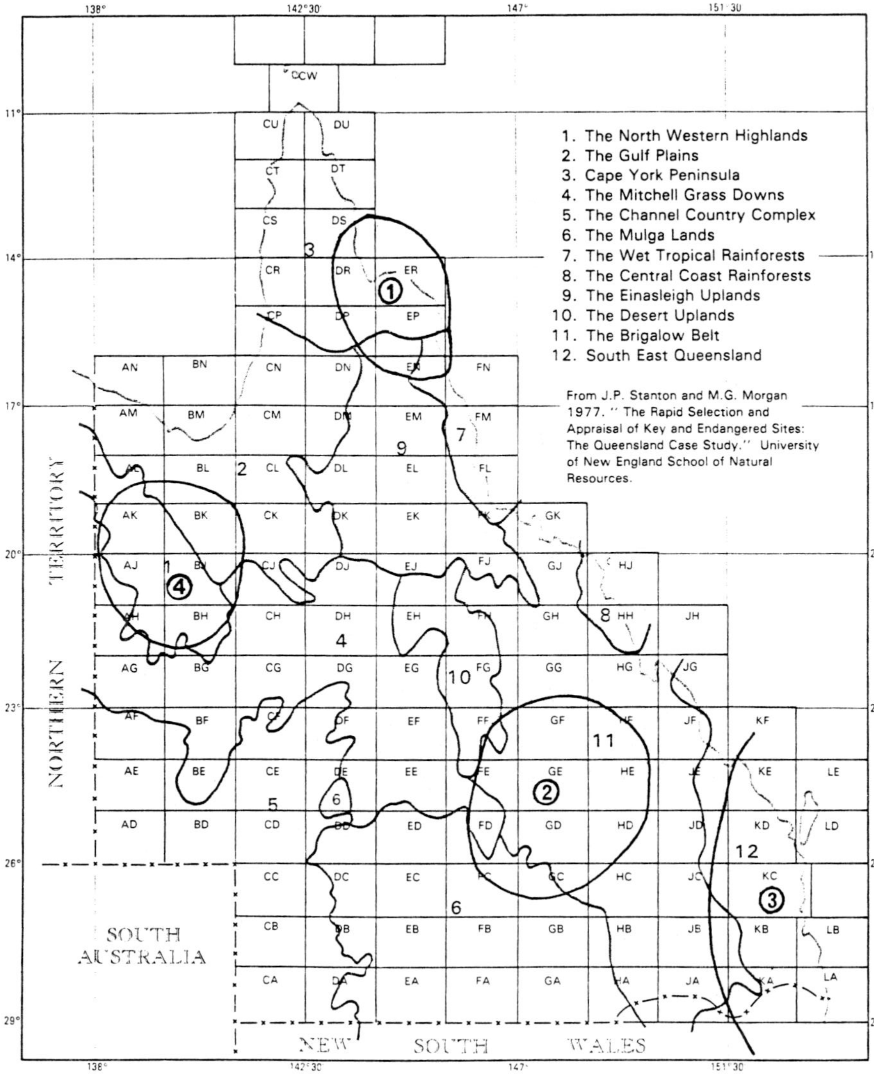

Figure 9. *Rock art regions defined by Morwood (1984):*
(1) Northeast Queensland: intra-regional differences. Figurative paintings of humans, mammals, fish and birds increasing north from Townsville.
(2) Central Queensland. Stencilling and painting or engraving of simple range of geometric designs and tracks.
(3) South-east Queensland. Intra-regional variation similar to New England region (N.S.W.) paintings and engravings comprise a restricted range of geometric motifs. Stencils absent. No major concentrations of art.
(4) Western Queensland. Extensive open engraving sites. High non-figurative component.

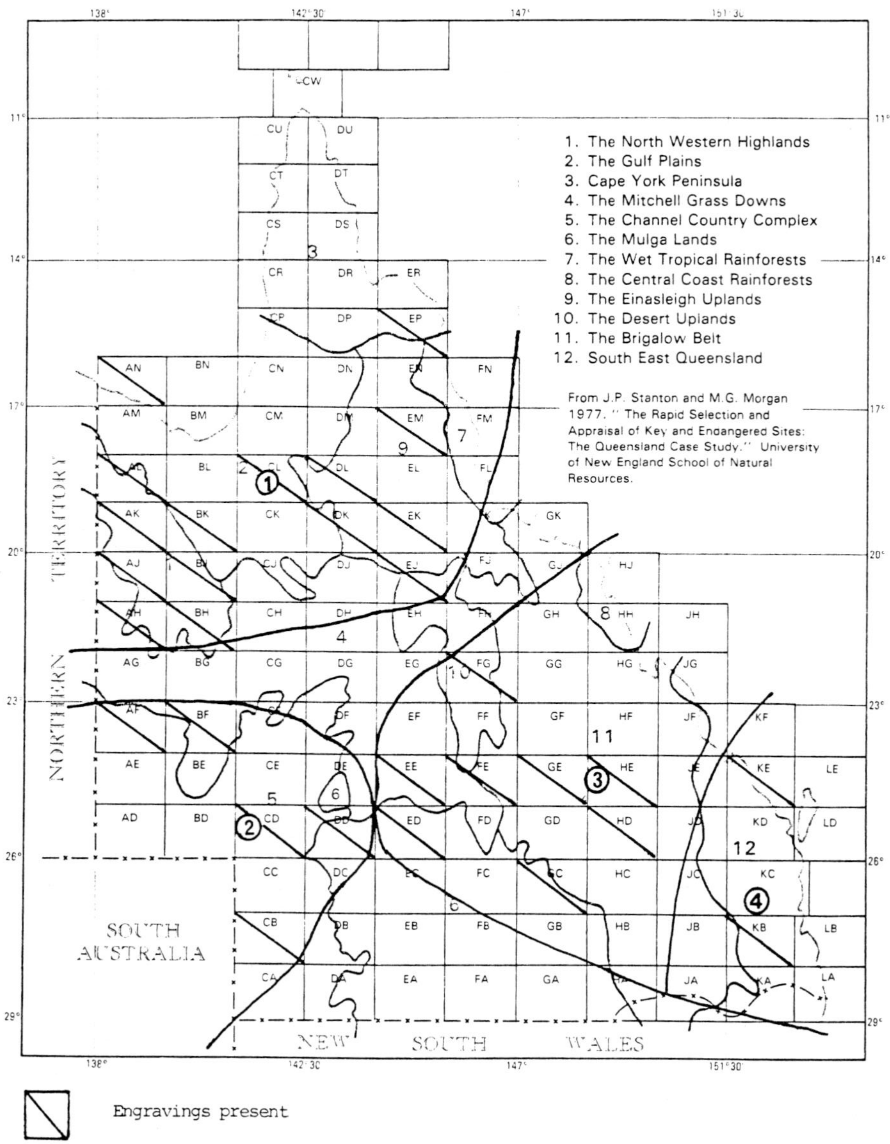

Figure 10. *Engravings – distribution of 'types'.*
(1) Figurative/geometric. Slight predominance of geometric.
(2) Predominance of geometric designs.
(3) Figurative/geometric. Slight emphasis on figurative designs.
(4) Emphasis on geometric.

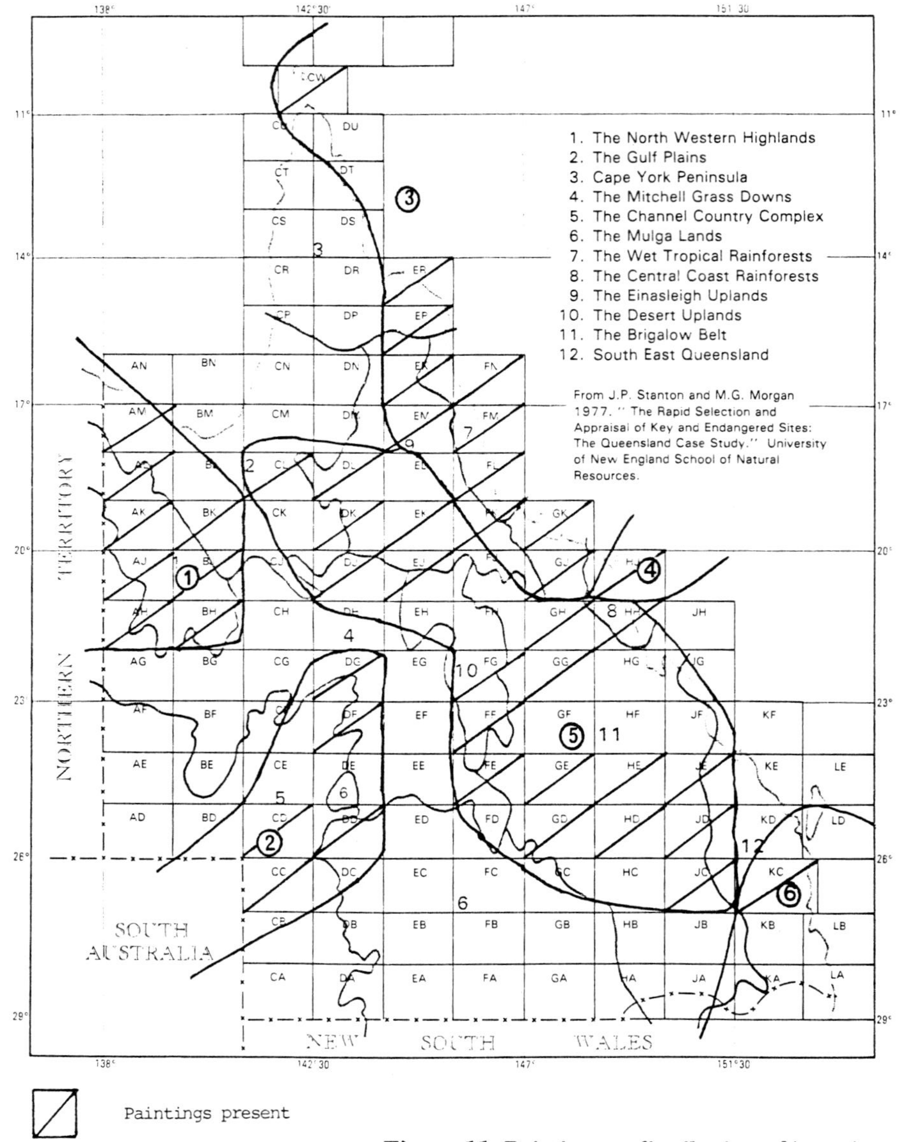

Figure 11. *Paintings – distribution of 'types'.*
(1) Figurative/geometric. Slight predominance of figurative. Snakes and boomerangs common.
(2) Geometric predominates, some figurative.
(3) Figurative/anthropomorphs and emphasis of marine motifs.
(4) Mainly geometric (restricted to Whitsunday Islands).
(5) Stencils predominate, particularly hands and weapons.
(6) Geometric.

25

A COMPARATIVE STUDY OF TWO PREHISTORIC ARTISTIC RECORDING LOCALITIES

Elanie A. Moore

Abstract. The difficulties in the recording of high murals at two sites in Baja California are discussed. In particular, methods of compensating for the foreshortening of tall figures are considered. The results of four different efforts of recording one large mural are compared, and the reasons for discrepancies are discussed.

Introduction

The purpose of this paper is to describe and evaluate the recording procedures at two sites in Baja California. The Baja California peninsula is located in north-western Mexico. It is formed by a mountain range that runs in a northwest-southwest direction. Between the 27th and 28th north parallels is the Sierra of San Francisco, an area cut through by deep and steep rugged canyons.

The rock art complex known as Cuesta del Palmarito is located on the south-east slope of these mountains, near the ranch of Higuajín, a few kilometres to the north-west of Rancho Santa Marta. It is situated on the west side of the arroyo of Cuesta del Palmarito, at a height of about 800 m above sea level and faces toward the east.

On the south-western slope of the same mountains is found the Serpiente complex. It is located in the upper reaches of the Arroyo Parral. The shelter faces north-east and is situated 3.5 m from the arroyo bed. The paintings on the walls and ceiling are clearly visible from the bottom of the arroyo.

The first non-indigenous reports of the paintings were by early Jesuit missionaries. León Diguet, a Belgian chemist, provided published descriptions (Diguet 1895, 1899: 34). In the 1970s Crosby and Hambleton located and compiled 200 sites (Crosby 1975; Hambleton 1979). In 1985 the first reproduction of the paintings at Palmarito was undertaken (Moore 1985). Smith (1983) discussed the cognitive aspects, a possible colour encoding system and the paintings' interpretations which were based on pertinent ethnological documents of the Cochimí Indians (Mission Period). Viñas and Sarría reported on the first topographical works of the Cuesta del Palmarito in 1981 (Viñas et al. 1985).

The large shelter known as Cueva del Palmarito I is 93 m in length, 10 m in depth and 13.7 m in height at the centre. Most of the figures are life-size, and are painted along the base of the wall up to and including the roof or ceiling of the shelter (Figs 1 and 2).

Figures are principally painted in earth tones with a combination of flat or stippled silhouettes. For the viewer, an array of colour meets the eye: reddish brown, purple, pink, orange, red, ochre, yellow, black and white. Close examination and comparative study indicates the artist saw these variations as only four colours: red, yellow, white and black (Moore 1986).

The majority of figures are monochrome, but many are painted in two or three colours, bisected vertically or horizontally, one side red and the other black, or the top half red and the bottom half black, then outlined in white. The artist often applied a flat colour and outline that is similar to the ornamental stripes painted on the trunk of the figure.

Objective

In 1982 I set out to record the major aboriginal painting in the Cuesta Del Palmarito complex, Palmarito I. At about the same time Ramón Viñas and Elisa Serría from Barcelona, Spain, began independent work on the same site. The work has continued intermittently since and still is incomplete. Methods of recording have evolved and improved. Because the majority of figures are out of reach, well above the head of the viewer, it is impossible to photograph or draw them directly from the front. The roughness of the shelter surface combined with multilayered applications of paint make deciphering the images slow and tedious, occasionally approaching the impossible. As a result, indications for new methods of recording and visual interpretation were utilised. By applying these methods at another site, which is accessible to recording by tracing, they can be verified. In this paper photography and measured grid drawings used in areas which are not flat nor at eye level are compared. Grid visualisation of Serpiente I has been undertaken, but not yet completed. Results so far indicate this method can be as effective as photography.

There are three methods which seem most commonly used in recording rock art: (1) photography (Brooks and Wakankar 1976), (2) copying from a linear grid which has been placed over the original (Anati 1960), and (3) tracing the actual images onto clear plastic (Johnson 1958) using acetate, acrylic or mylar. These three methods are being used and their results correlated at Palmarito I and Serpiente I. The recording methods will be discussed, then their correlation will follow.

Palmarito I

Because the walls upon which the images were painted consist of volcanic rock, mainly conglomerate and tuff (Chan 1980: 7), the painting is in extremely poor condition. The softer rock has eroded away, erasing much of the painted surface. The paintings are exposed to sandblasting from without and water seepage through the walls from within. The aboriginal artists painted the murals in layers. Such piecemeal disintegration, plus the poor rock surface, makes perception difficult to impossible.

The account which follows is a description of the methods being used by Moore for recording Palmarito I. All photographs arc made with 35 mm film, using Konica

Autoreflex and Minolta XDII cameras. Lenses used are 57 mm, 135 mm, 50 mm, 28 mm and 100-200 mm. Photographs were shot in existing light without filters. Several film types including Kodachrome 64 and Fujichrome 64 were employed, but Ektachrome 200 was primarily used because of avaiability and cheaper processing. Exposure and other shooting records are kept for each photograph. Flash is used when more light is needed. The camera, with few exceptions, is hand-held and positioned so the horizontal dimension of the picture is parallel to the natural horizon. This position keeps any lens distortion constant.

The mural is photographed first from a distance to provide an overview establishing major landmarks for grid shots. The latter are made by shooting close-ups that overlap at least one-fourth of the preceding photograph. This is done to compensate for distortion at the outer edge of each picture due to lens and wall shapes. Such grid shots are made with a 50 mm or 57 mm lens since it affords the least distortion. The entire mural is photographed in the same manner.

The camera is always positioned directly in front of the image so that the subject appears in the centre of the picture (Brooks and Wakankar 1976: 109). It is then tilted so the film plane is parallel to the wall upon which the image is painted. Keeping the two planes parallel is possible only along the horizontal axis of the painting. All shots along the vertical axis must be made from the shelter floor. In cases where the wall remains vertical, the plane of the wall recedes from the camera, causing perspective distortion. Most often, however, the shelter wall extends at an angle from the floor out over the proscenium so distortion can be minimised by changing the angle of the film plane (Heizer and Baumhoff 1962: 274).

My shooting stance is with both legs solidly planted slightly apart on the shelter floor, except when the paintings are on a low ceiling, in which case the shot is made in supine position. The camera is hand-held which keeps it 61 inches above the ground in my standing position. I maintain a constant distance between myself and the murals whenever possible, using varied focal lengths for compensation only when necessary.

All major figures which are within reach are measured on at least a horizontal and vertical axis. Vertical measurements for figures above head level are made with the surveyors pole when the angle of the shelter wall makes that possible.

Photographs include a 10-foot measuring tape and the surveyors' vertical pole. These photographs are then used to establish the vertical and horizontal axes, the scale for the mural sketches and references for the overlapping photographs and subsequent layout.

Measuring figures from head level when the surveyor's pole cannot be used presents another problem. Standard methods of triangulation are out of the question since the shelter floor is uneven and the outer edge steeply drops off to the canyon below. Positioning measuring posts of different heights and placing viewers at distances adequate for viewing the uppermost figures cannot always be done. Therefore the camera is substituted. Though not as accurate, it gives a credible approximation by using the following technique: the camera is set for a shallow depth of field and positioned at a point equidistant from the image above and a vertical measuring stick, or my guide, who measures exactly six feet from the ground to the peak of his hat. He stands directly under the figure to be measured. The camera is judged to be equidistant from both he and the figure when both are in focus in the view finder. These images are shot from the side, so horizontal axes are not accurate, but vertical axes are reasonably correct.

When budget allows, slides are printed to be used as references for sketching, then mounted to form a composite and marked with a grid. Each section refers to a drawing specifically designated for a particular mural. The 35 mm slides from which the composite photograph is made are next projected and traced for transfer onto a field sketch. These sketches serve as guides for proportions, details and placement of the images on the drawing page (Brooks and Wakankar 1976: 109). On each subsequent trip, the sketches are double-checked with each photograph and its corresponding image on the wall. The sketch book drawings are then corrected or modified as required. On at least two additional visits to the mural the same set of drawings is again carefully checked for any information previously missed.

The drawing method used for verifying proportions of figures and their positions on the wall is a technical approach used by professional artists which, in common usage, is referred to as 'sighting'. It was most recently publicised by Betty Edward in 1979. Moore calls it 'grid visualisation,' a term which is more specific to its definition because it is based on the concept of the grid used for enlarging and transferring drawings. The method works well as a substitute when the images are too distant to be reached.

The following is a description of how grid visualisation is implemented. Several parallel vertical and horizontal lines in the mural are chosen to serve as a frame of reference or landmark from which angles and layout proportions are sighted and measured. These lines are the first marks established in the preliminary sketches. All subsequent marks in the mural are measured visually by holding the pencil at arm's length, with elbow straight, and sighting along the top of the pencil toward the desired landmarks in the mural which is being copied. These marks are measured while sighting, by placing the thumb on one mark and the pencil end on the other. Holding this mark, the hand and pencil are moved to match each measurement to others in the mural or to sight on longer elements, counting the number of pencil measures they contain. Angles are taken in the same way, referring the measure to the established verticals and horizontals.

This method is tedious but provides comprehensive recording when combined with photography. It is the best approach at present for drawing areas of Palmarito I which cannot be traced. The highest image above the shelter's proscenium is 13.7 m away, the floor is uneven, and the site isolated, so the expense and awkwardness of taking additional equipment to the site for scaffolding to aid in tracing the murals is not feasible.

Sketching and photographing the upper portions of the mural creates some difficulty since the figures become distorted if viewed from directly below. In such instances, Moore works from positions outside the shelter, using the telephotographic lens for photography and binoculars for verifying details. In this way distortion is minimised.

All drawings are checked for layers of colour, extraneous lines, and miscellaneous marks whenever possible. Lines and dots are all counted. Variations in red are also noted. Layering is determined by a close study of the edges of paint strokes.

It is surprising how often the camera image is either deceptive or distorted. The importance of double-checking camera images with the original painting has been demonstrated many times. Several examples are described in detail by Moore (1986).

In the studio, the field sketches are completed and adjusted to scale, then transferred to the composite drawing for the whole mural. Scale is established by first projecting the original slides portraying the measuring tape and vertical pole, then adjusting overlapping images to fit. The information recorded in the field sketches is then transferred to the composite drawing.

Colour in the field sketches is noted last, along with layering, through a key established with coloured pencils This drawing serves as a preliminary sketch for the final painting, which is made with acrylic on canvas. Colour is then keyed to that of the original mural through the use of Munsell (1971) Color Chart for geologists.

Colours can be matched with the Munsell Color Chart by direct comparison only for the lower section, extending 2.5 m above the ground, since these areas are within reach. The remaining colours must be approximated by relating them to the known colours in the accessible parts of the mural. Shadow intensity and distance are matched for both sets colour records so that estimates can be in relation to the same variables. This is done by positioning oneself at equal distances from the images, tipping the chart to mimic the same planar angle as that of the wall, and finally double-checking by visual comparison to known colours.

Individual records are kept on each figure, noting layering, colour keying, size (if available), and the general rock surface which provides shape to the image. Notations are made on photographs and photocopies of the original drawings. Film type, exposure, lenses, time of day and camera type are also recorded.

Serpiente I

The methods discussed so far in recording Palmarito I, i.e. for photographing and drawing, were also used at Serpiente I. However, because the mural at Serpiente I is within reach and could also be traced, it was chosen as the focus for a comparative study of recording techniques. By comparing the recordings used at both sites and cross-referencing the results, the validity of each method reveals itself. These were then cross-referenced with illustrations already published.

Crosby (1984) published a rendition of Cueva de la Serpiente which was done as a generalised illustration and was not intended to be an accurate scientific recording (Fig. 3). Uriarte (1981) used a tracing of the Crosby painting when she attempted to catalogue the figures in 72 of the rock paintings. Smith (1983) made a drawing using Crosby's work and his own slides (Fig. 4). Hambleton (1979) also published a copy of the Crosby illustration which has been widely circulated in Mexico. Both Crosby's drawing and Hambleton's copy are now being used by researchers as reference material internationally. A team of Spanish archaeologists from Barcelona (Viñas, Sarría, Rubio and del Castillo 1985) did their own field recording and produced their own drawing (Fig. 5). Theirs differs in a number of ways from those of the Crosbys, Uriarte and Smith.

Palmarito I is too large to cover in a linear grid or to trace onto plastic. That leaves photography as the main tool for recording. Photography was employed in all three illustrations of Serpiente I: the Crosbys, Smith, and the Viñas team. Yet, comparing these illustrations at the site revealed disparities. One must then conclude that (1) photographic accuracy depends on the photographer and (2) that the human eyes are still more accurate than the camera.

All three methods, photography, tracing and grid visualisation have their drawbacks. Because of its accessibility to all three methods and because others have also used photography to record it, Serpiente I affords the best opportunity to test the relative merits of these methods. The data gleaned from field research was gathered by using two of them, tracing and photography. Averaging differences found during the cross-referencing of these two methods, a composite drawing was constructed. Tracing and actual painting was done on clear plastic drop cloths. The drawing was then reduced by photocopier (Fig. 6).

Two drawings were completed by tracing photographs, the final result was then compared to the Crosby, Uriarte, Smith and Viñas illustrations. The results indicated that all four had worked from photographs and that Uriarte had traced Crosbys' drawing. Subsequent letters from and conversations with Crosby, Smith and Viñas verified these observations. However, they did not use the same system for photographing, so differences occur. All work was then compared to the actual rock painting. Additional differences appeared, some of which are quite significant.

The most outstanding differences were the proportions. Size relationships (length as it relates to height) and distances between figures vary dramatically, depending on the method used for gathering visual information. In the case of photography, if the photographer is standing directly opposite the painting, and the camera is perpendicular to it, the only distortion with which he or she will have to contend will be produced by the camera lens (Martineau 1976: 189). If there are figures overhead and the photographer must take them from below, i.e. from ground level, then distortion created by foreshortening through linear perspective must be considered along with that created by the camera lens.

Figure 7 shows what happens to images in Serpiente I when drawings made by Viñas and Crosby, who both worked from photographs, are superimposed onto a tracing. A drawing corrected for vertical height proportion results.

Comparative study brings up another point. The rituals of the portrayed shamans were probably held at ground level so they were as accustomed to that view as is the modern viewer. When the upper drawings are extended to their correct proportions they look outsized to anyone accustomed to viewing from ground level. Murals of the Sierras of San Francisco have been rendered by established 'rules' of proportion based on a constant view.

It is unlikely that a hunter-gatherer society such as that of the painters would adjust a mural's perspective for a viewer on the ground. This takes an understanding of anamorphic drawing and linear perspective, and their application to vertical murals. An artist can achieve the same effect, without any sophisticated technical knowledge, by standing at ground level and painting with a long pole, an idea suggested by Smith (1985: 51).

Correlating the Serpiente I illustrations

Crosby omitted a few small figures in their painting. They are not near the area of image concentration and are faint on photographs. He photographed the mural using a 35 mm camera with a 50 mm lens. Employing a measuring tape, he kept the camera distance from each image consistent and the angle perpendicular. He overlapped each photograph by thirds. He shot nine overlapping photographs (Crosby 1985: 32). When he traced the resulting slides, he drew only the centre third connecting each. This method reconciled the distortions created by a curved lens and curved surface along the edges of the frame. He did not compare the final drawing to the actual rock painting.

Smith's drawing was taken from 35 mm slide transparencies which he took with a wide-angle lens. His stance was on the ground below and in front of the shelter. The camera-man's position and the wide-angle lens would combine to distort the mural's proportions considerably. Smith, like Crosby, was not trying to produce an accurate reading of the art (R. Smith, pers. comm. 1987).

By placing a tracing of Smith's drawing over Crosby's and Viñas (the tracing which Moore made from the original) it is evident that proportions have become distorted. Also, it is evident that Crosby and Smith came up with the same information. This may be due to the fact that both stood below the mural to shoot their photographs and did not double-check their results at the site. It is known that Smith cross-checked his drawing with Crosby's.

Of all illustrations published so far, the one of the Viñas team is the most accurate, though its proportions are slightly distorted. They used a Minolta, a Nikon and a Yashika with a 35 mm macroscopic lens for photographing Serpiente I. They kept the camera lens parallel to the painting. The camera was secured to a tripod. The centre three quarters of the photograph was used to minimise distortion (A. Rubio, pers. comm. 1988).

The group did not say exactly where the photographer or tripod stood, but the proportions of the drawing indicate that it may have been at the same place as Smith and Crosby. When viewing the mural from this position, outside and below the shelter, the upper images are perpendicular to the viewer's eye line, but the lower figures are shortened as they recede. This is the same distortion that appears in the Viñas drawing when it is compared to the actual tracing.

The Viñas group recording is thorough. They included all of the extraneous figures and left incomplete figures unfinished. The omissions were small details, for instance the number of fingers and toes on appendages. Since there is reference to the practice of digital mutilation by Cochimí shamans in the ethnology, such oversights could be significant.

When counting the black bands painted across a distinctive element, a horned serpent on the viewer's right in the Viñas drawing, the result is 35 bands. It is interesting that none of the researchers nor Moore's two methods of recording and site observation agree in number. Smith shows 38 in his illustration of the serpent; Crosby has 47. Moore's drawing, done from slides, produced 34; while her tracing on clear plastic produced only 25. Moore returned to the site and counted the black bands while carefully and closely studying the actual serpent, counting 48 bands. They are not clear nor complete enough to be *certain* that all were actually intended to be black bands.

Conclusion

It is interesting that four different researchers, using similar methods, Crosby, Smith, the Viñas team and Moore produced such different information on the same mural. The most intriguing example is the discrepancy between each researcher's total number of black bands across the serpent to the right of the viewer. Even when using three methods, each produced a different number.

If clarification of the individual researcher's, i.e. illustrator's objective for representing the rock mural is understood, the four recording projects all gave an adequate rendition of the same mural, Serpiente I. Crosby illustrated the major figures, which are also those of interest to a general public. Smith drew only what was needed to clarify his theories. Viñas' team made a black and white record of what they thought important. Moore has created adequate maps for future study.

The comparative studies of Serpiente I are continuing and include the yet to be completed grid visualisation method used at Palmarito I where figures are beyond reach. The hope is to validate this method as another tool for rock art study. Present indications are it will be a useful adjunct to photography, but like photography it will have the same difficulty with distortion of images above eye level. However, unlike photography, it can be useful for double-checking work on the site and as a tool for comparative analysis when correlating information gleaned by other methods.

Using more than one method for obtaining visual information at a recording site helps to validate as well as clarify the information recovered. Documenting methods can, subsequently, validate recordings made by only one or two approaches, when available, such as at Palmarito I. This study has shown the use of photography in rock art recording must be carefully documented. If information gleaned through photography and grid visualisation is documented then researchers other than the original recorder will have a valid yardstick by which to make further analysis.

The scientific method of research requires that hypotheses be verified. Illustrations are the visual evidence of hypotheses. The next step is to clarify what aspects of the various recording methods discussed herein can produce valid and useful information. The reverse is also necessary, that is to identify the aspects of each method that are detrimental to an accurate record and determine how to compensate for them.

REFERENCES

ANATI, E. 1960. La civilisation du Valcamonica. *Mondes Anciens*, No. 4, Artaud, Paris.

BROOKS, R. R. R., and V. S. WAKANKAR 1976. *Stone Age painting in India.* Yale University Press, New Haven.

CHAN, G. 1980. *Observations and notes on the Indian cave Murals of the San Francisco Mountains, Baja California.* College of Marin, Kentfield.

CROSBY, H. 1984. *The cave paintings of Baja California.* Copley Books, La Jolla.

DIGUET, L. 1895. Note sur la pictographic de la Basse California. *L' Anthropologie* 6: 160-75.

DIGUET, L. 1899. Repport sur une mission scientifique dans la Basse, California. *Nouvelles Archives des Missions Scientifiques* 9: 1-53.

EDWARDS, B. 1979. *Drawing on the right side of the brain.* J. P. Torcher Suc., Los Angeles.

HAMBLETON, E. 1979. *La pintura rupestra de Baja California.* Fomento Cultural Banamex, Mexico.

HEIZER, R. F. and M. A. BAUMHOFF, 1962. *Prehistoric rock art of Nevada and eastern California.* University of California Press,

Berkeley.
JOHNSON, T. 1958. Facsimile tracing and redrawing of rock paintings. *South African Archaeological Bulletin* 13: 67-70.
MARTINEAU, L. 1976. *The rocks begin to speak.* KC Publications, Las Vegas.
MOORE, E. A. 1985. A compositional analysis of two Baja California murals: an artist's point of view. In K. Hedges (ed.) *Rock Art Papers* 2: 19-32. San Diego Museum Papers No. 18, San Diego.
MOORE, E. A. 1986. The Cochimi ritual landscape. In 'The great murals of the Sierras of San Francisco, Baja California: Is Ron Smith's system valid?' *American Indian Rock Art* (in press). American Rock Art Research Association, El Toro.
MUNSELL, A. H. 1971. *Munsell Soil Color Charts.* Munsell Color Company, Inc. Maryland.
SMITH, R. 1983. Color encoding sequences and the pursuit of meaning in the great mural rock art of Baja California. In K. Hedges (ed.) *Rock Art Papers* 1: 17-24. San Diego Museum Papers No. 16, San Diego.
URIARTE, M. T. 1981. Pintura rupestre en Baja California. Instituto Nacional de Anthropologia e Historia, Departmento de Investigaciones Historica, *Collection Cientifica* 106, Mexico City.
VIÑAS, R., E. SARRIA, A. RUBIO and V. DEL CASTILLO 1985. Rupertonio tematico de la pictura rupestra de la Sierra de San Francisco, Baja California, (Mexico). *Ars Praehistorica* 314: 201-32. Barcelona.

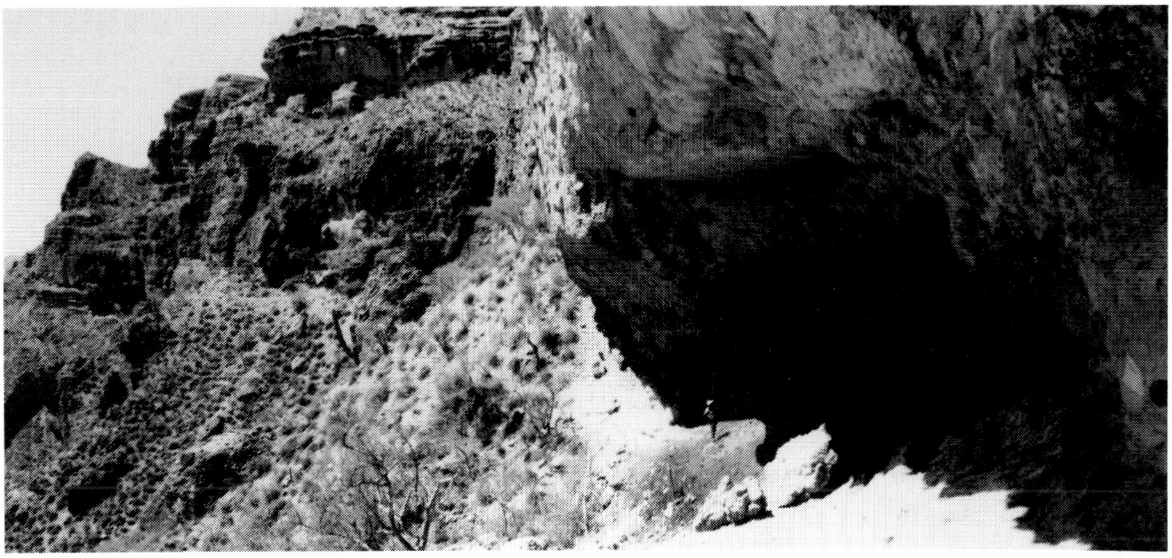

***Figure 1.** Palmarito I is of three sections totalling 93 m in length, 10 m in depth and 13.7 m in height at the centre of its stage-like form.*

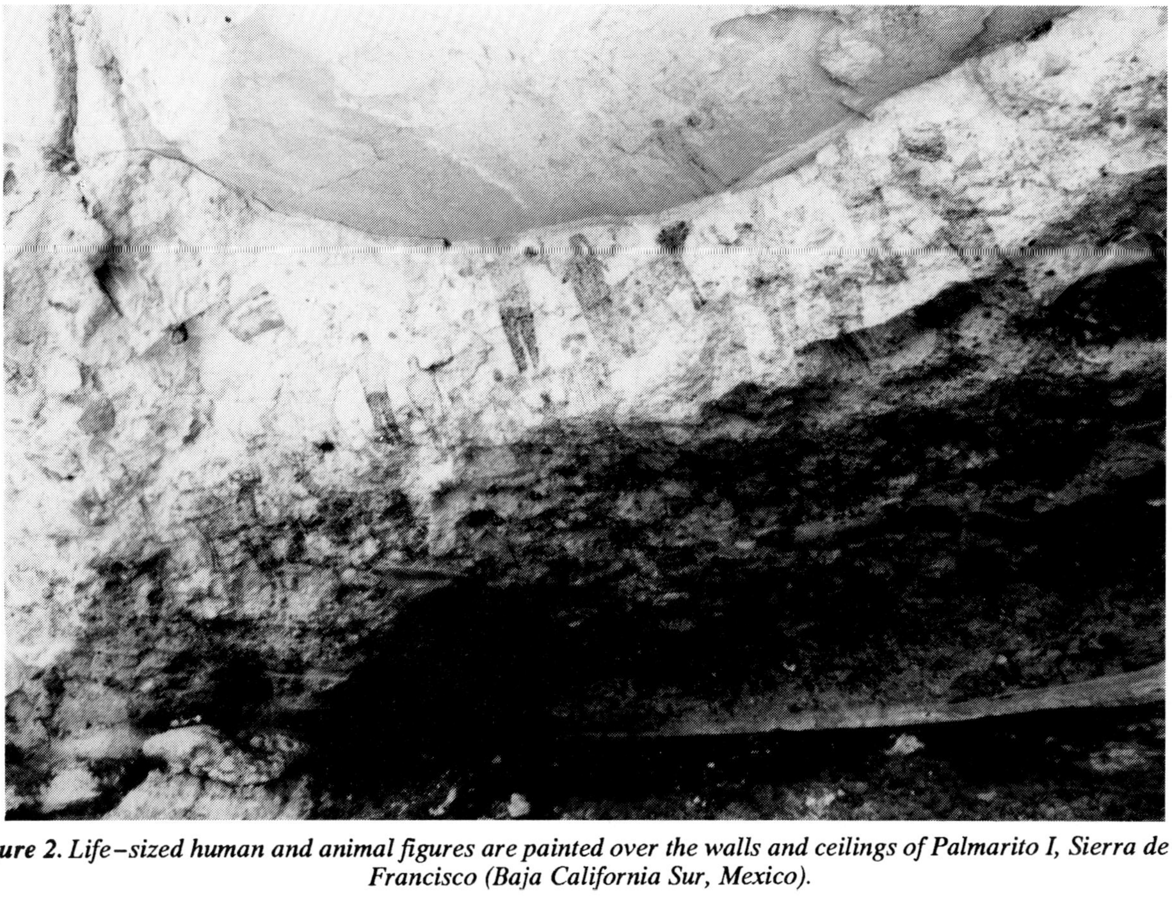

***Figure 2.** Life-sized human and animal figures are painted over the walls and ceilings of Palmarito I, Sierra de San Francisco (Baja California Sur, Mexico).*

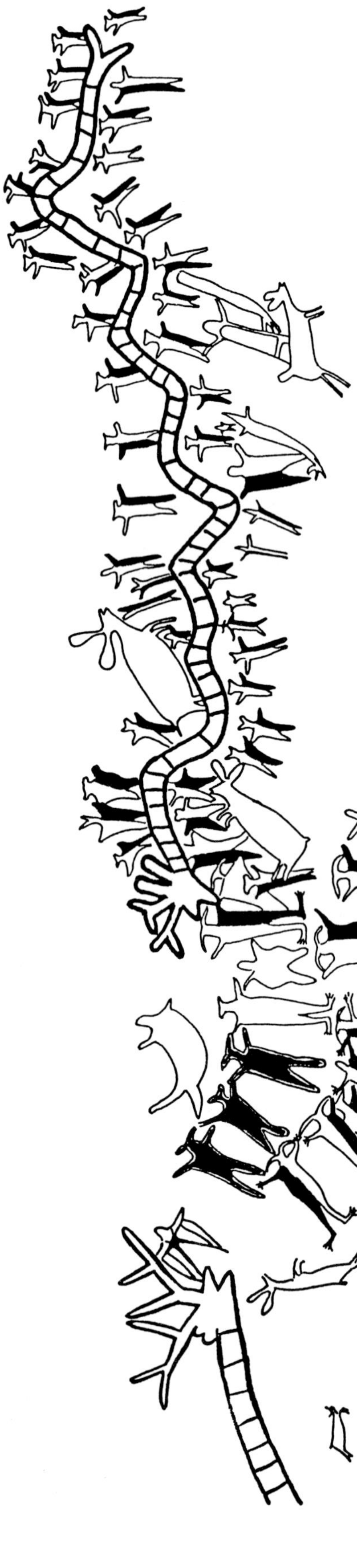

***Figure 3**. Moore's drawing of the Serpiente I painting executed by Joanne Crosby for Harry Crosby's book,* The cave paintings of Baja California.

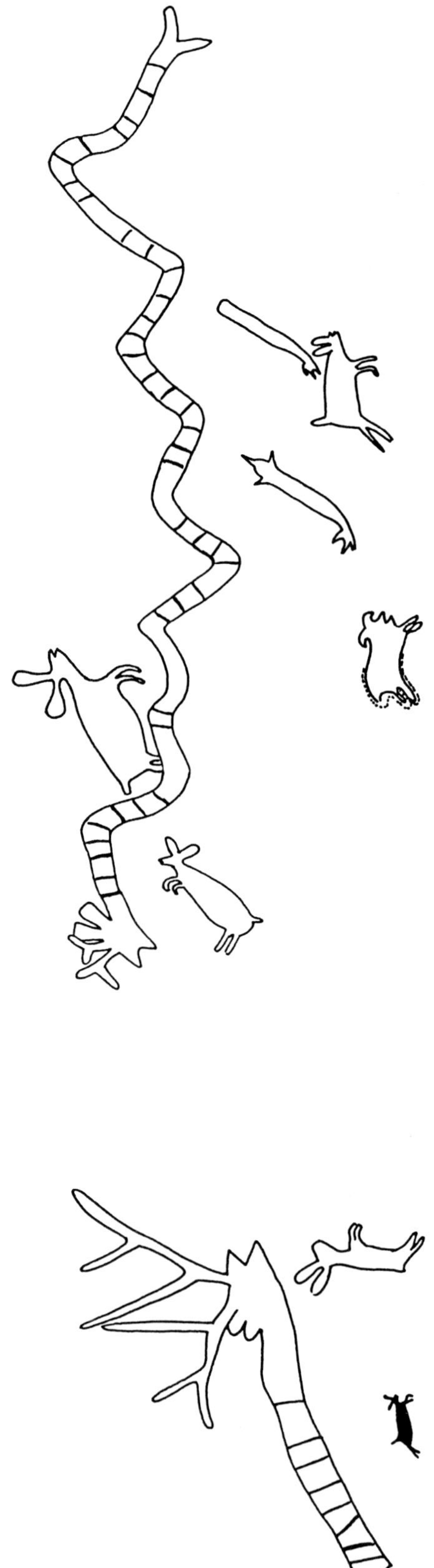

***Figure 4**. A copy of the Serpiente I drawing made by Ron Smith for his publication,* Serpent Cave.

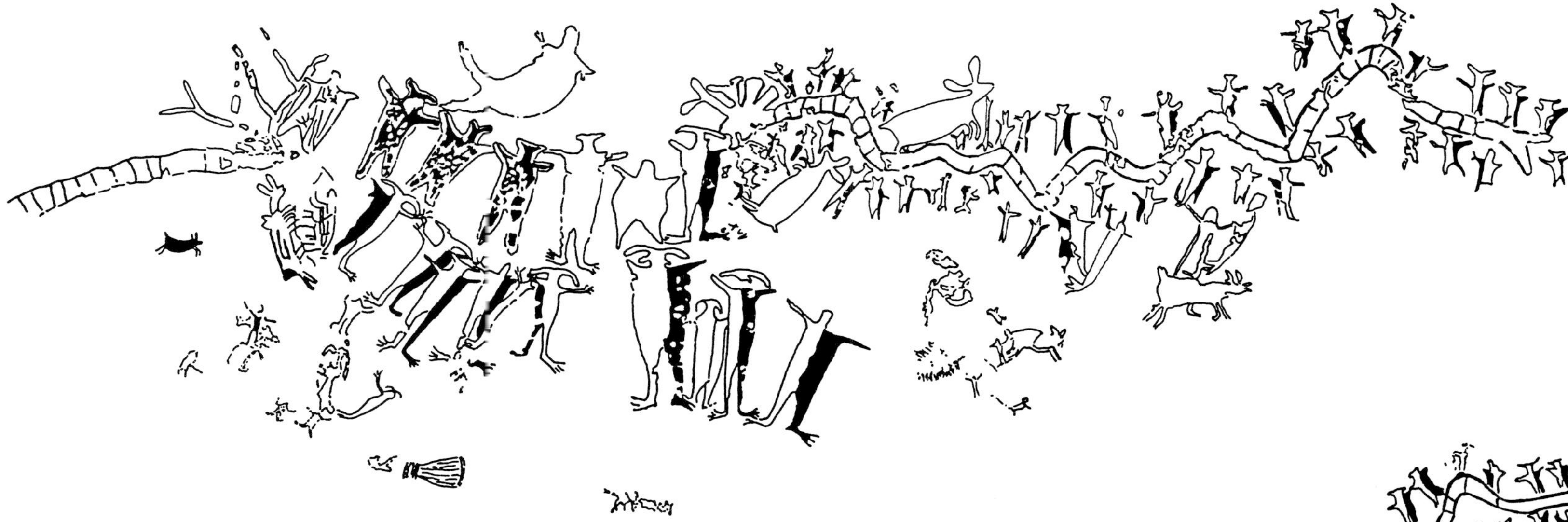

Figure 5. A copy of the Serpiente I drawing that was done by Albert Rubio for the Viñas team's publication, Cueva de la Serpiente and its painted murals.

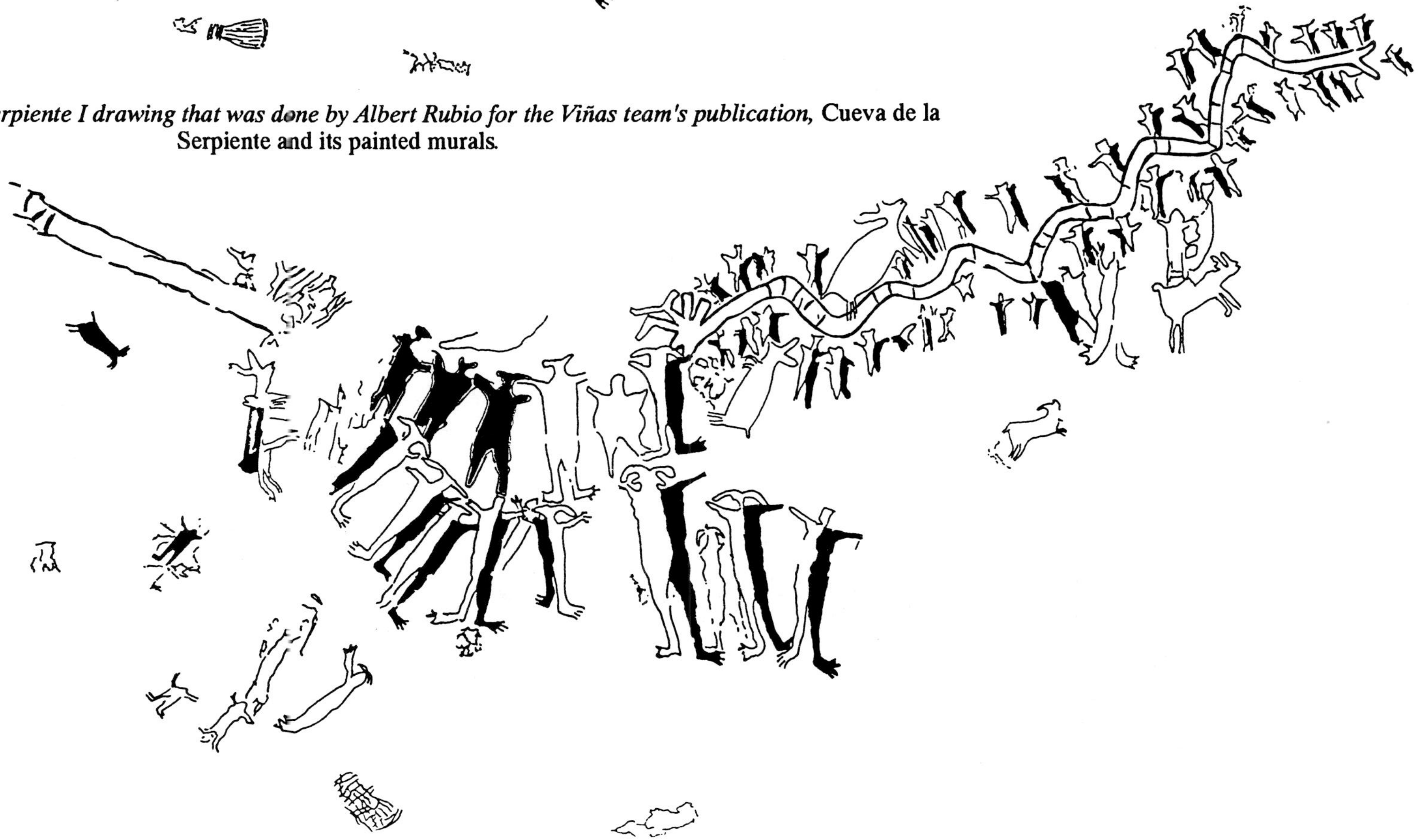

Figure 6.
This Serpiente I drawing was made from a copy machine reduction of a tracing of the original mural by Moore in 1986.

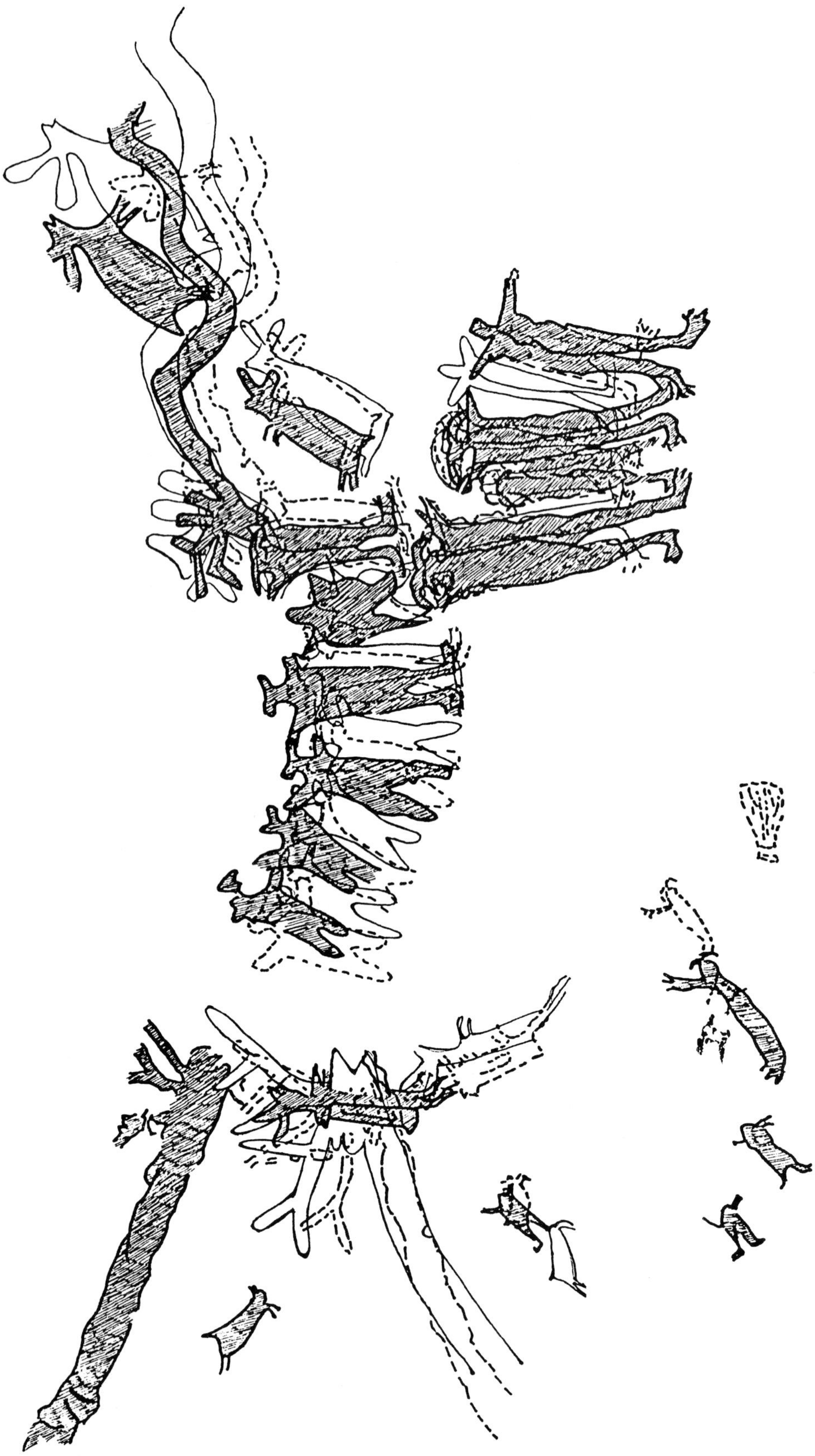

Figure 7. *The tracing (shaded area) of Serpiente I is superimposed over the Viñas drawing (dotted line) and the Crosby drawing (solid line) in order to compare distortions caused by photography and rolling out (tracing).*

26

A ROCK ART REPLICA AT THE SOUTHWEST MUSEUM: A STEP TOWARD PRESERVATION?

Paul Faulstich

Abstract. For the long-term preservation of rock art in its original setting it is crucial that the public is educated about its importance. Toward this end, the Southwest Museum in Los Angeles constructed a facsimile of a Chumash rock art site. This report examines the story behind the replica to determine whether it has fulfilled its goal of educating the public while doing justice to the magnificent art of the Chumash.

In 1985 the Southwest Museum in Los Angeles, U.S.A., opened the doors of its final exhibit renovation, *People of California*. This permanent exhibit presents, from an anthropological viewpoint, the historic and prehistoric cultures of California. The general cultural patterns of California Indian society - subsistence, social and political organisation, religion, and art - are portrayed. A central feature of the hall is a simulated Chumash painted rock shelter, modelled after a site in the Santa Susanna Mountains of southern California. This replicated rock art site, along with a simulated bedrock mortar, allows visitors, especially children, to experience features of California Indian ways of life that normally would not be presented in a museum exhibit. The installation, however, is not free from potential controversy.

The site and the model

Southern California contains many outstanding examples of historic and prehistoric rock art. Often executed with meticulous detail, Chumash pictographs are believed to have illustrated aspects of tribal myths, and to have expressed cultural values and norms. The Museum's curatorial staff, of which I was a member during the exhibit renovation projects, maintained that public understanding of the importance of rock art in Native American life would enhance the appreciation of the corresponding religious and artistic complexity of American Indians.

The Museum decided to include rock art as a major theme of the California hall in order to help convey aspects of the more enigmatic expressions of traditional Native American culture. Rock art is an irreplaceable cultural resource, and a compelling window onto the past. For the long-term preservation of rock art in its original setting it is crucial that the public is informed about the importance of this unique art form. To this end, the curatorial staff planned for a reproduction of a Chumash painted shelter to be incorporated into the exhibit. The project, it was felt, would be an important step toward the preservation and appreciation of one of the most significant elements of southern California's cultural heritage. The gradual but irreversible disappearance of Native American rock art, we felt, made it imperative that the public be informed about the fragility and importance of this ancient cultural expression. The actual site chosen for the reproduction, commonly referred to as Burrow Flats (Rozaire 1959), contains outstanding examples of Chumash rock paintings. Situated on private land, access to the site is strictly limited; it is not uncommon for qualified rock art researchers to be denied access. Largely because of its inaccessibility, the site is very well preserved, showing few signs of vandalism.

The Burrow Flats shelter functioned, at least partly, as an astronomical site, dramatically marking the winter solstice. On and around the solstice (December 21), the rising sun enters the shelter through a natural notch in the rock and casts a triangular beam of morning light onto a series of concentric circles. The circles are thought by some researchers to depict the Chumash conception of a five-tiered universe. The tip of the glowing triangle of light falls directly on the second concentric ring, which may have been representative of the Sun's world, that is, Earth.

The site replica was constructed at approximately 3/4 scale. No molds or tracings were made at the site, but photographs and field drawings were extensively used in constructing and painting the model. It was built to be as much *experienced* as *viewed*; children (and uninhibited adults) are encouraged to crawl inside to view the paintings on its ceiling and to feel contained within the space of this simulated rock shelter. Children love exploring the small but inviting shelter, and they delight in discovering the paintings which are tucked away in small nooks. They love pushing the button that activates the light simulating the winter solstice sunrise effect. They listen attentively as their parents read to them about this art which, in its fantastic imagery, stirs the imagination.

The Burrow Flats site is remarkable in many ways; it is in relatively pristine condition, it is a solstice marking site, it contains motifs unique to Chumash rock art, and it is undeniably and strikingly beautiful. Through the constructed replica at the Southwest Museum, it was hoped that these qualities, traditional to Chumash art, would be conveyed to the museum visitor.

An exhibit on the rocks?

Criticisms have been levelled against the display; one rock art student complained that it is a useless installation since it is not an accurate facsimile and therefore has no research potential. The curatorial staff conceived of the display as a replica, not as a facsimile. A facsimile is an exact copy or duplicate; a replica, on the other hand, is a meaningful reproduction, not necessarily accurate in every detail. The Burrow Flats replica was installed as an educational tool, not as a research tool. Although the replica was painted to look like the original, it is a schematised version. Its main goal is to express a feeling for Chumash artistic accomplishment and to convey the significance of rock paintings in traditional Chumash life. But the installation, being an artificial reproduction of prehistoric Native American art, can only hint at the vitality that the *original* work has *in situ*.

It could be argued that the rock art replica does not convey the artistic richness of Chumash culture. In this I agree. The installation is disturbingly out of context. It is surrounded with carpeting, and it is artificially lit. One cannot feel the mountain breeze or smell the chaparral. One cannot sense the time depth, or the enduring quality of the at-once old and fragile paintings. One cannot experience the *place*. One can, however, marvel at the array of colours that were gathered from the earth, and wonder about the meaning of the motifs.

It can also be questioned whether the exhibit does more harm than good by encouraging children to crawl in the shelter. Once inside, children are compelled to touch the paintings. This is acceptable, perhaps, if it is clearly communicated to the museum visitor that they can touch the replicated paintings but must never touch original pictographs. (Sadly, the label for this replica does not spell this out.) Rock art replications, if they are to be successful, must encourage constructive appreciation, not destructive inquisitiveness.

A well-written and informative museum label is essential. It provides an opportunity to convey the message that rock art is culturally *significant*. Through this simple but essential message we can dispel the widely held myth that rock art is simply graffiti. If the symbolic genius of the rock artists is conveyed, then the replication succeeds.

I am not certain that the replica has succeeded, although visitor response so indicates. Being the staff member who initiated and supervised the replica's installation, I feel an uneasy mix of delight and uncertainty. Given an extremely limited budget, and the usual deadlines, we did the best that we could. Three years after the unveiling of this exhibit, questions remain unanswered. Is rock art safer now from vandalism and unintentional destruction? Does the public now have a better appreciation of Native American rock art?

Replication: a step toward preservation?

At best, replication is an educational tool. Arguably, education is the biggest contribution we can make toward the long term preservation of rock art. It is certainly the biggest step that we can take toward appreciation and understanding.

The role of museums in Western society is expanding. The traditional museum functions of collecting, preserving and interpreting the world's cultural and natural histories are no longer sufficient. Museums are now being charged with a host of other functions that essentially present the achievements of one cultural group to others. The responsibility they have for authenticity is real. It is a responsibility that must be translated into educationally meaningful exhibits which both teach and 'touch' museum visitors.

Replication can never be a replacement for direct conservation measures; at best it can serve as a sensitive portrayal of aspects of another culture. In this portrayal lies the greatest hope of long-term preservation.

REFERENCE

ROZAIRE, C. E. 1959. Pictographs at Burro Flats. *Ventura Historical Society Quarterly* 4: 2-6.

27

MULTIDIMENSIONAL SCALING IN ROCK ART STUDIES: TWO CALIFORNIA EXAMPLES

William D. Hyder

Abstract. Multidimensional scaling has been used successfully in ceramic analysis for seriation studies and in reconstructing spatial relationships among distant archaeological sites. This study explores application of the technique using sites from three distinct areas in North America. In general it appears that multidimensional scaling techniques are ideally suited to the analysis of rock art data.

Recording and cataloguing images is just the first step in integrating rock art into archaeological studies. Despite advances in recording technologies and the ready availability of cheap computing power, we continue to be plagued with difficulties in translating graphic images into data suitable for statistical analysis. I propose that the problem has been made more difficult than necessary. In exploring this thesis, I first examine the question of style versus typology in rock art classification. As a way out of the impasse created by arguments over what constitutes style, I next propose one method of quantifying rock art data that can be adapted to a variety of analyses. Finally, I explore multidimensional scaling as one technique suitable to exploring rock art data with all its inherent noise and subjectivity.

Problems with style and typology

To analyse rock art data using quantitative methods, we must somehow classify and quantify graphic images. Maynard (1977: 387) notes that rock art categories are frequently called styles. Classification may or may not yield styles. For example, the definition of styles may be the goal of quantitative studies as Whitley (1982) demonstrates. Alternatively, the presence of different styles may be used to define variables to be measured in the recording process. Different situations require different solutions not unlike any archaeological study of artefacts. Just as there are types of types, there are styles of style. Too often style is used in aimless arguments about the validity of one or another classification scheme.

I must profess from the start that I am biased against most applications of the concept of style. All too often I cannot understand what different authors really mean by style. The problem is not that style is undefined. On the contrary, there are too many definitions of style. As Clegg (1986: 62) asserts, nearly everyone understands something different when encountering the term style. Some people recognise style as a set of defining attributes, others know it as a functional attribute. Both may be right, so style remains a useful - albeit confused - concept for many aspects of rock art study. But so far, style has rarely been useful in quantitative studies because it too often obscures important variation (cf. Hedges 1982). In coding data for quantitative analysis we must be careful to preserve sources of variation. That means that some traditionally accepted styles may encompass two or more types.

Spaulding (1982: 1-2) reminds us that 'thc primary task of archaeology is to discover and describe whatever structure there may be in the data of archaeology'. Rock art is just another class of archaeological data (cf. Clegg 1986), although a class with its own set of special problems. We can define a type as a group of internally cohesive items separated from other groups by one or more discontinuities (Maynard 1977: 387-88; Whallon and Brown 1982: xvii), but that just begins to tackle the problem of classifying rock art. What is our unit of analysis - the element, the panel, the site, a collection of sites, styles? The ideal rock art typology will lend itself to a hierarchical recording of data that can be used at several levels of analysis, across a wide variety of sites. Accomplishing that goal is not an easy task, however. Indeed, most classification schemes may be too cumbersome to easily transfer from one situation to another. Archaeology is replete with examples of incomparable data, but that lamentable fact has not stopped the business of archaeology.

A rock art typology should yield unambiguous, well-defined, and uniformly understood divisions (Maynard 1977: 387). Although this is a laudable goal, an element of subjectivity will always be present in rock art classifications. For example, Whitley (1982: 66) constructed a hierarchical classification key that he admits is not ideal, but 'results in the most consistent and replicable typology'. He presents no evidence to confirm his claim to consistency. Nevertheless, it stands as one of the best attempts to produce an objective classification key. Unfortunately, his key yields 30 element types, 29 of which are defined and one of which is the catch-all category called 'other'. We are never told the count for the other category as it is excluded from further analysis. In checking the count for the largest site in his sample, I discovered that only 60% of the elements had been included in valid categories (127 out of 207 total elements). In another case, only 40% of the elements were included (93 out of 256 elements). A 40-60% loss of data is too high a price to pay for objectivity.

How then can we approach rock art classification? The Herculean effort that led to Pager's (1971) Ndedema Gorge study in which every element is described according to an exhaustive set of precisely defined terms will rarely be repeated. Such dedication, no matter how laudable, comes only at a high price. Nissen (1982), on the other hand, defined some 999 designs as a shorthand approach to objective classification. Whenever a new discrete design was encountered, a new definition was added to the element catalogue. Bruder (1983: 113) followed a similar procedure in recording the rock art at Hedgpeth Hills. Wallace and Holmlund (1986: 64) similarly started from existing typologies and added categories as necessary. To control for recorder inexperience and subjectivity, all categorisation was performed in the field by the two principal investigators. Their assignments were cross-checked to achieve a reasonable degree of consistency.

The aforementioned examples exhibit a variety of shortcomings. Pager's typology is the most thorough, but it is also highly specific to the Ndedema sites. Nissen's typology preserves much detail about individual elements, but becomes too unwieldy for a recording crew to apply in the field. Whitley demonstrates one method that improves the objectivity of applying a typological scheme to rock art, but it does not cover a sufficient variety of elements to be useful even in the area where it was applied. The work of Bruder and Wallace and Holmlund begins to approach a workable compromise between detail, utility, and objectivity. It appears, however, that any approach to creating an element typology for rock art will be forced to make some compromises between objectivity and utility. With this caveat in mind, I offer my own imperfect contribution to rock art typology.

An approach to element typology

In constructing my typology for the study of Chumash rock art, I sought to combine the flexibility of schemes such as those of Nissen and Bruder with the consistency of the Whitley and Wallace and Holmlund approaches. Above all, I wanted to maintain the freedom of adjusting the definition of analytic categories as new information or new questions became available. An earlier effort by Horne (1976) to quantify Chumash rock art data relied on general categories defined by Heizer and Clewlow (1973) such as curvilinear, rectilinear, and animal. Horne's study demonstrates the potential of quantitative approaches, but his data cannot be adapted to other studies. Too much information about graphic images is lost in simplistic approaches to constructing element categories.

Maynard's (1977) proposed classification based on technique, form, motif, size and character offered a promising model for my own typology. Some changes were necessary, however. To begin with, Maynard (1977: 389) criticises one classification system organised by colour on the grounds that design characteristics are more important. That is not always true. Previous Chumash research had already demonstrated a pattern in sequences of overpainting. Since my interests are in discovering chronological and spatial dimensions in Chumash rock art, my classification scheme capitalises on patterns of overpainting. Elements are sorted first into the categories of paintings or petroglyphs (see Fig. 1). By maintaining parallel categories, the data can then be combined for a single analysis ignoring production techniques or it can be maintained as separate classes of data. Painted elements are further distinguished by whether they are drawn, painted in a single colour, or painted in two or more colours. These divisions were defined specifically to test the hypothesis that Chumash rock art evolved from monochrome drawing to painted polychromes. Since petroglyphs are rarely encountered in Chumash rock art, the petroglyph category has not been further defined for this study.

The next two divisions in my scheme were designed to maintain some compatibility with Heizer and Clewlow's earlier scheme. They record basic types, shapes and general classes such as curvilinear, aviforms, anthropomorphs and so on. Except for their use in helping to uniquely define particular shapes or motifs, they are not central to this discussion.

Maynard (1977: 393) next looks at form in assigning elements to mutually exclusive categories. I did not find this distinction useful in my study of Chumash rock art, although analogous distinctions have been made in other studies (cf. Swartz 1978) to detected changes over time. Motif, or the particular figure or shape of the element (Maynard 1977: 395), is defined by a letter and number combination. As new figures are encountered that do not easily fit into existing categories, new letter and number combinations can be defined. Together, form and motif define what Schaafsma (1985: 287) calls a figure type, an important component of style. As I suggested earlier, there are many similarities between the definition of a typology and a style. A style is best thought of as one of many possible typologies, although not all typologies define styles.

Maynard's (1977: 397-99) final two categories, size and character, are not used. Size was not felt to be a useful category in my particular study, although it can be important to other studies as Strange (1986) eloquently demonstrates. One might argue, however, about whether absolute or ordinal measures of size are most useful. Character was also felt to be extraneous to my study and in general reintroduces an element of subjectivity that Maynard tries to avoid. In their place, I include a colour category that can be used or ignored for different analytic purposes. In my studies, the colour designation helps distinguish between monochrome and polychrome paintings. Figure 1 shows a typical element, the corresponding category, and an explanation of its constituent parts.

The availability of microcomputers and the use of sophisticated spreadsheets serve to manage this cumbersome scheme, yet avoid the analytic problems raised by mostly zero data (Speth and Johnson 1976). As Nissen found and Whitley should have noted, too many zeros in a data matrix skews statistical results. The element categories as defined above are not used outside the spreadsheet format. Clegg (1986: 60) notes that his own 'knob and blob' approach of letter and number descriptive tags 'solves the problem of interpretation-free definition, but seems to maximise incomprehensibility'. My element naming convention is designed solely to take advantage of the database power of spreadsheet programs.[1]

It is important to note that no coding is done in the field. Instead, elements are assigned codes in the laboratory working from photographs or drawings and a master catalogue. Classifications are then field checked for accuracy. The coding conventions are too complex to be applied from memory or under the pressure of time constraints imposed by field work. Once a site has been coded, it is entered into a matrix with element types as rows, and sites as columns. The resulting matrix, or database, is composed largely of zeros, but it maintains a maximum level of detail about the rock art elements. The problem is to reduce this matrix to a manageable size suitable for further analysis.

Since my goal is to maintain a maximum level of detail while also producing useful data, I do want to permanently modify my database. Fortunately, I can reduce the matrix by defining composite motifs from collections of

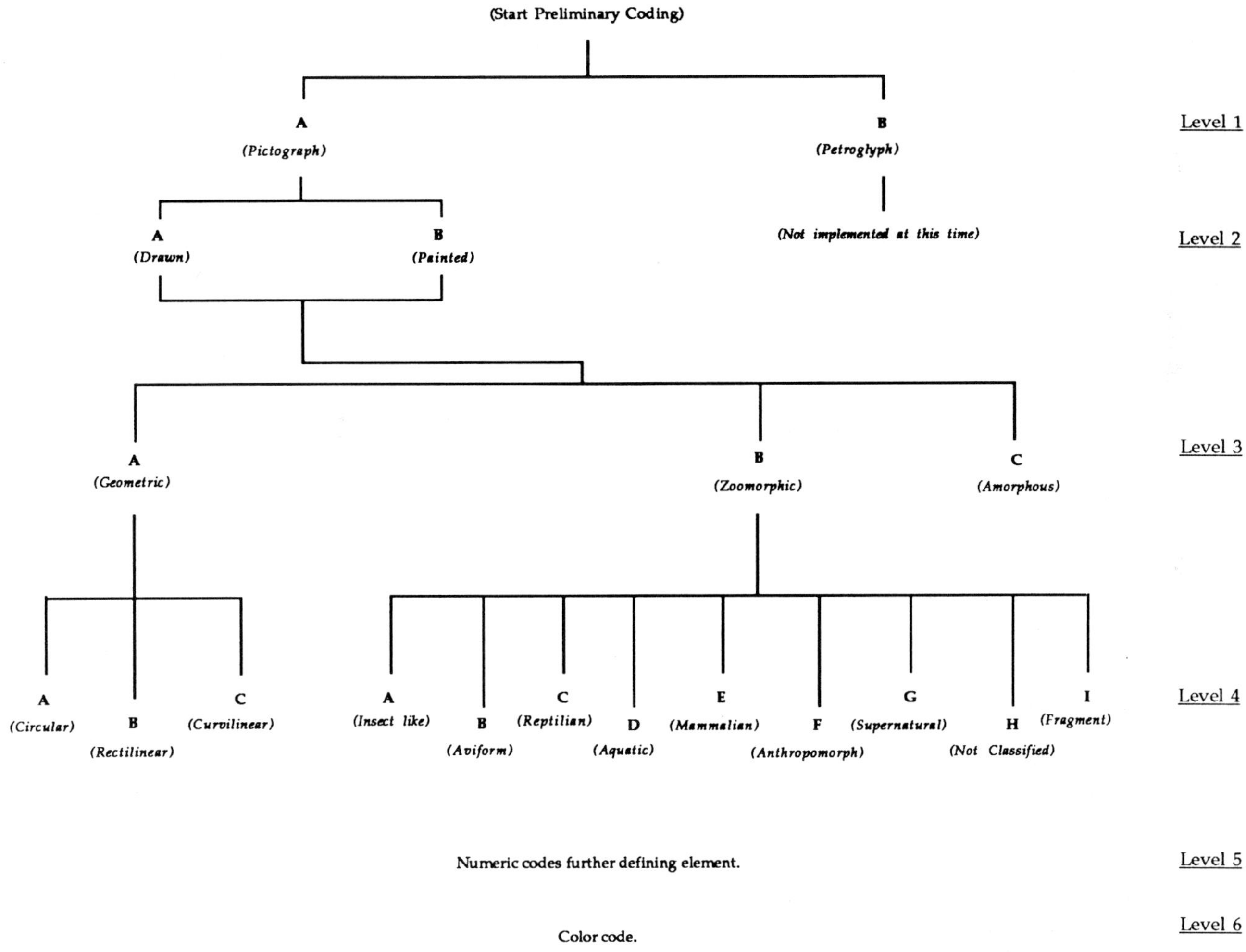

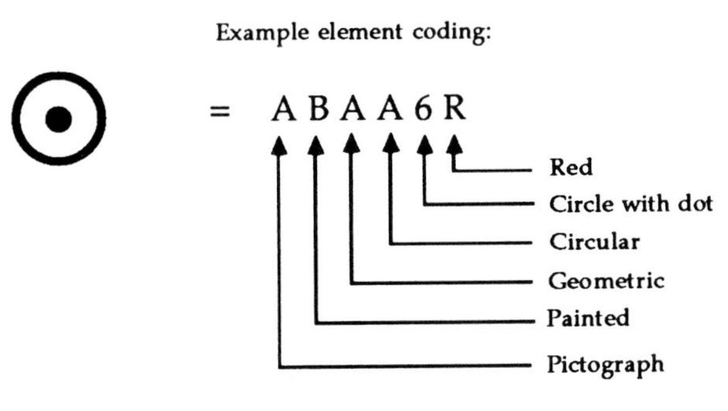

Figure 1. *Hierarchical coding scheme for Chumash rock art, and a sample coded element.*

similar elements. The terminology becomes confusing in that motif, element, category, figure type and so on usually refer to the same thing, but I am at a loss for an alternative term. So elements are basic figures, and motifs are collections of elements. Motifs are defined in the spreadsheet through a criterion reference list. The criterion reference list is then used to search the database and sum occurrences of appropriate element types on a site by site basis. To redefine or add a motif, all one has to do is redefine the criterion reference lists. The spreadsheet can be used to produce tables of motif counts, percentages, and colour distributions. These tables can then be output to text files for use in statistical packages or linked to relational databases for other analyses.

My typology does not solve all the problems of rock art classification, nor does it record all the data many consider relevant. It was designed only to produce data suitable to explore various dimensions of Chumash rock art. At this point, my assumptions about the meaning of superimpositions and production techniques come into play. Typically, in a motif analysis, sites are cases, motifs are variables, and the counts or proportions are the relevant attributes of motifs. If my assumptions are correct, then the site is not the proper unit of analysis. Instead, I must isolate sets of motifs by production techniques as if they were different site components or strata. I used a discriminate analysis to verify that the defined components were in fact composed of different motif clusters. The results are not central to this discussion and are reported elsewhere (Hyder 1989).

Multidimensional scaling in rock art analysis

If we accept for the moment that I have arrived at a reasonable set of motif types and that those motifs can be used as variables in quantifying attributes of sites, then we can turn to the question of analytic usefulness. Remember that our purpose as archaeologists is to discover or describe the structure in archaeological data. Clegg (1986: 62) identifies four sources of variability in rock art data: the artist's personality, the medium, function, and culture. These interact in ways that create distinct distributions in form, time and space. In undertaking a motif analysis, we assume that sites from different ages, cultures or functions will be composed of elements drawn from different design pools (cf. Kodack 1988). By using various measures of similarity, we can order these sites from most similar to most dissimilar.

Typically, similarity measurements are calculated using the Brainerd-Robinson coefficient (Robinson 1951), although several alternative measures are available depending on the characteristics of the data.[2] The similarity matrix is then sorted to order the cases from most similar to most dissimilar. Clegg (1977) was one of the first to use this procedure in rock art studies, but he did not go beyond the initial sorting of his similarity matrix. Until recently,[3] computer programs were not readily available to manage the task. In addition, many rock art similarity matrices do not yield an ideal ordering of sites. The data are too complicated for that.

Similarity measures are affected by the multiple sources of variation that might be present in rock art data. In my Chumash data, for example, I hope to have captured chronological variability, but functional variability muddies the analysis where it cuts across production techniques. The subjectivity inherent in all rock art recording introduces another source of noise that further reduces the power of statistical measures. It has been my experience, however, that multidimensional scaling can tolerate a reasonable degree of noise and still produce useful results.

Multidimensional scaling (MDS) is a quantitative technique that seeks to find the best placement of points in n dimensional space such that similar points are closer together and dissimilar points are farther apart. The configuration must then be examined for interpretable arrangements of points. These are sometimes referred to as the 'hidden structure' in data (Kruskal and Wish 1978). The goodness-of-fit, or more appropriately 'badness-of-fit,' is measured by the stress statistic. Lower values generally indicate better solutions. In archaeology, we typically assume that points will align themselves along chronological or spatial (i.e. cultural) dimensions. Outside its obvious use in seriation studies (Cowgill 1972), MDS has been used in ceramic studies to detect spatial and chronological distributions (Washburn and Matson 1985).

The most exhaustive use of MDS in rock art analysis published to date is Wallace and Holmlund's (1986) study of the Picacho Mountains petroglyphs. They explore a variety of combinations of data and sample sizes to address questions of chronology and spatial distributions in their study area. Although they found evidence for both chronological and spatial patterning in their data, the authors also found MDS difficult to work with (Wallace and Holmlund 1986: 173). In particular, collapsing data to remove empty cells and the generation of alternative combinations of element categories to test different solutions was time consuming. I have attempted to address some of these difficulties in the structure of my database.

Unaware of Wallace and Holmlund's work at the time, I produced my own attempt at using MDS to seriate a regional subset of my Chumash data (Hyder 1989). The results were encouraging, and they are the basis for work still in progress (Hyder 1988). One drawback to MDS, however, lies in interpreting multidimensional solutions. It is not so easy to detect what underlying dimensions are causing observed patterns. Two dimensions are easily displayed and interpreted, three dimensions can be difficult to visualise and interpret, while four or more dimensions are unmanageable. Again, computer programs[4] now allow us to display and rotate three-dimensional data in the search for patterns.

In the following two examples, I explore two- and three-dimensional solutions for disparate sets of data, in part to show the ability of MDS to tolerate ill-formed data. The first is a subset of my own Chumash data, coded specifically for this type of analysis. It should produce a distribution of points wherein drawn art clusters separate from monochrome art which is separate from polychrome art. The second example is drawn from Whitley's (1982) dissertation data. As noted earlier, the data include only a subset of elements present at each site, and no attempt was made to distinguish between possibly distinct site components. Since Whitley was seeking to explore ways to objectively define styles, I specifically explored his data with the hope of identifying spatial patterning.

My first example includes 30 site components drawn from three regional subsets of my Chumash data. The sites with the highest element counts and greatest diversity were selected as being best fitted to the requirements of MDS. Although two dimensions display an easily interpretable chronological pattern that I sought, Kruskal's stress drops from .10093 to .06361 as one moves from a two- to three-dimensional solution. The four dimension solution does not significantly improve the stress measure. Figure 2 shows one alternative display of the three dimension solution. As you move from the upper left corner to the lower right corner, you generally move from drawn to polychrome painted site components. This 'dimension' represents time if my earlier assumptions are correct. Moving from the lower left corner to the upper right corner, you move from coastal sites to interior sites. This pattern supports the ethnographic record which indicates cultural differences between the coastal and interior Chumash. These interpretations are over-simplistic and the patterns imperfect, but then so is the data. It does demonstrate, however, that MDS can be used

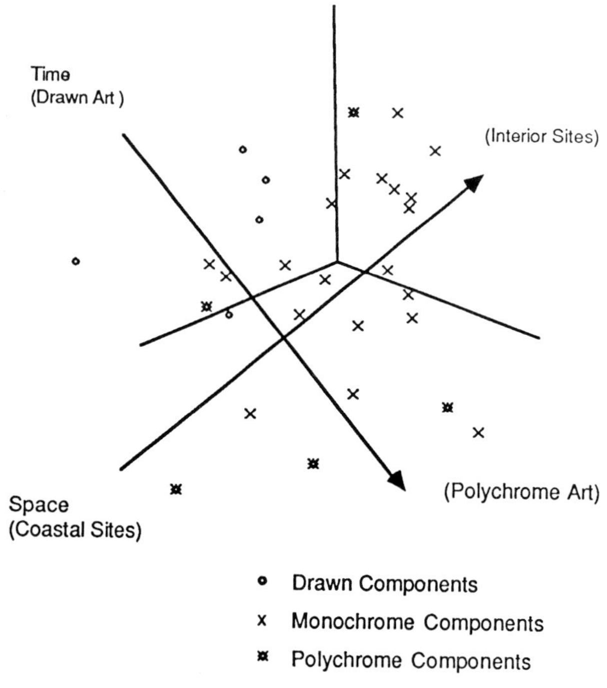

Figure 2. *Three dimension MDS solution for 30 Chumash site components showing chronological and spatial dimensions in rotated display.*

to detect important structure in rock art data.

The second example, Whitley's data, was also limited to a subset best fitted to the requirements of MDS. Of his original 90 sites, 18 with the highest element counts were selected for analysis. Because I had no expectations about chronological ordering, I only looked for spatial patterning. Three sites in particular were suspect at the outset. They are spatially quite distant from the other sites included in the analysis, and they include horse-and-rider elements unlike any of the other sites. Again, Kruskal's stress indicates the three dimension solution presents a better fit than the two dimension solution (a reduction in stress from .11873 to .07605), but I chose to present only the two dimension solution (see Fig. 3). With some exceptions, the two dimension solution roughly sorts the sites into sets that correspond to historic cultural patterns. Significantly, the three sites that are most distant geographically are most isolated on the graph. The patterns are not as clear as my Chumash data, but then nearly half of the elements are excluded from the analysis.

Conclusion

In her recent review of the state of North American rock art research, Schaafsma (1985: 239-41) notes the many problems inherent in recording rock art in the absence of a theoretical focus. We spin our wheels, expend a lot of energy, make pretty pictures, and sometimes produce useful records. But, Schaafsma's (1985: 245-246) own bias against non-stylistic classification schemes influences the rest of her review. Indeed, the field of rock art in general is biased against basic archaeological approaches to archaeological data. The classification of data is inherently related to the kinds of questions we ask. Style is no better or no worse a classification scheme than any other scheme that reveals structure in rock art data. It just addresses different kinds of questions than those I often want to ask.

How we translate our records of rock art images into data plays an important role in its future usefulness. The more flexible our typological scheme, the more useful our classification might be for other studies. In this paper I have presented examples of one exploratory technique that can make use of non-stylistic classification schemes to address particular kinds of archaeological questions. These questions are neither better nor worse than other kinds of questions, such as the pursuit of meaning. They are, however, basic to the understanding of all archaeological data including rock art. That alone makes them worth asking.

NOTES

1 I happen to use Microsoft Excel on an Apple Macintosh, but the same functions can be performed using Lotus 1-2-3 on an IBM PC or a variety of other compatible machines and software packages.

2 Kintigh's (1988) *Archaeologist's Analytical Toolkit* includes a program that will produce a similarity matrix from count data using a variety of different measures. All the similarity matrices analysed for this paper were produced using Kintigh's program.

3 The analyses for this paper were performed on IBM XTs, a Compaq 386/20, and an Apple Macintosh using SYSTAT V3.0 and V4.0 (Wilkinson 1988).

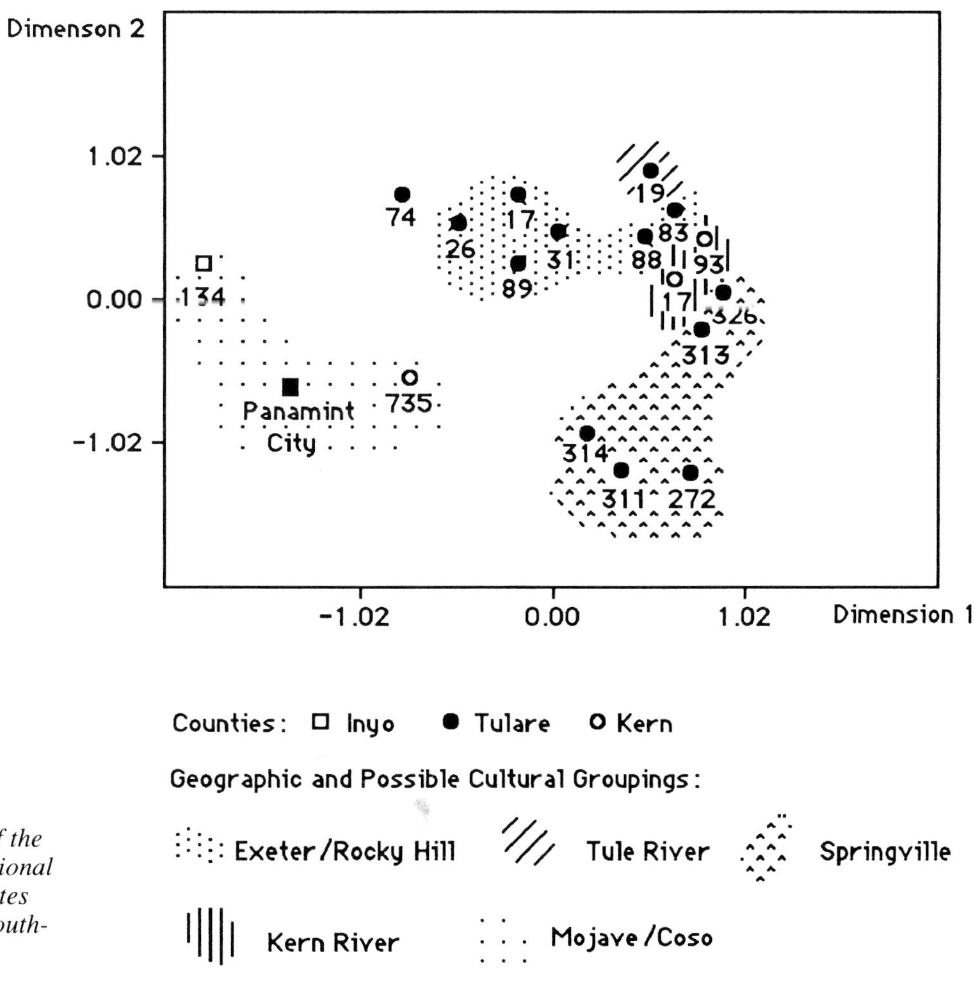

Figure 3.
Two-dimensional plot of the results of a multidimensional scaling analysis of 18 sites from Whitley's (1982) south-central California data.

4 MacSpin (Donoho et al. 1985) duplicates many of the statistical display features previously available only on mainframes. With creative use of filters, you can partially view four-dimensional data.

REFERENCES

BRUDER, J. S. 1983. Archaeological Investigations at the Hedgpeth Hills Petroglyph Site. *Research Paper* 28, Museum of Northern Arizona, Flagstaff.

CLEGG, J. 1977. A method of resolving problems which arise from style in art. In P. J. Ucko (ed.), *Form in indigenous art: Schematisation in the art of Aboriginal Australia and prehistoric Europe*, pp. 260-76. Australian Institute of Aboriginal Studies, Canberra.

CLEGG, J. 1986. The archaeological approach to prehistoric pictures in Australia. In G. N. Bailey and P. Callow (eds.), *Stone Age prehistory. Studies in memory of Charles McBurney*, pp. 55-65. Cambridge University Press, Cambridge.

COWGILL, G. L. 1972. Models, methods and techniques for seriation. In D. L. Clarke (ed.), *Models in archaeology*, pp. 381-424. Methuen, London.

DONOHO, A. W., D. L. DONOHO, M. GASKO and C. W. OLSON 1985. *Macspin Graphical Data Analysis Software*. D2 Software, Austin.

HEDGES, K. 1982. Great Basin rock art styles: A revisionist view. In F. G. Bock (ed.), *American Indian Rock Art*, Vols 7-8, pp. 205-11. American Rock Art Research Association, El Toro.

HEIZER, R. F., and C. W. CLEWLOW, Jr. 1973. *Prehistoric rock art of California*, (2 vols.) Ballena Press, Ramona.

HORNE, S. 1976. Analysis of Chumash rock art from Sierra Madre Ridge, California. In K. Sutherland (ed.), *American Indian Rock Art*, Vol. 2, pp. 115-25. El Paso.

HYDER, W. D. 1989. Rock art and archaeology in Santa Barbara County, California. *Occasional Paper* 13. San Luis Obispo County Historical Society, San Luis Obispo.

KINTIGH, K. W. 1988. *The archaeologist's analytical toolkit.* Privately published, Tempe, Arizona.

KODACK, M. 1988. Design pools. MS in possession of author.

KRUSKAL, J. B., and M. WISH 1978. Multidimensional scaling. *Sage University Paper Series on Quantitative Applications in the Social Sciences*, Series no. 07-011. Sage Publications, Beverly Hills.

MAYNARD, L. 1977. Classification and terminology in Australian rock art. In P. J. Ucko (ed.), *Form in indigenous art: Schematisation in the art of Aboriginal Australia and prehistoric Europe*, pp. 387-402. Australian Institute of Aboriginal Studies, Canberra.

NISSEN, K. M. 1982. *Images of the past: An analysis of six western Great Basin petroglyph sites*. Ph.D. thesis, University of California, Berkeley. University Microfilms, Ann Arbor.

PAGER, H. 1971. *Ndedema*. Akademische Druck- und Verlagsanstalt Graz, Austria.

ROBINSON, W. S. 1951. A method for chronologically ordering archaeological deposits. *American Antiquity* 16: 293-301.

SCHAAFSMA, P. 1985. Form, content, and function: Theory and method in North American rock art studies. In M. B. Schiffer (ed.), *Advances in archaeological method and theory*, Vol. 8, pp. 237-77. Academic Press, New York.

SPAULDING, A. C. 1982. Structure in archaeological data: Nominal variables. In R. Whallon and J. A. Brown (eds), *Essays on archaeological typology*, pp. 1-20. Center for American Archaeology Press, Evanston.

SPETH, J. D., and G. A. JOHNSON 1976. Problems in the use of correlation for the investigation of tool kits and activity areas. In C. E. Cleland (ed.), *Cultural change and continuity. Essays in honor of James Bennett Griffith*, pp. 35-57. Academic Press, New York.

STRANGE, W. C. 1986. Child of the Water, Child of the Wall: Essay in the semiotics of Barrier Canyon, Utah. Paper presented at the Thirteenth Annual Meeting of the American Rock Art Research Association, Flagstaff, Arizona.

SWARTZ, B. K., Jr. 1978. Klamath Basin petroglyphs. *Anthropological Papers*, No. 12. Ballena Press, Socorro.

WALLACE, H. D., and J. P. HOLMLUND 1986. Petroglyphs of the Picacho Mountains South Central, Arizona. *Anthropological Papers*, No. 6. Institute for American Research, Tucson.

WASHBURN, D. K., and R. G. MATSON 1985. Use of multidimensional scaling to display sensitivity of symmetry analysis of patterned design to spatial and chronological change: Examples from Anasazi prehistory. In Ben A. Nelson (ed.), *Decoding prehistoric ceramics*, pp. 75-101. Southern Illinois University Press, Carbondale.

WHALLON, R., and J. A. BROWN (eds) 1982. Preface in *Essays on archaeological typology*. pp. xv-xix. Center for American Archaeology Press, Evanston.

WHITLEY, D. S. 1982. *The study of North American rock art: A case study from south-central California*. Ph.D. thesis, University of California, Los Angeles. University Microfilms, Ann Arbor.

WILKINSON, L. 1988. *SYSTAT: The system for statistics*. SYSTAT, Inc., Evanston.

List of Contributors (Part 1, in alphabetical order)

Robert G. Bednarik - AURA, P.O. Box 216, Caulfield South, Vic. 3162, Australia

Tony Blanks - C/o Steve Brown

Steve Brown - 151 Warwick Street, West Hobart, Tas. 7000, Australia

J. Brunet - Laboratoire de Recherches des Monuments Historiques, Rue de Paris, Château de Champs, 77420 Champs sur Marne, France

Elizabeth Caldicott - 180 Upper Sturt Road, Hawthorndene, SA 5051, Australia

J. Clarke - Lot 188, Sexton Road, Mount Helena, WA 6555, Australia

Dr Paul Faulstich - 5824 Middle Crest Drive, Agoura Hills, CA 91301, U.S.A.

Dr J. Flood - Australian Heritage Commission, G.P.O. Box 1567, Canberra, ACT 2601, Australia

Ivan P. Haskovec - Kakadu National Park, P.O. Box 71, Jabiru, NT 0886, Australia

P. Haydock - 15 Henry Street, Shenton Park, WA 6008, Australia

D. Lambert - National Parks and Wildlife Service of NSW, P.O. Box 1393, Gosford South, NSW 2250, Australia

Dr Colin Pearson - National Centre for Cultural Heritage Science Studies, University of Canberra, P.O. Box 1, Belconnen, ACT 2606, Australia

N. North - 216 Balcatta Road, Balcatta, WA 6021, Australia

Peter Randolph - Aboriginal Sites Department, West Australian Museum, 17 Emerald Terrace, West Perth, WA 6005, Australia

Dr Eric W. Ritter - 238 Wilshire Drive, Redding, CA 96002, U.S.A.

Nicholas P. Stanley Price - The Getty Conservation Institute, 4503 Glencoe Avenue, Marina del Rey, CA 90292-6537, U.S.A.

Hilary Sullivan - Kakadu National Park, P.O. Box 71, Jabiru, NT 0886, Australia

Andrew Thorn - 2/43 Coppin Grove, Hawthorn, Vic. 3122, Australia

Gajendra S. Tyagi - 590/13 Indrani Dershan, Dehu Road, Pune 412-101 (M.S.), India

P. Vidal - 7 Rue de la Cité, 24000 Perigueux, France

J. Vouvé - C/o J. Brunet

Joseph Wallam - Aboriginal Sites Department, West Australian Museum, 17 Emerald Terrace, West Perth, WA 6005, Australia

Grahame L. Walsh - Takarakka Rock Art Research Centre, Carnarvon Gorge, via Rolleston, Qld 4702, Australia

Information about AURA

The Australian Rock Art Research Association (AURA) was established in October 1983 by Robert G. Bednarik. Its principal objectives are to provide a forum for the dissemination of research findings in the fields of rock art, palaeoart studies and cognitive archaeology; to promote Aboriginal custodianship of sites externalising indigenous Australian culture; to co-ordinate studies concerning the significance, distribution and conservation of rock art, both nationally and with individuals and organisations abroad; and to generally promote awareness and appreciation of Australia's indigenous cultural heritage.

AURA produces two periodicals, the international journal *Rock Art Research*, and the *AURA Newsletter*. *RAR* appears bi-annually (May and November), and is devoted to developing theory and methodology for the systematic and rigorous understanding of palaeoarts and related phenomena. Emphasis is given to communication across the various disciplines related to the study of global rock art, and to synthesising related subjects around the journal's focus: the surviving externalisations of early world views. The journal is read in over 50 countries, and is designed to encourage international debate. Most major articles are followed by debates of specialists and the authors' responses. The *AURA Newsletter* is an ad hoc communication organ which is received by full members of AURA.

In addition, AURA publishes a series of *Occasional AURA Publications*, of which the present book is Volume No. 4. All AURA publications are produced by volunteers, who receive no compensation whatsoever. This includes writing, editing, typesetting, artwork and formatting. *RAR* is produced without any subsidisation, and has been published at a loss since 1984.

Membership with AURA is open to all persons and institutions professing a commitment to the study, appreciation or conservation of rock art. Within the first five years of operation, the membership of AURA grew to about 700.

The AURA Congress is the largest international event in the discipline of palaeoart studies/cognitive archaeology. It has been called the 'world congress of rock art studies' in overseas publications. It is a quadrennial event, coinciding with the years of the Olympic Games. The First AURA Congress was held in Darwin, Northern Territory, from 29 August to 2 September 1988. It was attended by 343 specialists from throughout the world, and immediately established the AURA Congress as the world's premier event in this discipline. The First AURA Congress was chaired by Robert G. Bednarik and George Chaloupka, who were elected as Secretary and President respectively, with Professor Jack Steinbring as Vice-President. It consisted of eleven subject-oriented symposia. This present volume comprises most of the papers in two of these symposia.

The Second AURA Congress will be held from 30 August to 4 September 1992 in Cairns, Queensland. It will be chaired by Robert G. Bednarik and Percy Trezise, and will also consist of approximately ten symposia.

AURA is involved in numerous other projects and activities. Annual AURA Meetings are held in various parts of Australia, consisting of weekend seminars and field trips. In 1992/93, AURA will launch a major international exhibition in collaboration with The Australian Museum, Sydney. AURA has established a major archive of specialised literature, and the association is a founding member of the International Federation of Rock Art Organisations. Through the multiple roles of its Secretary and Editor, Robert G. Bednarik, who also serves as Convener of IFRAO and National Co-ordinator of the Comité International ICOMOS pour l'art rupestre, AURA co-operates most closely with almost twenty similar organisations abroad. AURA's journal *Rock Art Research* also serves as the Federation's official journal.

Enquiries concerning membership, subscriptions and other relevant matters can be addressed to:

The Editor
AURA
P.O. Box 216
Caulfield South, Vic. 3162
Australia